The Ocean Life of Atlantic Salmon

Environmental and Biological Factors Influencing Survival

Edited by
Derek Mills
MSc., PhD., FIFM, FLS

Atlantic Salmon Trust, Pitlochry

Fishing News Books
An imprint of Blackwell Science

Blackwell Science

Copyright © 2000 by
Fishing News Books
A division of Blackwell Science Ltd
Editorial Offices:
Osney Mead, Oxford OX2 0EL
25 John Street, London WC1N 2BL
23 Ainslie Place, Edinburgh EH3 6AJ
350 Main Street, Malden
  MA 02148 5018, USA
54 University Street, Carlton
  Victoria 3053, Australia
10, rue Casimir Delavigne
  75006 Paris, France

Other Editorial Offices:

Blackwell Wissenschafts-Verlag GmbH
Kurfürstendamm 57
10707 Berlin, Germany

Blackwell Science KK
MG Kodenmacho Building
7-10 Kodenmacho Nihombashi
Chuo-ku, Tokyo 104, Japan

The right of the Author to be identified as the Author of this Work has been asserted in accordance with the Copyright, Designs and Patents Act 1988.

All rights reserved. No part of this publication may be reproduced, stored in a retrieval system, or transmitted, in any form or by any means, electronic, mechanical, photocopying, recording or otherwise, except as permitted by the UK Copyright, Designs and Patents Act 1988, without the prior permission of the publisher.

First published 2000

Set in 10/13 pt Times
by DP Photosetting, Ltd, Aylesbury, Bucks
Printed and bound in Great Britain by
MPG Books Ltd., Bodmin, Cornwall

The Blackwell Science logo is a trade mark of Blackwell Science Ltd, registered at the United Kingdom Trade Marks Registry

DISTRIBUTORS

Marston Book Services Ltd
PO Box 269
Abingdon
Oxon OX14 4YN
(*Orders:* Tel: 01865 206206
        Fax: 01865 721205
        Telex: 83355 MEDBOK G)

USA
Blackwell Science, Inc.
Commerce Place
350 Main Street
Malden, MA 02148 5018
(*Orders:* Tel: 800 759 6102
            781 388 8250
        Fax: 781 388 8255)

Canada
Login Brothers Book Company
324 Saulteaux Crescent
Winnipeg, Manitoba R3J 3T2
(*Orders:* Tel: 204 837-2987
        Fax: 204 837-3116)

Australia
Blackwell Science Pty Ltd
54 University Street
Carlton, Victoria 3053
(*Orders:* Tel: 03 9347 0300
        Fax: 03 9347 5001)

A catalogue record for this title is available from the British Library

ISBN 0-85238-271-5

Library of Congress
Cataloging-in-Publication Data
The ocean life of Atlantic salmon: environmental and biological factors influencing survival/edited by Derek Mills.
    p.   cm.
  'Proceedings of a workshop organised by the Atlantic Salmon Trust and held at the Freshwater Fisheries Laboratory, Pitlochry, on 18th and 19th November, 1998.'
  Includes bibliographical references (p.   ).
  ISBN 0–85238–271–5
  1. Atlantic salmon Congresses.   I. Mills, Derek Henry.
  QL638.S2023   1999
  597.5'6—dc21   99-32067
                                      CIP

For further information on
Fishing News Books, visit our website:
www.blacksci.co.uk/fnb/

Contents

Contributors   v

Chapter 1   Introduction   1
Derek Mills

Chapter 2   Executive Summary   7
Peter Hutchinson and Derek Mills

Chapter 3   A Perspective on the Marine Survival of Atlantic Salmon   19
E.C.E. Potter and W.W. Crozier

Chapter 4   Description of Marine Growth Checks Observed on the Scales of Salmon Returning to Scottish Home Waters in 1997   37
J.C. MacLean, G.W. Smith and B.D.M. Whyte

Chapter 5   Tracking Atlantic Salmon Post-smolts in the Sea   49
A. Moore, G.L. Lacroix and J. Sturlaugsson

Chapter 6   Distribution and Possible Migration Routes of Post-smolt Atlantic Salmon in the North-east Atlantic   65
Jens Christian Holst, Richard Shelton, Marianne Holm and Lars Petter Hansen

Chapter 7   Distribution and Migration of Atlantic Salmon, *Salmo salar* L., in the Sea   75
Lars Petter Hansen and Jan Aarge Jacobsen

Chapter 8   Survival of Atlantic Salmon (*Salmo salar* L.) Related to Marine Climate   88
D.G. Reddin, J. Helbig, A. Thomas, B.G. Whitehouse and K.D. Friedland

Chapter 9   The NAO: The Dominant Atmospheric Process Affecting Oceanic Variability in Home, Middle and Distant Waters of European Atlantic Salmon   92
R.R. Dickson and W.R. Turrell

Chapter 10   Changes in Ocean Climate and its General Effect on Fisheries: Examples from the North-west Atlantic   116
K.F. Drinkwater

Chapter 11 Historical and Potential Long-term Climatic Change in the
            North Atlantic   137
            Grant R. Bigg

Chapter 12 Long-term Planktonic Variations and the Climate of the
            North Atlantic   153
            Philip C. Reid and Benjamin Planque

Chapter 13 Feeding Habits of Atlantic Salmon at Different
            Life Stages at Sea   170
            Jan Aarge Jacobsen and Lars Petter Hansen

Chapter 14 The Food and Feeding of Atlantic Salmon (*Salmo salar* L.)
            During Feeding and Spawning Migrations in Icelandic
            Coastal Waters   193
            Johannes Sturlaugsson

Chapter 15 Problems Facing Salmon in the Sea – Summing Up   211
            A.D. Hawkins

Index   223

Contributors

Grant R. Bigg
School of Environmental Sciences, University of East Anglia, Norwich, England, UK, NR4 7TJ.

W.W. Crozier
Department of Agriculture for Northern Ireland, Agriculture and Environmental Sciences Division, Belfast, Northern Ireland, UK, BT9 5PX.

R.R. Dickson
Centre for Environmental, Fisheries and Aquaculture Research, Pakefield Road, Lowestoft, Suffolk, England, UK, NR33 0HT.

K.F. Drinkwater
Department of Fisheries and Oceans, Bedford Institute of Oceanography, Box 1006, Dartmouth, Nova Scotia, Canada, B2Y 4A2.

K.D. Friedland
UMass/NOBA CMER Program, Blaisdell House, University of Massachusetts, Amherst, MA 01003, USA.

Lars Petter Hansen
Norwegian Institute for Nature Research, Dronningensgate 13, PO Box 736, Sentrum, N0105, Norway.

A.D. Hawkins
Fisheries Research Services, Marine Laboratory, PO Box 101, Victoria Road, Aberdeen, Scotland, UK, AB11 9DB.

J. Helbig
Department of Fisheries and Oceans, Box 5667, St John's, Newfoundland, Canada, A1C 5X7.

Marianne Holm
Institute of Marine Research, Nordnesgaten 50, P.b. 1870, Nordnes, N-5024, Bergen, Norway.

Jens Christian Holst
Institute of Marine Research, Nordnesgaten 50, P.b. 1870, Nordnes, N-5024 Bergen, Norway.

Peter Hutchinson
NASCO, 11 Rutland Square, Edinburgh, Scotland, UK, EH1 2AS.

Jan Aarge Jacobsen
Fishery Laboratory of the Faroes, PO Box 3051, F0-110, Tórshavn, Faroe Islands.

G.L. Lacroix
Department of Fisheries and Oceans, St Andrews Biological Station, St Andrews, New Brunswick, Canada, E0G 2X0.

J.C. MacLean
Fisheries Research Services, Freshwater Fisheries Laboratory, Field Station, 16 River Street, Montrose, Angus, Scotland, UK, DD10 8DL.

Derek Mills
Atlantic Salmon Trust, Pitlochry, Moulin, Perthshire, Scotland, UK, PH16 5JQ.

A. Moore
CEFAS, Pakefield Road, Lowestoft, Suffolk, England, UK, NR33 0HT.

Benjamin Planque
Centre for Environmental, Fisheries and Aquaculture Sciences, Pakefield Road, Lowestoft, Suffolk, England, UK, NR33 0HT.

E.C.E. Potter
Centre for Environmental, Fisheries and Aquaculture Sciences, Pakefield Road, Lowestoft, Suffolk, England, UK, NR33 0HT.

D.G. Reddin
Department of Fisheries and Oceans, Box 5667, St John's, Newfoundland, Canada, A1C 5X7.

Philip C. Reid
Sir Alister Hardy Foundation for Ocean Science, 1 Walker Terrace, The Hoe, Plymouth, England, PL1 3BN.

Richard Shelton
Fisheries Research Services, Freshwater Fisheries Laboratory, Pitlochry, Perthshire, Scotland, UK, PH16 5LB.

G.W. Smith
Fisheries Research Services, Freshwater Fisheries Laboratory, Field Station, 16 River Street, Montrose, Angus, Scotland, UK, DD10 8DL.

Johannes Sturlaugsson
Institute of Freshwater Fisheries, Vagnhöfda 7, 112 Reykjavik, Iceland.

A. Thomas
School of Marine Sciences, University of Maine, 5741 Libby Hall, Orono, ME 04469-5741, USA.

W.R. Turrell
Fisheries Research Services, Marine Laboratory, PO Box 101, Victoria Road, Aberdeen, Scotland, UK, AB11 9DB.

B.G. Whitehouse
Alliance for Marine Remote Sensing, Suite 620, 1550 Bedford High Way, Bedford, Nova Scotia, Canada, B4A 1E6.

B.D.M. Whyte
Fisheries Research Services, Freshwater Fisheries Laboratory, Field Station, 16 River Street, Montrose, Angus, Scotland, UK, DD10 8DL.

Chapter 1
Introduction

DEREK MILLS

> And then one day he's off to sea.
> And only six inch long
> Into the Black Hole, under the Ocean,
> Rows himself along, my dears,
> He rows himself along.
> To Hell with Russian, Viking, Hun!
> This great-hearted simpleton
> Takes the whole Atlantic on.
>
> He hauls his trawl from Scilly Isles
> To the Subarctic shore
> No overheads, no crew to pay
> Whose wives will cry for more, my dears
> Wives always cry for more,
> Through storm and freeze, with cheerful grin,
> Candelfish and Capelin,
> He crams the Ocean's goodness in.
>
> (Ted Hughes)*

Since the advent of the high seas fisheries off West Greenland in 1957 and off the Faroes and in the Norwegian Sea some years later, much more is now known about the salmon's ocean life. This knowledge, accumulated from the results of research vessel cruises to study salmon distribution and migration to and on their feeding grounds, feeding habits and survival, was disseminated at a major international Atlantic salmon symposium in 1992 and subsequently published a year later (Mills 1993a).

One of the important points made at the end of this Symposium was the need to understand what is happening in the ocean. To this end it was recommended that: techniques be investigated to incorporate environment data in assessment and management of salmon stocks; that hydroacoustic gear be developed and, if practical, used to determine the distribution and abundance of salmon in the sea; and that research vessel investigations be continued to determine relationships between salmon abundance, mortality and environment. Furthermore, it was agreed that a multi-national workshop, to be organised jointly by the Atlantic Salmon Trust and the Atlantic Salmon Federation, should be held to discuss ways to increase further our

---

* Reproduced from *The Best Worker in Europe* (1985). Atlantic Salmon Trust. © Ted Hughes 1985.

knowledge of salmon in the sea. Such a workshop was held in Edinburgh on 9 and 10 December 1992, and its proceedings were reviewed by Potter and Moore (1993). A number of studies were recommended, including the development of survey and tracking methods and environmental studies including the extraction and collation of various satellite data and their relationship to salmon abundance. To this end, a three-year project, funded by the Ministry of Agriculture, Fisheries and Food and the Molson Family Foundation, was started in 1995 involving the Atlantic Salmon Trust, the Atlantic Salmon Federation, the European Space Agency and the Atlantic Centre for Remote Sensing of the Oceans in Nova Scotia to improve our knowledge of the effect of ocean conditions on Atlantic salmon using information from the European ERS-2 earth observation satellite. This involves a comparison of salmon stock abundance with 15 years of accurate sea surface temperature data from satellite observation and, if a correlation can be established, the aim is to be able to use current data to forecast salmon abundance and thus plan appropriate salmon management measures. Some of the results are presented in Chapter 8.

One of the major problems confronting marine salmon studies has been one of integration of research projects and of communication between the various disciplines. This lack of integration between the various disciplines and co-ordination of investigations has been mentioned regularly at the Anadromous and Catadromous Fish Committee of the International Council for the Exploration of the Sea over a number of years. Consequently, it was thought timely to redress this situation and bring together climatologists, oceanographers, planktologists and salmon biologists to update our information on various aspects of the salmon's marine life and those environmental and biological factors likely to have effects on the growth and survival of salmon in the sea, and furthermore, to formulate proposals for further research and collaboration in this field.

A simplified flow diagram depicting marine environmental and biological factors influencing salmon survival (Fig. 1.1) will help to focus on the issues that require consideration.

The climate and climatic changes have a direct effect on the ocean environment. Storms and strong winds drive the oceanic gyres and influence sea temperature. Strong winds may also delay the onset of the spring bloom of phytoplankton and may therefore affect zooplankton production (Dickson *et al.* 1988). Climate will also govern the melting of arctic ice and iceberg abundance. The volume of melting ice will inevitably affect sea surface temperature and salinity.

Major changes in climate may result in significant faunal oscillations, and Dunbar (1993) illustrated such occurrences in Greenland involving cod and capelin.

A relationship between salmon abundance and distribution and sea-surface temperature has been demonstrated in important work by Reddin (1988) and Reddin & Shearer (1987). Furthermore, Reddin, Friedland and co-workers have demonstrated a relationship between post-smolt growth and survival and sea-surface temperature (Friedland & Reddin 1993, Friedland *et al.* 1998). Sea temperature also appears to influence maturation and growth (Scarnecchia 1983, Martin & Mitchell

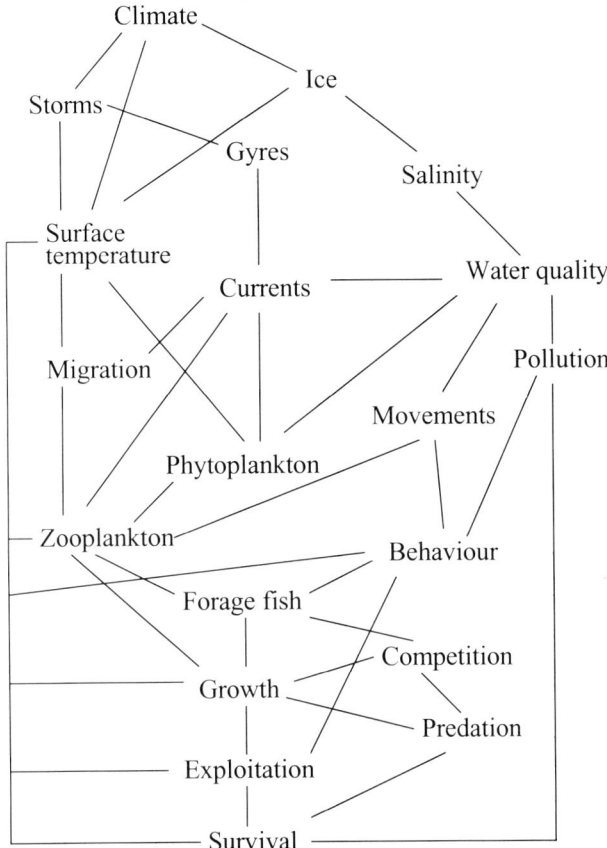

Fig. 1.1 Simple flow diagram depicting marine environmental and biological factors influencing salmon survival.

1985). An important question is whether it is the direct effect of sea-surface temperature that influences survival, abundance, mortality and growth of salmon or whether it is the effect of ocean productivity that is affected by sea-surface temperature. For example, it is known that there is a relationship between sea-surface temperature and phytoplankton distribution and abundance. The abundance of phytoplankton may have an effect on zooplankton production.

Sea-surface temperature may also influence salmon behaviour and migration, which, in turn, may have an effect on the exploitation level through altering catchability. For example, in 1983 and 1984 the Greenland salmon catches were well below average and Greenland salmon fishermen attributed this to the colder than average surface water (Drinkwater 1996) keeping the salmon at a much greater depth than normal and below the depth at which their drift nets were set. Lower water temperatures may also decrease fish activity and affect growth rates, and any consequent reduction in average fish size may also affect catchability.

Gyres and currents may have some influence on salmon migration patterns, as they have on other fish such as herring, while their temperature, salinity and water quality, particularly in the form of silicate and phosphate levels, will affect phytoplankton production. The manner in which salmon use the gyres and currents in their feeding and homing migrations is very poorly understood, although some important results have been collected from post-smolt surveys undertaken by Norwegian and Scottish scientists (Shelton *et al.* 1997, Holm *et al.* 1998). Work on tracking post-smolts at sea and the movements of homing salmon in coastal waters has intensified (Potter & Moore 1993, Sturlaugsson 1995). The oceanic phase of salmon migration is less well-known and has been reviewed by Hansen (1993).

One aspect of the water quality of ocean current systems which appears to have received little attention, in so far as it affects salmon, is that of pollutants and their effect on salmon movements, growth and survival. A number of organic contaminants enter the oceans and include polychlorobiphenyls (PCBs), benzofurans, dioxins, aromatic hydrocarbons (PAHs) and organochlorines (Livingstone *et al.* 1994, Koistinen 1990, D. Scott pers. comm.). It has been found that very low level contamination by PAHs experienced by salmon embryos caused a reduction by up to 60% of the return rate of adults of the exposed cohort to their breeding site. Other substances which could be of importance are mutagens such as heavy metals and radionuclides (Dixon & Pascoe 1994). Such substances may also affect salmon behaviour and survival.

Variations in the physical and chemical environment of the ocean are transferred through phyto- and zooplankton to pelagic fish including salmon and their prey species such as capelin, sprats, sandeels and herring. To date little attempt has been made to investigate a relationship between zooplankton and salmon abundance. However, this workshop brings planktologists and salmon biologists together and this situation will no doubt be redressed.

Information on the salmon's diet is available for a number of oceanic regions including Greenland, the Faroes, the Norwegian Sea and the Labrador Sea/Davis Strait. Although fish form the main part of the diet, various zooplankton organisms, such as copepods, euphausids and amphipods, are also of importance (Hislop & Shelton 1993). It is not known whether or not there is any serious competition for these forage fish with other predators such as man. Concern over the effects of industrial fishing have already been raised and some investigations into the effects of sandeel fishing have been made (Hawkins *et al.* 1998). However, other forage fish species, such as capelin, blue whiting, lantern fish and barracudinas, are also exploited, and it is not known what effects this has on salmon growth and survival. Furthermore, do salmon, at some stage, particularly as post-smolts, share feeding grounds with other pelagic fish such as herring and mackerel?

Hislop & Shelton (1993) listed known records of marine predators of salmon. However, these were mainly casual records and predation on salmon in the sea remains very much an unknown quantity, and, as Hislop and Shelton point out, predation on salmon by fish in the open ocean would have to be investigated by means of dedicated mid-water sampling programmes targeted at likely predators such as sharks (Ritter, 1989). Squid might also be found to be salmon predators.

An evaluation of marine exploitation of salmon in Europe was made by Potter & Dunkley (1993) and in Canada by Chadwick (1993). Most of the recognised high seas salmon fisheries have now either been curtailed (Faroes and the Norwegian Sea) or significantly reduced (Greenland). Practically all Canadian coastal salmon fisheries have now been closed and the unauthorised fisheries in international waters (Mills 1993b) would also appear to have either stopped or be intermittent. Potter (1996) estimated the increases in returns of salmon to home waters following the suspension of commercial salmon fishing at Faroes and Greenland. However, the remaining sea fisheries for salmon off the west of Ireland and the north-east of England could be having a serious effect on some salmon stocks.

The survival rates of salmon at various times in the salmon's ocean life have not been estimated fully, although return rates are better known. Much more information is required on this subject.

This workshop will fill in many of the gaps in our knowledge of the salmon's ocean life and will help investigators to resolve some of the problems. The recommendations proposed for future studies will be invaluable.

References

Chadwick, E.M.P. (1993) Measuring marine exploitation of Atlantic salmon in Canada. In *Salmon in the Sea and New Enhancement Strategies* (ed. D. Mills), pp. 184–202. Fishing News Books, Oxford.

Dickson, R.R., Meincke, J., Malmberg, S.-A. & Lee, A.J. (1988) The 'Great Salinity Anomaly' in the northern North Atlantic 1968–82. *Progressive Oceanography*, **20**, 103–51.

Dixon, D.R. & Pascoe, P.L. (1994) Mussel eggs as indicators of mutagen exposure in coastal and estuarine marine environments. In: *Water Quality and Stress Indicators in Marine and Freshwater Ecosystems: Linking Levels of Organisation* (ed. D.W. Sutcliffe), pp. 124–37. Freshwater Biological Association Special Publications, Kendal.

Drinkwater, K.P. (1996) Climate and oceanographic variability in the northwest Atlantic during the 1980s and early-1990s. *North Atlantic Fisheries Organisation Science Council Studies*, **24**, 7–27.

Dunbar, M.J. (1993) The salmon at sea – oceanographic oscillations. In: *Salmon in the Sea and New Enhancement Strategies* (ed. D. Mills), pp. 163–70. Fishing News Books, Oxford.

Friedland, K.D. & Reddin, D.G. (1993) Marine survival of Atlantic salmon from indices of post-smolt growth and sea temperature. In: *Salmon in the Sea and New Enhancement Strategies* (ed. D. Mills), pp. 118–38. Fishing News Books, Oxford.

Friedland, K.D., Hansen, L.P. & Dunkley, D.A. (1998) Marine temperatures experienced by post-smolts and the survival of Atlantic salmon, *Salmon salar* L., in the North Sea area. *Fisheries Oceanography*, **7**(1), 22–34.

Hansen, L.P. (1993) Movement and migration of salmon at sea. In: *Salmon in the Sea and New Enhancement Strategies* (ed. D. Mills), pp. 26–39. Fishing News Books, Oxford.

Hawkins, A.D., Christie, J. & Coull, K. (1998) *The Industrial Fishery for Sandeels*. Atlantic Salmon Trust, Pitlochry.

Hislop, J.R.G. & Shelton, R.G.J. (1993) Marine predators and prey of Atlantic salmon (*Salmo salar* L.). In: *Salmon in the Sea and New Enhancement Strategies* (ed. D. Mills), pp. 104–18. Fishing News Books, Oxford.

Holm, M., Holst, J.C. & Hansen, L.P. (1998) Spatial and temporal distribution of Atlantic salmon post-smolts in the Norwegian Sea and adjacent areas – origin of fish, age structure and relation to hydrographical conditions in the sea. International Council for the Exploration of the Sea. Theme Session on Ecology of Diadromous Fishes During the Early Marine Phase. CM 1998/N:15.

Koistinen, J. (1990) Residues of polychloroaromatic compounds in Baltic fish and seal. *Chemosphere*, **20**, 1043–8.

Livingstone, D.R., Forlin, L. & George, S.G. (1994) Molecular biomarkers and toxic consequences of impact by organic pollution in aquatic organisms. In: *Water Quality and Stress Indicators in Marine and Freshwater Ecosystems: Linking Levels of Organisation* (ed. D.W. Sutcliffe), pp. 154-71. Freshwater Biological Association Special Publication, Kendal.

Martin, J.H.A. & Mitchell, K.A. (1985) Influence of sea temperature upon the numbers of grilse and multi-sea winter Atlantic salmon (*Salmo salar*) caught in the vicinity of the River Dee (Aberdeenshire). *Canadian Journal of Fisheries and Aquatic Science*, **42**, 1513–21.

Mills, D.H. (ed.) (1993a) *Salmon in the Sea and New Enhancement Strategies.* Fishing News Books, Oxford.

Mills, D.H. (1993b) Control of marine exploitation. In: *Salmon in the Sea and New Enhancement Strategies* (ed. D. Mills), pp. 233–48. Fishing News Books, Oxford.

Potter, E.C.E. (1996) Predicted increases in the returns of multi-sea winter salmon to home waters following the reduction in fishing at Faroes and Greenland. In: *Enhancement of Spring Salmon* (ed. D. Mills), pp. 67–86. Atlantic Salmon Trust, Pitlochry.

Potter, E.C.E. & Dunkley, D.A. (1993) Evaluation of marine exploitation in Europe. In *Salmon in the Sea and New Enhancement Strategies* (ed. D. Mills), pp. 203–19. Fishing News Books, Oxford.

Potter, E.C.E. & Moore, A. (1993) *Surveying and Tracking Salmon in the Sea.* Atlantic Salmon Trust, Pitlochry.

Reddin, D.G. (1988) Sea surface temperature and distribution of Atlantic salmon. In: *Present and Future Management of Atlantic Salmon* (ed. R.H. Stroud), pp. 25–35, Savannali, GA:NCMC Inc.

Reddin, D.G. & Shearer, W.M. (1987) Sea surface temperature and distribution of Atlantic salmon (*Salmo salar* L.) in the northwest Atlantic. *American Fisheries Society Symposium on Common Strategies in Anadromous/Catadromous Fishes*, **1**, 262–75.

Ritter, J.A. (1989) Marine migration and natural mortality of North American Atlantic salmon (*Salmo salar* L.) Canadian Manuscript Report of Fisheries and Aquatic Science, No. 2041.

Scarnecchia, D.L. (1983) Age at sexual maturity in Icelandic stocks of Atlantic salmon (*Salmo salar*). *Canadian Journal of Fisheries and Aquatic Science*, **40**, 1456–68.

Shelton, R.G.J., Turrell, W.R., MacDonald, A., McLaren, I.S. & Nicoll, N. (1997) Records of post-smolt Atlantic salmon, *Salmo salar* L., in the Faroe-Shetland Channel in June, 1996. *Fisheries Research*, **31**, 159–62.

Sturlaugsson, J. (1995) Migration study on homing of Atlantic salmon (*Salmo salar*) in coastal waters in west Iceland – depth movements and sea temperature recorded at migration routes by data storage tags. International Council for the Exploration of the Sea CM 1995/M:17.

Chapter 2
Executive Summary

PETER HUTCHINSON and DEREK MILLS

The basis of concern

In recent years, increasing concern has been expressed about the decline in abundance of the North Atlantic salmon. The total nominal catch has declined from a peak in excess of approximately 12 700 t in 1973 to approximately 2300 t in 1997, with a particularly marked decline in catch since the late 1980s (Fig. 2.1). While much of this decline can be accounted for by the introduction of restrictive management measures, specifically intended to reduce fishing effort, estimates of pre-fishery abundance developed by the International Council for the Exploration of the Sea (ICES) through run-reconstruction models indicate marked declines in the abundance of both maturing and non-maturing one-sea-winter salmon of North American origin, of multi-sea-winter salmon and one-sea-winter salmon from southern European stocks and of multi-sea-winter salmon from Northern European stocks. While there are wide confidence limits on these estimates, if the information for mean recruits is used, there has been a decline of about 46% (from about 4.7 million fish to about 2.6 million fish) in the abundance of one-sea-winter stocks and of about 65% (from about 4.2 million fish to about 1.5 million fish) in the abundance of multi-sea-winter stocks in the North Atlantic region from the early 1970s to the mid 1990s.

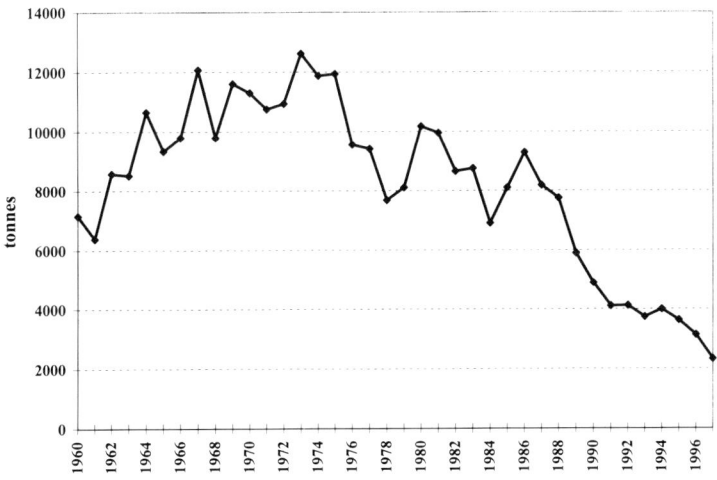

Fig. 2.1 Total declared catch of Atlantic salmon in the North Atlantic region.

It is widely believed that the factors responsible for the decline in abundance have been operating during the marine phase of the life cycle. However, there are considerable difficulties, and not inconsiderable expenses, involved in studying salmon at sea. Indeed, this phase of the life cycle has been described as a 'black-box' because, while it is possible to monitor the number of smolts leaving the river and the number of adults returning, little is known about the factors influencing returns. The problem is, in part, related to the fact that limits on production in fresh water mean that the density of salmon at sea is low and until the 1960s there were no directed fisheries throughout much of the salmon's marine range. The development of fisheries on the salmon's feeding grounds at West Greenland, in the Northern Norwegian Sea and at Faroes provided 'fishery windows' for the study of salmon in the sea and much valuable information has been obtained. These studies provided the first insight into the biology of salmon at sea. Initially these fisheries were implicated in the decline in abundance, but despite the major sacrifices that have been made by Greenland and the Faroe Islands through internationally agreed regulatory measures, and the complete cessation of the Northern Norwegian Sea fishery, stocks have continued to decline. Attention has, therefore, shifted away from the fisheries and is now focusing on the biological and environmental factors likely to affect marine survival.

Evidence of problems facing Atlantic salmon in the sea

Coherence in trends in returns of salmon over wide geographical areas has led scientists to conclude that common events operating in the marine environment are responsible for these trends. Certainly, the marine phase is a period of high mortality. Estimates of natural mortality of salmon in the sea provided by ICES vary from about 70% (River Bush, Northern Ireland, wild salmon) to over 97% (Drammen River, Norway, hatchery-reared salmon) for European stocks, and from about 95% (Western Arm Brook, Newfoundland, wild salmon) to over 99% (Penobscot River, USA, hatchery-reared salmon) for North American stocks. However, relatively few data sets exist which allow trends in marine survival to be assessed. Data from a small number of rivers discharging into the North Atlantic that are monitored in detail but in which patterns of change in the salmon stocks are not necessarily representative of larger groups of stocks (i.e. those rivers referred to by ICES as 'monitored rivers') reveal that:

- For North American stocks, marine survival in recent years has remained low relative to historic levels, despite reduced levels of exploitation, although freshwater production levels are thought to have been maintained. Concomitant with the decline in abundance of North American stocks has been an increase in the proportion of smolts maturing after one sea-winter from less than 50% in the early 1970s to more than 70% in the mid 1990s.
- For European stocks, marine survival has generally been lower in the 1990s compared with earlier years. There appears to have been a widespread and sudden

decline in marine survival of the one-sea-winter stocks at the end of the 1980s and survival has remained low for some, but not all, of these stocks. For two-sea-winter stocks there has been a more gradual decline that has been continuing for at least 25 years.

Survival rates of salmon from the Western Arm Brook and the Conne River (Newfoundland) and for the River North Esk (Scotland) have halved in recent years compared with rates in the 1970s and 1980s.

Relative importance of freshwater and marine survival

There is, therefore, evidence of a reduction in marine survival in recent years, at least in some of the monitored rivers in both North America and Europe, while freshwater production levels in these rivers appear to have been maintained. However, in one of the monitored rivers, the River Bush (Northern Ireland), survival from egg to smolt (1.03%) is considerably less than survival from smolt emigration to coastal return (31.3%). In this river freshwater survival has declined to one-third of its value since the late 1980s as a result of habitat degradation and avian and mammalian predation, but marine survival has remained relatively constant. A further complication is that factors in fresh water may also have important influences on the subsequent performance of salmon in the sea. For example, in acidified rivers, accumulation of aluminium during the freshwater phase may lead to an impaired ability to osmoregulate and, therefore, to adapt to conditions in the sea. Recent studies in New Brunswick also suggest that 4-nonylphenol, a commercial compound widely used in industrial, commercial and domestic chemicals, including pesticides and cleaning products, may adversely affect the later stages of smolting and lead to delayed mortality at sea. Such effects may be manifested as a reduction in return rate and the problem attributed to factors during the marine phase when it clearly lies in the freshwater environment. In these circumstances inappropriate management action may be taken to address the problem. It is, therefore, desirable that survival estimates for the entire life cycle are obtained. In this regard, studies of growth patterns in historical scale collections could provide valuable information about the relationship between growth, ocean climate and freshwater conditions. Such collections exist for some Scottish rivers (back to 1963) and for the River Wye (back to 1910). A clearer understanding of how factors in fresh water may subsequently affect marine survival would also be desirable. It is clear, however, that for some salmon stocks there is a real problem linked to conditions at sea.

Stage at which marine survival is affected

Marine survival of the Atlantic salmon is thought to be heavily influenced by events during the first few months at sea since it is often assumed that predation pressure declines as the fish grow (the so-called 'inverse weight hypothesis'). There is, however, conflicting evidence concerning the validity of this hypothesis in relation to Atlantic salmon. In the north-east Atlantic, a correlation has been demonstrated between salmon catch and an index of spring thermal habitat thought to be preferred by post-smolts. Studies in Canada also suggest that sea surface temperatures inshore around the coast of Newfoundland may influence post-smolt survival and ultimately the number of salmon produced. Furthermore, significant positive relationships between recapture rates of tagged one-sea-winter and two-sea-winter adults from the North Esk (Scotland) and the River Figgjo (Norway) have been reported. These findings suggest that the dominant factors affecting survival operate early in the marine phase.

However, a preliminary analysis presented at the workshop of the relative marine survival to homewaters of fish of differing sea ages, suggests that marine survival of one-sea-winter and two-sea-winter stocks are not related within individual stocks. This finding could indicate that factors affecting the fish later in the marine phase are an important source of marine mortality. In this regard salmon returning to Scottish home waters in 1997 showed a higher than previously recorded incidence of summer checks on the marine zone of their scales. Such checks could be the result of periods of reduced growth, which may be associated with reduced marine survival. The incidence of checks was significantly higher in one-sea-winter salmon than in two-sea-winter salmon but the checks were laid down in the same calendar year (1996). However, while the marine survival indices for North Esk salmon returning to home waters in 1997 as one-sea-winter and two-sea-winter fish were the lowest on record, no correlation has been demonstrated between marine survival and the incidence of growth checks for the years examined. The extent to which the summer checks were present varied with both month and sea-age at return, suggesting that the causal effect may not be uniform throughout the marine environment and that run timing differences are related to different marine migration patterns.

Further studies are clearly necessary in order to improve our understanding of the stage at which the increased marine mortality occurs, and whether fish from different populations migrate to different areas of the North Atlantic where they are exposed to different conditions influencing their survival. In this regard much recent progress has been made in understanding the oceanic distribution and migration of the Atlantic salmon.

Oceanic distribution and migration of Atlantic salmon

Knowledge of the biology and distribution of post-smolts at sea has been limited by an inability to locate salmon once they leave the river. Miniaturised acoustic tags have been used to track post-smolts in the sea, but because of the short life of the tags and the limited detection range in coastal waters tracks to date have been relatively short (up to 60 km). In Iceland, sampling using drift-nets has been used to study the coastal migration of ranched post-smolts microtagged prior to release. Most of the information gathered to date suggests that post-smolts move relatively quickly into the ocean close to the water surface, and that the patterns of movement are strongly influenced by surface water currents, wind direction and tidal cycle. However, in some years post-smolts have been caught in the nearshore zone of the northern Gulf of St Lawrence, throughout their first summer at sea, and in Iceland some post-smolts, mainly maturing males, foraged along the shore following release from ranching stations. These results indicate that the migratory behaviour of post-smolts can vary among populations.

In recent years, scientific research fishing has been conducted over wide geographical areas of the North Atlantic. The first studies were carried out in the north-west Atlantic in 1988, 1989 and 1991 using surface drift-nets. The highest catch rates were made in relatively warm waters (4–8°C) and the results suggest that frontal zones may play some role in the marine phase of the salmon's life cycle. Low catches occurred at water temperatures below 4°C. Since 1991, major scientific research fishing programmes have been conducted in the north-east Atlantic. Pelagic trawl surveys conducted in the Iceland/Scotland/Shetland area during May and June and in the Norwegian Sea from 62°N–73°N in July and August, have shown that post-smolt salmon are widely distributed throughout the sampled area, although they do not reach the Norwegian Sea until July and August. Salmon from Southern Europe (British Isles, France, Iberian Peninsula) dominate the western samples while post-smolts from Northern Europe (principally Norway) are better represented in the eastern samples. Over much of the area of study, catches of post-smolts are closely linked to the main surface currents although north of about 64°N, where the current systems are less pronounced, post-smolts appear to be more diffusely distributed. Use of the northerly surface currents may allow migration to the northern feeding areas with minimum expenditure of energy. However, while post-smolt migration can be rapid, as demonstrated by a smolt tagged in southern England which was recaptured 2000 km to the north only 3 months later, swimming in pursuit of food or entrainment in eddies may lead to migration at speeds below those of the currents.

This research fishing has shown that there is spatial and temporal overlap of post-smolts and large scale pelagic fisheries in the Norwegian Sea and there is concern that there may be a significant by-catch of post-smolts, particularly in the mackerel fishery. Screening of the catches in the pelagic fisheries through an observer programme is a high priority.

The only opportunity there had previously been to study salmon at sea was in the fisheries conducted on the feeding grounds. It can be concluded from the studies conducted at Greenland, Faroes and in the northern Norwegian Sea that salmon of North American origin tend to remain in the north-west Atlantic area, although some fish move into the north-east Atlantic. However, a larger proportion of the European multi-sea-winter salmon stock moves far into the western Atlantic to feed. Salmon from Canada and Scotland are thought to be major contributors to the stocks off West Greenland, with southern European rivers contributing more than northern European rivers. Salmon from Norway, Scotland and Russia are thought to be the major contributors to stocks to the north of Faroes, and these countries, together with Ireland, are also thought to be the main contributors to stocks in the south of the Faroese zone. A major change in recent years has been the occurrence of farmed fish (up to about 40% of fish sampled) in the Faroese zone. A limitation of the studies based on sampling in the fisheries is that they provide little information on the distribution in time and space of salmon in the marine environment.

More detailed knowledge of the spatial and temporal distribution of salmon at sea, of their behaviour, migration routes and the environmental conditions experienced should lead to a clearer understanding of the factors influencing their survival. Further research fishing is planned in both the north-east and north-west Atlantic. The application of tracking and telemetry technology also promises to be a valuable tool for increasing our understanding of this stage of the salmon's life cycle. Archival or data storage tags (DSTs) could provide information on geographical position, behaviour of salmon at sea and marine environmental conditions if they can be sufficiently miniaturised. In the meantime, pilot studies involving DSTs applied to kelts (as substitutes for smolts) are continuing. The techniques for the live capture of post-smolts during their first summer at sea are now well established and DSTs might also be used to study the movement of these fish, and the environmental conditions they experience, following release. These tags require a procedure for their recapture, which is a concern given recent reductions in exploitation in the North Atlantic. Information is obtained only for those fish that survive and are recaptured, whereas the most valuable information would be derived from those fish that have died. The use of miniaturised radio tags fitted with release mechanisms ('pop-up tags') set to allow the tag to surface at pre-set times and transmit stored information to satellites, and collaboration with the military with a view to the utilisation of existing acoustic systems for salmon tracking, might also be considered.

Links between ocean environment and abundance of salmon stocks

There is now considerable evidence that changes in the survival, growth and sea-age at maturation of salmon are significantly correlated with changes in sea surface temperature. Research in Canada suggests that marine survival is influenced by the sea surface temperature in the nearshore environment through which post-smolt and adult salmon pass on their migration from, and return to, rivers. Furthermore, the pre-fishery abundance of North American origin (potential two-sea-winter) salmon has been shown to be correlated with a temperature-based ocean habitat index (using data for the month of February) and is one of the few examples of the use of environmental information to predict abundance for management purposes. This index has declined since the 1970s although it has increased since 1996, reaching a value in 1998 as high as those experienced in the 1970s. Despite the increase in thermal habitat index in recent years, the pre-fishery abundance of North American origin salmon has remained low. The breakdown of this correlation would have implications for the present management system but it may also offer an opportunity to obtain a clearer understanding of the mechanisms underlying the relationship.

Quantitative thermal habitat concepts have also been applied to European salmon, although they have not, so far, been used to provide management advice. The abundance of European salmon estimated from catch data has been shown to be positively correlated with changes in the ocean area enclosed by the 7°C and 13°C isotherms in the spring.

It has been concluded from the literature relating growth, maturity and survival of salmon to climatic factors that the increased mortality of salmon in recent years may, in part, be climate-related. However, the underlying mechanisms linking ocean climate to survival remain unclear.

Climate-related changes in the ocean environment

The large-scale atmospheric pressure patterns over the North Atlantic Ocean are dominated by the Icelandic Low and the Azores High. The strength of these systems varies from year to year with a tendency for both systems to intensify (or weaken) in the same year. This tendency is known as the North Atlantic Oscillation (NAO). The difference in the sea surface pressure at the Azores and that at Iceland is known as the NAO index. A high index corresponds to a deep Icelandic low and an intense Azores high. While the NAO index is subject to annual and decadal change, the 1960s were characterised by an extreme negative phase of the NAO index, while the late 1980s and early 1990s experienced the most prolonged and extreme positive phase of the index.

Associated with these changes is a wide range of physical and biological responses in the North Atlantic, including effects on wind speed, ocean circulation, sea surface temperature, prevalence and intensity of Atlantic storms and changes in production of zooplankton. For example, during years of positive NAO index, the eastern and western North Atlantic display increases in temperature while temperatures in the central North Atlantic and in the Labrador Basin decline. The extremely low temperatures in the Labrador Sea during the 1980s and early 1990s coincided with marked changes in the fish stocks in the area, including the decline in abundance of Atlantic salmon. Similarly, in Europe, years of low NAO index were associated with high catches, while stocks have declined dramatically during the high index years. Periods of high NAO index also result in stronger winds over most of the central and eastern North Atlantic during winter months.

There is evidence that the climatic events in the North Atlantic region may, through teleconnections, be linked to the El Niño Southern Oscillation (ENSO). This is of particular interest because it has been shown that changes in the size and distribution of Pacific salmon stocks are associated with important changes linked to El Niño and possibly climate change. There could, therefore, be benefits from closer collaboration between scientists working on Atlantic and Pacific salmon.

Impacts of long-term climate change

The North Atlantic is one of the most climatically sensitive regions of the world and given the available information relating growth, age at maturity and survival of salmon to climatic factors, there is inevitably concern about the possible effects of enhanced greenhouse warming on Atlantic salmon. Models of the ocean–atmosphere climate system based on a 1% increase per annum in greenhouse gases predict a global temperature increase of 1–2°C by the middle of the next century. However, most models suggest that the temperature change over the North Atlantic will be less than elsewhere in the world. It is predicted that there will be an increase in precipitation of about 5% for a 2.5°C increase in temperature and that this will contribute to a slowing of thermohaline circulation by up to 30% by 2050. On land, temperatures around the North Atlantic are predicted to warm more in winter than summer, with increased precipitation in the winter but a reduction in summer. While there are considerable uncertainties in the climate models, the predicted changes in temperatures, rainfall and ocean circulation could have major implications for the Atlantic salmon. For example, studies in the Girnock Burn, Aberdeenshire, Scotland, have shown that the lack of snow melt water in recent years has led to elevated water temperatures in the spring and as a result earlier migration of smolts into the main river. This behaviour could have consequences for subsequent survival by, for example, increasing exposure to avian predators. Future warming could lead to an extension of salmon-bearing rivers into the Arctic but a loss of populations at the southern limit of the range in Spain, France and the United States. There may also be an increase in productive sub-polar waters in which salmon could feed, although changes in coastal waters might make it harder for post-smolts to reach these feeding grounds. Salmon have had to cope with major changes of climate in the past, and it has been speculated that they will be able to adjust to the coming climate change provided that management is also able to adapt to the new conditions.

Possible causal factors for the increased marine mortality

The physical environment of the ocean (temperature, salinity, mixing and transport processes) can affect the growth, maturation, distribution and abundance of fish populations through four main processes, direct physiological effects, feeding, diseases and predators.

Physiological effects

Evidence from research fishing in the north-west Atlantic indicates that salmon concentrate in waters in the temperature range 4–8°C, and any change in ocean temperature could therefore lead to changes in distribution or to salmon experiencing sub-optimal conditions with possible effects on growth and survival. In the 1980s, it was hypothesised that in periods when there were warmer conditions in the sub-arctic, salmon would migrate further north away from the more productive frontal regions, consuming more energy in the process, and that this would lead to a higher proportion of fish returning to home waters as multi-sea-winter salmon. Conversely, colder sub-arctic temperatures would lead to an increased proportion of fish maturing as one-sea-winter salmon. The increase in the proportion of North American origin salmon maturing after one sea winter and recent changes in ocean temperature associated with the high NAO index are consistent with this hypothesis. However, other studies have failed to find a link between sea age at maturity and sea surface temperature.

Feeding

Studies have shown that the size of the growth increment in the first year at sea is positively correlated with the survival rate of the North Esk smolts. Similarly, growth and survival have been shown to be linked for other pelagic fish species. However, limited knowledge exists of the factors influencing growth and survival of post-smolts. One obvious factor is the availability of adequate feeding. Density-dependent mortality due to food limitation has been reported for Pacific salmon, but no such relationship has been demonstrated for salmon in the north-east Atlantic.

Information from the Continuous Plankton Recorder (CPR) surveys indicates that over the last 50 years the plankton of the northern North Atlantic and the north-west European shelf seas has shown substantial and regionally widespread changes. Because the CPR surveys are conducted offshore in waters regarded as being relatively pristine, it is believed that the major cause of the observed variation in the plankton is environmentally induced. These changes are correlated with the NAO index. *Calanus finmarchicus*, a good indicator of zooplankton as a resource for higher trophic levels, has been shown to be inversely and highly significantly correlated with the NAO index.

Studies in Icelandic coastal waters indicated that the feeding intensity of post-smolts released from ranching stations was initially very low, although this increased with increasing distance (and time) from release and was not thought to affect survival. Recent research fishing at sea has shown that growth in the early weeks at sea is rapid, at least in the southern part of the range. The fish captured were in good condition and there was no evidence of post-smolt starvation.

During the period 1992–1995, a major study was undertaken of the diet of approximately 4000 salmon caught by long-lines to the north of the Faroes. The results of this study confirm the previously expressed view that salmon forage opportunistically, but that they demonstrate a preference for fish rather than crustaceans when both are available. Salmon were also found to be selective when feeding on crustaceans, preferring hyperiid amphipods to euphausiids. Feeding intensity and the feeding rate of salmon north of the Faroes have been shown to be lower in the autumn (November–December) than in the spring (February–March), which might suggest that limited food is available at this time of year. Similar results have been found for salmon in the Labrador Sea and in the Baltic. Salmon in the North Atlantic also seemed to rely on amphipods in the diet in the autumn, whereas at other times fish dominate the diet. The higher sea surface temperatures during the autumn, combined with the apparent reduction in abundance of fish prey, might result in suboptimal growth and, in extreme conditions, starvation. This finding is of considerable interest, given the report of late summer growth checks on the scales of salmon returning to Scottish rivers, and sampling of post-smolts at the end of the first summer at sea is a priority. A recent hypothesis developed in relation to natural regulation of coho salmon in the Pacific has proposed that mortality occurs at two stages, immediately after entry to the sea, when mortality is predation-based, and in the autumn and winter of the first year at sea, when juveniles that are not of a critical size are unable to maintain minimum metabolic requirements, resulting in mortality.

Predation

Predation on smolts by cod, pollack, seals and avian predators can be high and may be increasing. Studies in Canada have shown that while salmon were rarely recorded in the diet of gannets in the late 1970s and 1980s, they increased in the 1990s when the birds' diet shifted from warm water species such as mackerel to cold water species such as capelin and salmon. Shifts in diet, to include salmon, have also been observed in herring gulls and greater black-backed gulls in Newfoundland following closure of the fisheries for ground fish (and associated processing plants) and because of delays in the arrival of spawning capelin. Research suggests that seal predation may be a significant factor in the decline of large spring-running Pacific salmon, and it clearly deserves further study in relation to Atlantic salmon in view of the rapid growth of seal populations in the North Atlantic. Concern was expressed that predation on returning adult salmon might be particularly high during dry summers. In wet summers, such as 1998 in the United Kingdom, there can be an apparent 'spontaneous generation' of salmon which may be due to ease of river entry.

Other factors

Research fishing in the north-east Atlantic has revealed the occurrence of cataracts in wild post-smolts. It was speculated that these may be linked to water quality problems, but further studies are needed to increase understanding of their cause and possible significance. There is also concern that sea lice from fish farms may be adversely affecting smolt survival in areas where migration routes pass in close proximity to sea cages. Preliminary results from tracking studies in the Bay of Fundy, the centre of the Canadian east coast aquaculture industry, have revealed differences in the survival of smolts leaving two different river systems and following different migratory routes out of the Bay. Those tagged smolts that migrated past aquaculture cages experienced higher levels of mortality than those that migrated directly out of the Bay. However, further studies are needed to identify the factors influencing this differential mortality.

Concern was also expressed that there may be a loss of salmon at sea due to their inability to successfully navigate back to their parent river or due to straying.

Future research

The workshop served to highlight the very rapid progress that is now being made in understanding the problems facing salmon in the sea, particularly with regard to the relationship between the physical environment and trends in salmon abundance. The challenge ahead is to identify factors responsible for the observed reduction in marine survival and to improve our understanding of how these are linked to changes in the physical environment. A number of recommendations for future research were identified in the workshop papers and the discussions that followed their presentation. These included the following:

- The available data sets should be re-examined to evaluate the 'inverse weight hypothesis' and to determine whether current estimates of natural mortality of adult salmon in the sea are appropriate.
- Studies on monitored rivers that provide long-term survival and productivity data should be continued and expanded as resources permit. Particular effort should be made to develop monitoring programmes in rivers with multi-sea-winter stocks. Studies on wild smolts would be desirable but there are considerable financial implications. Data on freshwater survival should be obtained so as to facilitate a total life cycle approach.
- Studies of growth patterns in historical scale collections should be conducted in order to test hypotheses about the relationship between growth and ocean and riverine temperatures.
- The geographical coverage of research fishing should be extended as resources permit. Computer simulations, based on existing knowledge of currents and rates of migration, could be used to assist in the planning of future research fishing.
- Given the presence of late summer growth checks on the scales of some returning adult salmon, research fishing for post-smolt salmon later in the year, at the end of the first summer at sea, is desirable.
- Systematic sampling of the catch in fisheries for pelagic species, where these overlap spatially and temporally with the distribution of post-smolts, should be considered.
- Further studies of feeding and metabolic state of salmon at sea and studies to identify the predators of salmon at sea are desirable.
- Further studies should be conducted using data storage tags on adult salmon, kelts and post-smolts and, when feasible, smolts.
- Collaboration should be encouraged between researchers involved in tracking

Atlantic salmon with regard to aspects such as the costs and benefits of different approaches and the possible standardisation of identification codes.
- Closer collaboration between biologists working on Pacific salmon and Atlantic salmon should be encouraged.

In view of the costs involved in studies of salmon at sea future research will need to be carefully planned so as to determine the most practicable and appropriate use of the funds available. There are likely to be benefits from international co-operation between salmon biologists, oceanographers, planktologists and climatologists. The workshop represented the first step in this process.

Chapter 3
A Perspective on the Marine Survival of Atlantic Salmon

E.C.E. POTTER and W.W. CROZIER

Introduction

Recently, great attention has been given to the apparent decline of wild Atlantic salmon (*Salmo salar* L.) stocks throughout much of the North Atlantic range of the species. Such trends, if real and sustained, are of major concern, because of the economic dependence of commercial and recreational fisheries interests on the resource and the potential longer-term effects on bio-diversity. We hardly need reminding that bio-diversity is more easily lost than restored, as it is not simply a matter of numerically replacing the fish that have disappeared.

In view of the widespread nature of the decline in stocks, scientists have been quick to blame changes in the marine environment for these trends (ICES 1998). Various studies have demonstrated correlations between environmental parameters and stock numbers or survival rates (Martin & Mitchell 1985, Friedland *et al.* 1993), although none has shown a clear causal relationship. Accordingly, less emphasis has been given to the possibilities that problems in fresh water may be implicated, or that stocks may have been affected by a range of adverse factors across both environments. It is, however, essential that we have a clear understanding of all the factors that may be contributing to these declines if managers are to introduce appropriate stock protection or conservation measures.

The aims of this chapter are therefore to review the existing scientific evidence for declines in stock status of wild Atlantic salmon (especially in the north-east Atlantic area) and the part reduction in marine survival has played in this, and from this to conclude what (if any) additional monitoring is needed to support continued conservation and management activity. The chapter will focus on the broad qualitative/ quantitative evidence for declines in stock status. It will largely leave recommendations for specific research into some of the possible causes for the decline to subsequent chapters dealing with specialist areas.

Methods of estimating the natural mortality of salmon are first described, and then changes in the estimated abundance of North American and European salmon stocks derived from generalised modelling methods are considered and data on marine survival in some monitored river stocks in the north-east Atlantic are examined to determine whether this might account for the major changes that have been observed in stock abundance. Finally the evidence from one of these monitored stocks, the River Bush (UK), is considered in order to illustrate the possible role of freshwater factors. This information will be used to determine the future requirements for monitoring to support scientific advice to managers.

Marine survival and natural mortality

Marine survival for salmon is commonly expressed as the proportion of emigrating smolts that return to homewaters or to their river of origin as one-sea-winter (1SW) or two-sea-winter (2SW) adults (ICES 1998). Strictly speaking, it is not possible to measure the survival of the different sea-age groups in this way because changes in these 'return rates' reflect the effects of both mortality and maturation. Changes in the age at maturation may affect the relative proportions of fish from the same smolt year class returning as 1SW or 2SW salmon, but this may also be caused by changes in natural mortality in different areas of the ocean.

Survival (S) over a time period (t) is also expressed as:

$$S = e^{-Zt}$$

where $Z = M + F$
and M = instantaneous rate of natural mortality in unit time; and
 F = instantaneous rate of fishing mortality in unit time.

Ricker (1976) described various methods for estimating total mortality rates of salmon during the marine phase. Many of these depend upon simple marking experiments to estimate the numbers of fish alive at different life stages. These approaches may also be used to estimate the numbers of fish from a particular stock taken in mixed stock fisheries and thus to permit differentiation of the total mortality into that caused by fisheries and that which may be attributed to natural causes. Determining how the natural mortality changes during the marine phase is more complex. For many marine species, the estimates of natural mortality that are used in the assessments conducted for management purposes have little basis in observed data, but attempts have been made to investigate this in more detail for Atlantic salmon (Doubleday *et al.* 1979).

The above equation suggests that 20% survival from smolt to 1SW should equate to an average monthly mortality rate of about 0.1 for 16 months (equivalent to losses of about 10% per month). However, it is reasonable to assume that the natural mortality of salmon does not remain constant throughout the marine phase of the life cycle. If, for example, natural mortality in the sea is mainly caused by predation, it may be expected to decline as the fish grow. It may therefore be appropriate to apply the 'inverse weight hypothesis' proposed for coho salmon (*Oncorhynchus kisutch*) by Matthews & Buckley (1976). Doubleday *et al.* (1979) used an inverse weight model to obtain estimates of the natural mortality of salmon from the River Bush, Northern Ireland, and the Sandhill River, Labrador. Further estimates were subsequently calculated for the River Bush assuming no growth after the first sea winter (ICES

1986) and for salmon from the North Esk, Scotland (Shearer 1984). The results of these studies are given in Table 3.1. If the model is correct, it implies that over 90% of the marine mortality occurs in the first 6–8 months in the sea.

Table 3.1 Estimates of natural mortality at sea from 14 to 24 months for Atlantic salmon from Sandhill River, Canada; River Bush, Ireland; and Nork Esk, Scotland.

River	Method*	Estimated percentage mortality 14–24 months	Estimated monthly instantaneous mortality rate	Authors
Sandhill	1	10–14%	0.010–0.015	Doubleday et al. (1979)
Bush	1	2.3–3.4%	0.002–0.003	Doubleday et al. (1979)
Bush	2	5.9–8.6%	0.006–0.009	Doubleday et al. (1979)
North Esk	1	6.2–13.6%	0.006–0.015	Shearer (1984)
North Esk	2	1.6–3.7%	0.002–0.004	Shearer (1984)

* 1 = inverse weight method
2 = inverse weight method with constant M after one sea winter.

On the basis of these data, it has generally been accepted that the natural mortality of Atlantic salmon after the first sea winter is likely to be low, and a monthly mortality rate of 0.01 (about 1%) (or a range of 0.005–0.015) has been widely used in the models developed to assess salmon abundance and exploitation rates (e.g. Potter & Dunkley 1993, Rago et al. 1993a, Potter et al. 1998). However, it must be noted that both Doubleday et al. (1979) and ICES (1986) observed that owing to the lack of data describing the abundance of salmon between smolt emigration and adult return, the inverse weight hypothesis had not been proved.

The use of these values has reinforced the widely held view that the majority of the natural mortality of salmon in the sea occurs in the first few months, and similar assumptions have been made for Pacific salmon. For example, Furnell & Brett (1986) concluded that 90% of the mortality of sockeye salmon (*Oncorhynchus nerka*) occurred during the first 4 months in the sea. However, it must be emphasised that this is based almost entirely upon the assumption that the inverse weight hypothesis is valid. This has also led to the belief that variation in marine survival is most likely to be the result of factors affecting mortality in the early months in the sea. There is again little firm evidence to support this, although it is a logical conclusion because (assuming the inverse weight hypothesis is valid) a similar percentage variation in the level of mortality will have a greater effect when the rate is high (i.e. when the fish are small). Nevertheless, while it may not make a major difference to our overall understanding of the status of salmon stocks, we should not lose sight of the possibility that natural mortality rates might alter as the fish return from oceanic areas to the continental shelf and as they migrate through coastal and estuarine waters. In addition, conditions delaying the entry of salmon into fresh water, such as low flows, might result in increased mortality in coastal waters, from natural causes or fishing.

Similarly, the fact that unexpectedly good runs of salmon are sometimes observed when river conditions are particularly favourable suggests that mortality may be reduced when fish can move rapidly into fresh water.

Assessment models

The models currently used to estimate the pre-fishery abundance of North American and European salmon (Rago *et al.* 1993b, Potter *et al.* 1998) and levels of exploitation in the marine salmon fisheries (e.g. Potter & Dunkley 1993, Rago *et al.* 1993a) all depend upon a run-reconstruction approach, employing an assumed level of mortality after the first sea winter of about 0.1. The models back-calculate the stock size at points earlier in the life cycle from the numbers of fish returning to fresh water. In order to identify fish from particular river stocks in mixed stock fisheries, samples of emigrating smolts are marked, for example by microtagging. Back-calculation approaches have been developed in preference to forward-running models because of the belief that mortality after the first year in the sea is quite low and relatively stable. Thus more reliable estimates can be obtained of the numbers of adult fish alive at the time of the fisheries (pre-fishery abundance) and consequently of exploitation rates. Contingent upon these assumptions, the back-calculation models also allow for the separate estimation of the recruitment to the marine fisheries (as immature fish) of salmon that would ultimately return as mature 1SW or 2SW fish and the survival rate of the total stock from smolt to this recruitment stage.

These models have provided much of the information which is now fundamental to our current thinking about the changing status of Atlantic salmon stocks over the past quarter century. Thus Figs 3.1 and 3.2 show the estimated pre-fishery abundance

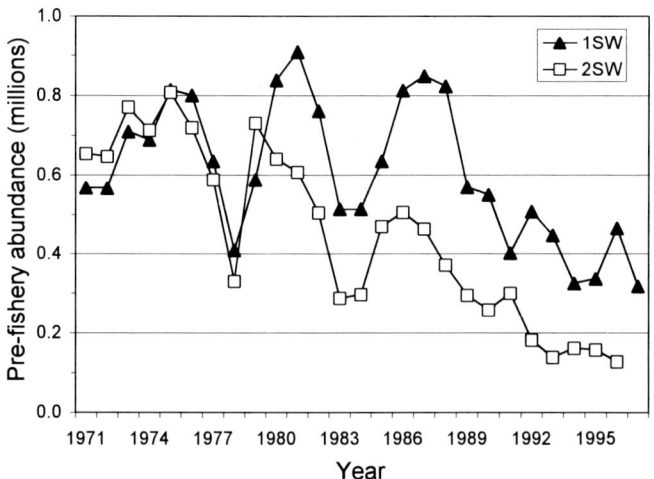

Fig. 3.1 Pre-fishery abundance of potential 1SW and 2SW salmon from North America estimated as numbers of immature salmon in the sea in 1971–97.

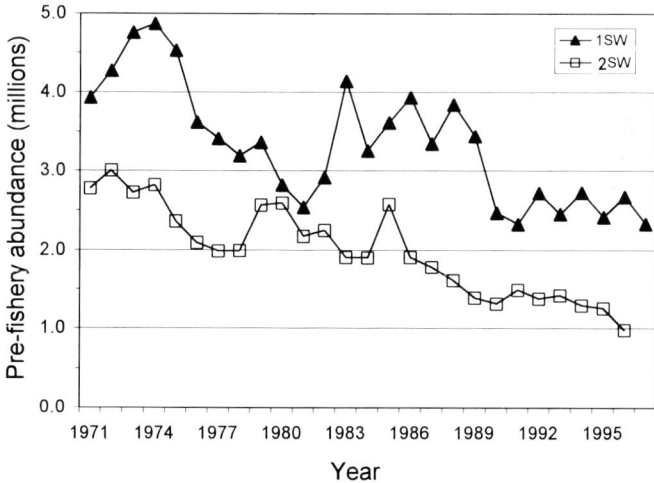

Fig. 3.2 Pre-fishery abundance of potential 1SW and 2SW salmon in the north-east Atlantic estimated as numbers of immature salmon in the sea in 1971–97.

of salmon originating from North American and North East Atlantic countries (ICES 1998). These figures both show significant declines, particularly in the multi-sea-winter (MSW) components of the stocks, despite significant reductions in the exploitation of MSW salmon in the distant water and home water fisheries.

Changes in marine survival

In order to assess changes in natural mortality in the sea we need to exclude as far as possible the effects of exploitation from the survival estimates. Many salmon stocks are exploited, in some cases quite heavily, in coastal waters and estuaries, and the assessments of marine survival in this chapter are therefore based mainly upon the estimated returns to home waters prior to these fisheries. However, it must be noted that no account has been taken of exploitation in the distant water fisheries, which will be small on most 1SW stocks but may have been significant on some MSW stocks in some earlier years before the substantial reductions in the quotas and the implementation of compensation arrangements (ICES 1998).

North American stocks

Data from a number of river stocks in Canada and the USA show a decline in the return rates as 1SW and 2SW salmon over the past 30 years (ICES 1998). Data for Western Arm Brook, for example, are shown in Chapter 8. In recent years, return rates have remained low relative to historical levels, even in areas where commercial exploitation has been reduced or banned (ICES 1998). This has been taken to indicate that the decline in stock status is largely due to reduced survival of salmon at sea, rather than over-exploitation. Canadian scientists have indicated that they believe that, with the exception of some acidification losses in east Nova Scotia, freshwater production rates are being maintained (ICES 1998). Generally, juvenile densities and smolt production have been normal or high in areas of good spawning deposition and have fallen to unsatisfactory levels only in areas where targets are not being met. However, data on freshwater survival trends, rather than just numbers and densities, are needed to fully assess whether freshwater factors may be affecting stock status.

Studies of North American stocks have demonstrated a correlation between the numbers of fish recruiting to the fisheries and a temperature-based oceanic habitat index (Reddin *et al.* 1993). The causal mechanisms for this relationship are not known, but it has led scientists to speculate that poor marine habitat conditions may have been limiting the survival of wild salmon in the sea in recent years, possibly through poor feeding and lower growth rates. Indeed, habitat indices are being used to predict numbers of salmon available for harvest in West Greenland and Canadian fisheries (e.g. ICES 1998). However, a recent upturn in the habitat index to the higher values of the 1970s has not yet led to an improvement in salmon marine survival, and is prompting more intensive examination of other potential contributory reasons for declines in stock status.

North-east Atlantic stocks

Most data on the status of stocks in Europe still come from a small number of monitored rivers, which provide information on smolt production, exploitation rates, adult returns and spawning escapement (ICES 1998). There are few long-term data sets, but results from the North Esk (Scotland) suggest that there was a marked reduction in return rates between the 1970s and the 1980s (Chapter 8). Shorter time series are available from nine European rivers and these data have been analysed by ICES (1998), who failed to find significant overall trends (based on 5- and 10-year periods) in return rates of wild smolts to 1SW or 2SW salmon or in production from fresh water. This is despite the fact that there had been a relatively sudden decline in catches and freshwater escapement in 1990 and 1991, believed to be associated with poor survival of the 1989 smolt cohort (ICES 1992). This presents a confusing picture, as it leaves open the possibility, for example, that declines in marine survival in some areas or in some monitored stocks are being offset by relatively good marine survival in other areas or among other monitored stocks. Such patterns would not have been picked up in an overall analysis. To address this point, we have carried out a simple re-analysis of marine survival among the monitored river stocks in the north-east Atlantic area.

Data on the return rates for the nine monitored stocks have been re-examined (Table 3.2 and Figs 3.3 and 3.4)(ICES 1998) in an attempt to answer the following simple questions:

Table 3.2 Estimated survival of wild smolts (%) to return to home waters (prior to coastal fisheries) for various monitored rivers in the NE Atlantic area. (From ICES 1998)

Smolt migration year	Iceland[1] Ellidaar	Iceland[1] R. Vesturdalsa[4]		Norway[2] R. Imsa		Ireland River Corrib		UK (N. Ireland)[3] R. Bush	UK (Scotland)[2] North Esk		France Oir[5]	France Nivelle[6]	France Bresle
	1SW	1SW	2SW	1SW	2SW	1SW	2SW	1SW[7]	1SW	2SW	All ages	All ages	All ages
1980	—	—	—	—	—	9.4	1.6	—	—	—	—	—	—
1981	—	—	—	17.3	4.0	11.8	3.8	—	—	—	—	—	—
1982	—	—	—	5.3	1.2	15.6	2.7	—	13.7	6.9	—	—	—
1983	—	2.0	—	13.5	1.3	10.6	1.2	—	12.6	5.4	3.2	—	8.5
1984	—	—	—	12.1	1.8	19.8	1.7	—	—	—	7.7	—	16.3
1985	9.4	—	—	10.2	2.1	15.4	1.4	—	10.0	4.1	7.5	—	12.2
1986	—	—	—	3.8	4.2	—	—	31.3	26.1	6.4	3.9	15.8	19.4
1987	—	—	—	17.3	5.6	12.0	1.0	35.1	—	—	9.3	2.6	—
1988	12.7	—	—	13.3	1.1	12.4	0.5	36.2	13.9	3.4	2.3	2.4	—
1989	8.1	1.1	2.0	8.7	2.2	5.3	1.0	25.0	—	—	2.4	3.5	—
1990	5.4	1.0	1.0	3.0	1.3	4.4	0.6	34.7	7.8	4.9	6.1	1.8	—
1991	8.8	4.2	0.6	8.7	1.2	5.6	0.1	27.8	7.3	3.1	13.2	9.2	—
1992	9.6	2.4	0.8	6.7	0.9	5.9	—	29.0	11.2	4.5	4.4*	8.9	6.9*
1993	9.8	—	—	15.6	—	9.0	0.2	—	—	—	8.3*	8.3*	10.3*
1994	9.0	—	—	—	—	7.8	0.1	27.1	17.2	2.3	3.7	7.1*	7.5*
1995	9.4	1.6	1.2	1.8	1.5	6.7	—	—	11.5	5.1	—	1.8	—
1996	4.6	1.4	n/a	3.4	n/a	4.1	n/a	31.0	10.7	n/a	n/a	n/a	n/a
Mean 1992–1996	8.5	1.8	1.0	6.9	1.2	6.7	0.2	29.0	13.1	3.7	3.7	5.4	8.2

[1] Microtags.
[2] Carlin tags, not corrected for tagging mortality.
[3] Assumes 30% exploitation in trap fishery.
[4] Assumes 50% exploitation in rod fishery.
[5] Minimum estimates.
[6] From 0+ stage in autumn.
[7] Microtags, corrected for tagging mortality.
* Incomplete returns.

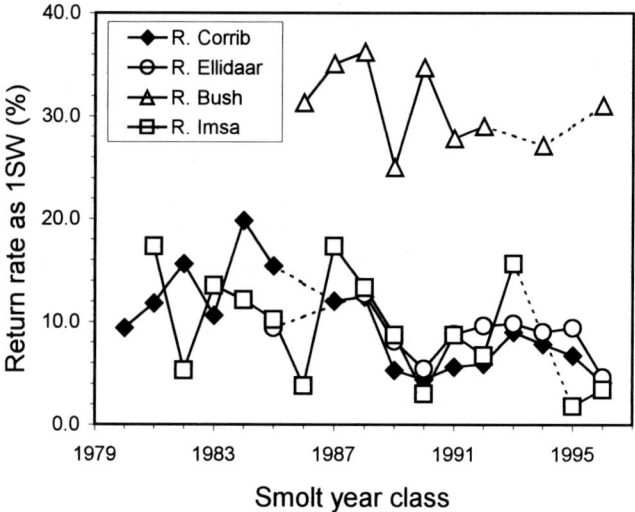

Fig. 3.3 Rates of return as 1SW salmon to coastal waters for four monitored stocks in the north-east Atlantic that showed a decline in survival after 1989. (Dashed lines indicate missing points.)

- Could a decrease in marine survival be responsible for the substantial reduction in catches in the north-east Atlantic after 1990?
- Do marine survival trends operate differently in different stocks/areas?
- Do marine survival trends operate differently between different components of the same stock (i.e. 1SW/MSW)?

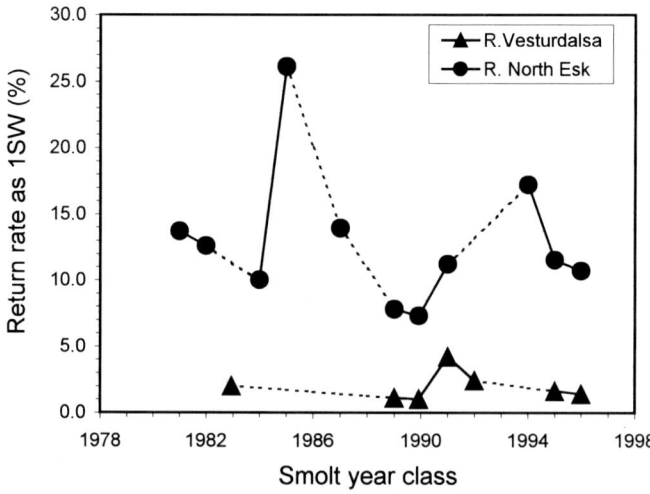

Fig. 3.4 Rates of return as 1SW salmon to coastal waters for two monitored stocks in the north-east Atlantic that showed no trend in survival after 1989. (Dashed lines indicate missing points.)

A non-parametric randomisation test (Rago 1993) has been used to determine whether there has been a significant change in the return rates for smolt cohorts before 1989 (the baseline period) and from 1989 onwards (the treatment period). For four of the six stocks return rates to home waters as 1SW salmon remained lower after the 1989 event compared with the period before it (Ellidar, Iceland, R_{crit}, $P < 0.05$; Corrib, Ireland, R_{crit}, $P < 0.001$; Bush, N. Ireland, R_{crit}, $P < 0.05$; Imsa, Norway, R_{crit}, $P < 0.05$), while for two stocks there was no significant difference in the return rates as 1SW fish between the periods before and after the 1989 event (Vesturdalsa, Iceland, R_{crit}, $P > 0.1$; North Esk, Scotland, R_{crit}, $P > 0.05$).

These results suggest that, for the stocks examined, a reduction in marine survival at the end of the 1980s started a period of poorer survival at sea that has been maintained in some, but not all, stocks. Further evidence of this is available from three stocks in France (Fig. 3.5), where return rates (all sea ages combined, but thought to be > 50% 1SW) suddenly declined in the late 1980s. The randomisation test indicates that return rates for the Oir (R_{crit}, $P > 0.5$) and Nivelle (R_{crit}, $P > 0.1$) stocks recovered after 1989, while that for the Bresle remained lower after 1989 than before (R_{crit}, $P < 0.05$). The marked decline in return rates for the 1989 smolt cohort is consistent with the sharp drop in the overall stock abundance estimates for north-east Atlantic 1SW fish in the sea in 1990 (Fig. 3.2).

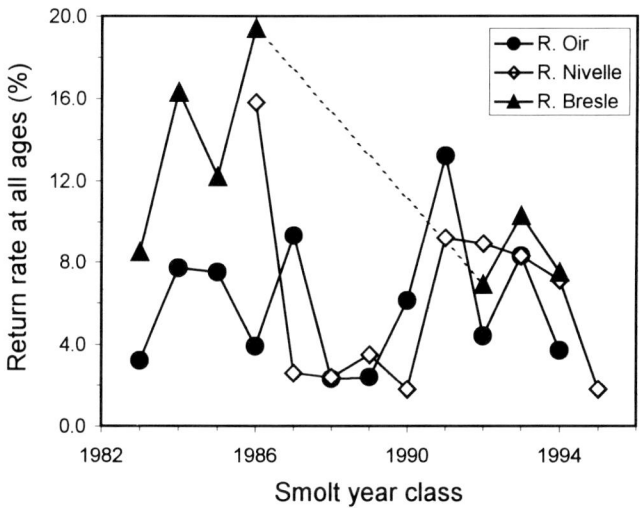

Fig. 3.5 Rates of return of salmon (all ages) to coastal waters for three French monitored stocks. (Dashed lines indicate missing points.)

Return rates as 2SW fish (Fig. 3.6) are available for only four monitored stocks and there are virtually no data on older age groups for any river. A more continuous decline in survival is evident for the 2SW salmon compared with the 1SW fish. The 1989 smolt year class does not appear to have suffered particularly poor survival,

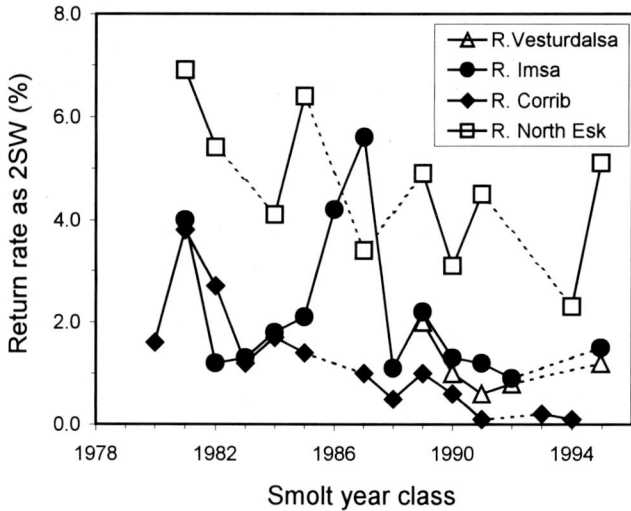

Fig. 3.6 Rates of return as 2SW salmon to coastal waters for four monitored stocks in the north-east Atlantic. (Dashed lines indicate missing points.)

indeed for that cohort return rates increased in all cases compared with the previous year. This is supported by the results of the randomisation test analysis, which indicates that the return rates for the North Esk (R_{crit}, $P > 0.05$) and Imsa (R_{crit}, $P > 0.05$) stocks were not significantly different before and after the 1989 event. However, survival in the Corrib stock was higher before 1989, possibly because of two particularly high values in 1981–1982 (R_{crit}, $P < 0.01$). The data from the Vesturdalsa only begin at 1989 and so could not be statistically tested, but for that stock 1989 had the highest return rate in the short time series available. These results suggest that the 1989 depression in marine survival was largely a grilse event and are consistent with the estimates of pre-fishery abundance which also show a longer-term declining trend for MSW salmon than is the case for 1SW fish (Fig. 3.2).

The apparent differing marine survival trends noted here between 1SW and MSW stock components are of interest because they may provide insight into the factors affecting marine survival. For example, if maturation was tending to occur at an earlier sea-age, this could cause a decline in returns as MSW fish, while grilse returns would remain stable or increase. Alternatively, a positive correlation between the return rates of 1SW and MSW salmon from the same smolt cohorts might indicate that factors affecting the fish relatively early in the marine phase had the greatest effect on marine survival or that there was little relationship between survival and age at maturity. Regressions of the proportion of smolts returning as 2SW on the proportion returning as 1SW for four of the stocks indicate that there is no consistent relationship between return rates of different sea-age groups within individual stocks. Non-significant positive trends were found in three stocks examined (N. Esk,

$r = 0.313$, $P > 0.1$; Corrib, $r = 0.503$, $P > 0.05$; Imsa, $r = 0.440$, $P > 0.1$) and a non-significant negative trend in the fourth (Vesturdalsa, $r = -0.693$, $P > 0.1$).

The conclusions from this analysis are that marine survival of salmon in the north-east Atlantic has generally been lower in the 1990s than in the 1980s. For grilse stocks, this appears to have begun with a widespread and sudden decline in marine survival during a short period at the end of the 1980s. Marine survival among grilse stocks has remained low for some but not all stocks. Marine survival of 2SW stock components (based on a very limited number of rivers examined here) has been lower in the 1990s than previously, but this appears to reflect a more gradual decline rather than a sudden event. The pre-fishery abundance analysis suggests that this decline may have been continuing for at least 25 years (Fig. 3.2).

The tentative conclusion from the analysis of relative marine survival to home waters of fish of differing sea ages is that return rates as between 1SW and 2SW fish do not appear to be statistically related within individual stocks. However, Hansen *et al.* (1995) have reported significant positive relationships between recapture rates (total return rates were not available) of returning tagged 1SW and 2SW adults from both the North Esk and the River Figgjo (Norway). In these cases it appears that the dominant factors affecting survival in the sea have operated relatively early in the marine phase. Clearly, more analyses of 1SW/MSW marine survival relationships across a range of stocks are required.

Relative importance of freshwater and marine survival

The analysis of data from monitored rivers in the north-east Atlantic has also failed to assess the effects of changes in freshwater production on the status of salmon stocks (ICES 1998). This is because these analyses have been based only on changes in the abundance and densities of juvenile salmon rather than on information on freshwater survival. We have therefore examined results from the long term monitoring of the wild salmon population in the River Bush as an example of a river stock for which we have both freshwater and marine survival indices.

Marine survival of River Bush salmon

As noted earlier, the return rates to coastal waters as 1SW fish for wild River Bush smolts have generally been lower since 1989 than in earlier years (Fig. 3.3). Unfortunately, no time series of data are available for returns of wild 2SW fish to coastal waters, as numbers of tag returns have been too small for reliable estimation. The 1SW/2SW ratio among wild adults returning to the River Bush from each smolt release has not shown any long-term trend with time (Fig. 3.7; $r = -0.204$; $P > 0.1$), in contrast to reports of changing grilse/salmon ratios elsewhere, where MSW components of population have been declining (ICES 1998)

From these data, it is possible to conclude that natural marine survival of wild River Bush 1SW fish has been lower on average since the 1989 downturn, but is not

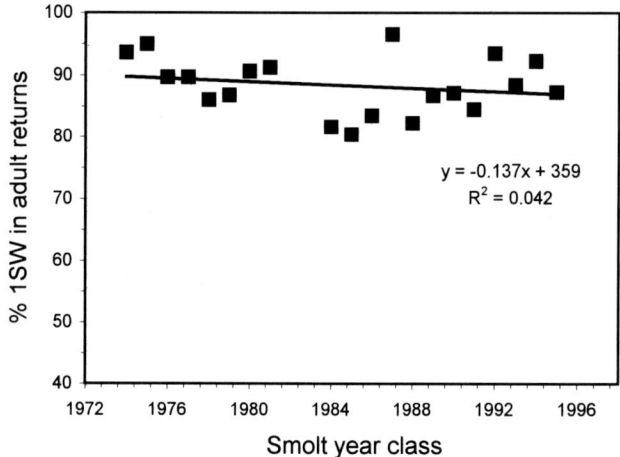

Fig. 3.7 Proportion of surviving adult salmon that returned as 1SW fish from each smolt cohort, 1973–95.

showing a significant downward trend and is again approaching levels typical before 1989. Returns to fresh water have remained high, especially in recent years, which have included some of the highest returns in the time series. Indeed, survival of 1SW fish to fresh water in 1997 (12.1%) was higher than the average for the previous 10 year period (11.0%), while returns of 2SW fish in 1997 (2.4%) were the highest in the time series to date and considerably higher than the average for the previous 10-year period (1.4%). This is heavily dependent upon changes in the relatively high levels of exploitation in coastal waters which has an overriding influence on realised returns into fresh water; exploitation in coastal waters has been decreasing in recent years, averaging 70% during the period 1987–91 and 56% since 1992.

In contrast to the wild stocks, returns of ranched River Bush fish (of two smolt ages) to home waters have been very variable, ranging from 1.1% (1993) to 15.4% (1985) for 1+ smolts and from 2.3% (1995) to 23.2% (1989) for 2+ smolts, with lower return rates generally being observed in recent years. The differing pattern of returns of hatchery fish, both on a single year basis (when wild returns can increase and hatchery returns decrease) and over a long time series, have previously been described by Crozier & Kennedy (1993). Contributing factors are believed to include variable condition at release of hatchery smolts, method and timing of releases and river conditions after release. In addition, returns to fresh water may be affected by differing migratory behaviour of wild and hatchery fish, leading to different levels of exploitation on returning adults. A lack of correlation of returns to fresh water of wild and hatchery fish for both 1SW and 2SW fish (Crozier & Kennedy 1993) contrasts with results reported from the Rivers Burrishoole and Imsa (ICES 1992) and suggests that more research is needed into the extent to which hatchery fish can be used as indicators of the marine survival of wild salmon.

Freshwater survival of River Bush salmon

Data on survival from the estimated egg deposition to smolt emigration for the River Bush population (Fig. 3.8) indicates that ova-to-smolt survival has varied from around 0.16% to 2.02% during the time series, with a long term average of 1.03%. Observed variability in survival during the freshwater phase is believed to reflect additive effects of density-dependent and density-independent mortality, the former defining the shape of the hypothesised domed stock/recruitment curve described for this population (Kennedy & Crozier 1995), and the latter representing a wide scatter of points around the fitted relationship, particularly at low to moderate ova depositions. While density-dependent mortality during a period of high ova deposition during 1985–1988 is believed to have contributed to lower freshwater survival of several cohorts at that time, there appears to have been a downturn in overall freshwater survival during the latter half of the 1980s, continuing into the early 1990s. Freshwater survival of the most recent fully recruited egg cohort showed an improvement relative to the very low values of the previous four years, but remains below the average for the 1970s and early 1980s. It would clearly be valuable to examine this and other data sets to investigate the relative roles of density-dependent and density-independent mortality in fresh water in more detail.

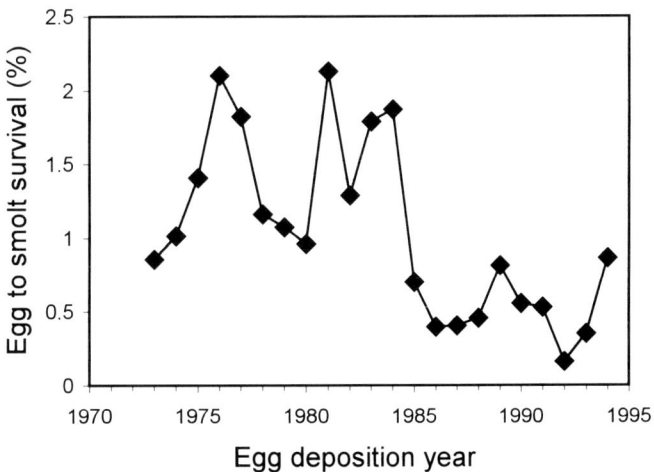

Fig. 3.8 Estimated egg to smolt survival for salmon eggs deposited in the River Bush, 1973–94.

Partitioning of the data into survival to 1+ and 2+ smolts, together with additional data from annual electrofishing surveys of the river has implicated a number of contributory factors. These include habitat degradation, with increased siltation of spawning gravels thought to be caused by land use changes, coupled to increased weed growth. For example, a recent all river survey (O'Connor 1997) examining particle composition of spawning gravels showed that 87% of samples contained more than 10% fines (i.e. particles <2mm), and this was a particular problem in the lower river. Continuing avian and mammalian predation is also thought to have significantly reduced survival, particularly of 2+ parr/pre-smolts (Kennedy & Greer 1988).

Relative importance of freshwater and marine mortality

An examination of the relative variability of marine and freshwater survival for the River Bush wild population provides insight into the relative importance of changes in mortality during the two main life cycle phases in driving population fluctuation. The coefficient of variation for marine survival of River Bush salmon (smolt to return to home water as 1SW fish, c.v. = 14.16%), is considerably less than that for natural survival in fresh water (egg to smolt migration) (c.v. = 56.6%) (Kennedy and Crozier 1993, 1995; W.W. Crozier and G.J.A. Kennedy unpublished data).

Average survival in this population during the freshwater phase from egg to smolt (1.03%) has been about one-thirtieth of that from smolt emigration to coastal return (31.3%) from the River Bush stock. Hence it is concluded that the potential for variation in natural marine survival to influence total life cycle variation is somewhat less than might be imagined. This contrasts with other Atlantic salmon populations in which marine survival variability has been shown to exceed that during the freshwater phase (Chadwick 1991), and illustrates the need for a total life cycle approach to understanding population fluctuation and differences between stocks. It also illustrates the point that trends in indices of abundance observed at the end of the marine phase of the salmon life cycle (such as net catches or counter data) are all too easily attributed solely to events in the marine environment.

Is surviving stock specific?

A major issue surrounding the debate on salmon marine survival is the extent to which changes in survival occur over geographical areas and across stock types. Apart from possibly providing insights into what influences survival at sea, this also affects the extent to which survival data from a monitored river or based upon hatchery stocks can be used as an index of what is happening to larger groups of stocks or stock complexes.

Kennedy & Crozier (1995) reviewed the limited extent to which factors controlling survival and production in fresh water had been examined for parallel trends, noting that ova to smolt survival rates in the River Burrishoole and River Bush catchments changed in parallel in 12 out of 13 years between 1973 and 1985. It was hypothesised that the mechanism was environmental, possibly related to similar climate (rainfall) patterns and consequent effects of river flow (the catchments are separated by about 250 km). The lack of a complete correlation was taken to indicate that in some years some of the climatic factors influencing annual production may be local in their effects.

Similar correlations between marine survival indices among various rivers

(Friedland et al. 1993, Hansen et al. 1995) have led to the view that marine conditions experienced in common by these stocks have similar effects on their survival. This conclusion has been reinforced by retrospective studies indicating correlations between marine growth and returns to fresh water for specific populations (Friedland et al. 1993). In addition, correlations between trends in marine habitat indices in the North Atlantic and abundance of salmon have provided a tentative basis for driving a growth/survival model, whereby habitat limitation may affect cohort survival by restricting growth during a critical post-smolt phase (Friedland et al. 1993). While correlations between the marine survival of wild River Bush and River Burrishoole grilse have been reported (Crozier & Kennedy 1993), a recent examination of marine growth rates of returning River Bush grilse found no relationship between variability in marine growth and marine survival to home waters (Crozier & Kennedy in press).

It is evident that trends in marine survival may be operating in parallel in some but not all of the 1SW north-east Atlantic stocks examined, while trends appear more uniform among 2SW stock components. Clearly further comparisons between stocks (in both freshwater and marine survival indices) are warranted to help resolve the extent to which survival trends operate in parallel between stocks. This will assist in determining the extent to which monitored rivers can act as true 'index' rivers for a region and whether it is appropriate to group stocks when considering management measures. Such studies should not just focus on trends in survival indices but should also attempt to identify underlying factors that might influence survival.

The relationships between the survival (marine and fresh water) and other stock parameters among monitored rivers with differing stock characteristics and from differing geographical areas should be examined further. This will help determine whether changes in survival are related over geographical areas and may provide insights into possible causal mechanisms. This will also help to determine the extent to which a small number of monitored rivers can act as true 'index' rivers for geographical regions or stock groupings.

Conclusions and recommendations

The main conclusions arising from our review of the data available to ICES is that marine survival of wild Atlantic salmon in the north-east Atlantic has been lower in the 1990s than in the 1980s and that, for grilse stocks, this appears to have begun with a widespread and sudden decline in marine survival around 1989 or 1990. Marine survival among grilse stocks has remained low, but not universally so, as it has recovered in some stocks. Marine survival of 2SW stock components (based on a very limited number of rivers examined) has been lower in the 1990s than previously, but this appears to reflect a steadier decline rather than a sudden event.

These patterns are consistent with the changes in the estimated pre-fishery abundance for north-east Atlantic salmon stocks for both 1SW and MSW fish, but the extent to which changes in freshwater survival may also have played a part is unclear. Also not known is the relationship between survival trends and hypothesised longer term cycles of abundance (both absolute and relative between sea-ages) and whether present trends are part of this or caused by specific recent events (Salmon Advisory Committee 1994, Summers 1995).

Given the general lack of long-term data on survival of freshwater stages of wild salmon, we are predisposed to focus on the marine survival issue as the main explanation for the declines in stock abundance while tending to ignore factors affecting freshwater productivity. While there appears to have been real changes in marine survival, this emphasises the need to consider all phases of the life cycle when assessing the status of stocks. Where such an approach is possible, it allows us to examine where factors have most impact in the life cycle and model how changes in those factors (marine or freshwater) can affect productivity and abundance (Kennedy & Crozier 1997).

Arising from the issues examined in this paper are several specific recommendations to address data deficiencies and the ways in which these data are utilised:

(1) Available data sets should be re-examined to evaluate the 'inverse weight hypothesis' and determine whether current estimates of the natural mortality of adult salmon in the sea are appropriate.
(2) The use of monitored rivers to provide accurate long-term survival and productivity data should be continued and expanded to provide a more representative coverage of the 1600 salmon river stocks in the north-east Atlantic area. Where marine survival estimates are developed, emphasis should be given to survival to home waters, prior to coastal or estuarine exploitation. Long-term time series of marine survival of MSW salmon are available for very few stocks and particular effort should be made to develop monitoring programmes on stocks with significant MSW runs.
(3) Hatchery stocks frequently show patterns of marine survival quite different from those of wild stocks in the same river, and in recent years there has been a significant decline in returns from hatchery releases. This should prompt a critical examination of the value of using information derived from studies of artificially reared fish to support management advice on wild salmon stocks.
(4) There is a serious lack of data on long-term trends in survival and productivity in fresh water. Without this, changes in stock status cannot be properly interpreted and advice to managers may be flawed. Wherever possible, existing data sets on freshwater survival should be examined in more detail. In addition, further freshwater survival estimates should be collected for the monitored rivers stocks for which marine survival data are already being obtained, thus affording a total life cycle approach. Alternative methods should also be explored for obtaining such information from a wide range of rivers, e.g. counters operating in conjunction with juvenile abundance indices or smolt production estimates (Crozier & Kennedy 1995).
(5) The relationships between the survival (marine and freshwater) and other stock parameters among monitored rivers with differing stock characteristics and

from differing geographical areas should be examined further. This will help determine whether changes in survival are related over geographical areas and may provide insights into possible causal mechanisms. This will also help to determine the extent to which a small number of monitored rivers can act as true 'index' rivers for broader geographical regions or stock complexes.

(6) The PFA (Pre-fishery Abundance) model has provided estimates of stock abundance that are broadly consistent with the observed trends in survival in monitored rivers. This modelling approach should be continued and refined, with particular attention being paid to the quality of the input data. Estimates of unreported catches, exploitation rates, and sea-age structure of stocks used in the model should be based on scientific sampling wherever possible. Independent methods of verifying the model outputs should also be investigated, including comparison with forward running models based upon smolt production estimates (perhaps from multiple-site semi-quantitative electrofishing surveys).

(7) The observation that trends in the marine survival of grilse do not appear to be uniform across all rivers suggests that a diversity of management regimes might be necessary in different geographical areas, reflecting the differing performance of various stocks/stock complexes. This presents a strong argument for recommending that exploitation at sea should be restricted to single stocks or as small groups of stocks as possible.

Acknowledgement

The authors express their thanks to the General Secretary of ICES for permission to cite the reports of the ICES Working Group on North Atlantic Salmon.

References

Chadwick, E.M.P. (1991) Stock-recruitment of Atlantic salmon. Working Paper for the *Joint Atlantic Salmon Trust/Royal Irish Academy Workshop on the Measurement and Evaluation of the Exploitation of Atlantic Salmon*, Dublin, 8–10 April, 1991.

Crozier, W.W. & Kennedy, G.J.A. (1993) Marine survival of wild and hatchery reared Atlantic salmon (*Salmo salar* L.) from the river Bush, Northern Ireland. In: *Salmon in the Sea and New Enhancement Strategies* (ed. D. Mills), pp. 139–62. Fishing News Books, Oxford.

Crozier, W.W. & Kennedy, G.J.A. (1995) Application of a fry (0+) abundance index, based on semi-quantitative electrofishing, to predict Atlantic salmon (*Salmo salar* L.) smolt runs in the R. Bush, Northern Ireland. *Journal of Fish Biology*, **47**, 107–14.

Crozier, W.W. & Kennedy, G.J.A. (in press) Relationships between marine growth and marine survival of 1SW Atlantic salmon, *Salmo salar*, from the R. Bush, Northern Ireland. *Fisheries Management and Ecology*, **5**.

Doubleday, W.G., Rivard, D.R., Ritter, J.A. & Vickers, K.U. (1979) Natural mortality rate estimates for North Atlantic salmon in the sea. International Council for the Exploration of the Sea CM 1979/M 26.

Friedland, K.D., Reddin, D.G. & Kocik, J.F. (1993) Marine survival of North American and European Atlantic Salmon: effects of growth and environment. *ICES Journal of Marine Science*, **50**, 481–92.

Furnell, D.J. & Brett, J.R. (1986) Model of monthly marine growth and natural mortaliy for Babine Lake sockeye salmon (*Oncorhynchus nerka*). *Canadian Journal of Fisheries and Aquaculture Science*, **43**, 999–1004.

Hansen, L.P., Friedland, K.D. & Dunkley, D.A. (1995) Examination of survival rates of Atlantic salmon (*Salmo salar* L.) from Norway and Scotland and the possible influence of marine habitat area. International Council for the Exploration of the Sea CM 1995/M9.

ICES (1986) Report of the Working Group on North Atlantic Salmon, Copenhagen, 17–26 March, 1986. International Council for the Exploration of the Sea CM 1986/Assess:17.

ICES (1992) Report of the Working Group on North Atlantic Salmon, Dublin, 5–12 March 1992. International Council for the Exploration of the Sea CM 1992/Assess:15.

ICES (1998) Report of the Working Group on North Atlantic Salmon, Copenhagen, 14–23 April, 1998. International Council for the Exploration of the Sea CM 1998/ACFM:15.

Kennedy, G.J.A. & Crozier, W.W. (1993) Juvenile Atlantic salmon (*Salmo salar* L.) – Production and Prediction. In: *Production of juvenile Atlantic salmon,* Salmo salar, *in Natural Waters* (eds R.J. Gibson & R.E. Cutting), pp. 179–87. Canadian Special Publication of Fisheries and Aquatic Sciences 118.

Kennedy, G.J.A. and Crozier, W.W. (1995) Factors affecting recruitment success in salmonids. In: *The Ecological Basis for River Management* (eds D.M. Harper & A.J.D. Ferguson), pp. 349–62. John Wiley and Sons, Chichester.

Kennedy, G.J.A. & Crozier, W.W. (1997) What is the value of a wild salmon smolt (*Salmo salar* L.)? *Fisheries Management and Ecology*, **4**, 103–10.

Kennedy, G.J.A. & Greer, J.E. (1988) Predation by cormorants (*Phalacrocorax carbo* L.) on the salmonid populations of an Irish river. *Aquaculture and Fisheries Management*, **19**, 159–70.

Martin, J.H.A. & Mitchell, K.A. (1985) Influence of sea temperature upon the numbers of grilse and multi-sea-winter Atlantic salmon (*Salmo salar*) caught in the vicinity of the River Dee (Aberdeenshire). *Canadian Journal of Fisheries and Aquaculture Science*, **42**, 1513–21.

Matthews, S.B. & Buckley, R. (1976) Marine mortality of Puget Sound Coho Salmon (*Oncorhynchus kisutch*). *Journal of the Fisheries Research Board of Canada*, **33**, 1677–84.

Potter, E.C.E. & Dunkley, D.A. (1993) Evaluation of marine exploitation of salmon in Europe. In: *Salmon in the Sea and New Enhancement Strategies* (ed. D. Mills), pp. 203–19. Fishing News Books, Oxford.

Potter, E.C.E., Hansen, L.P., Gudbergsson, G., Crozier, W.C., Erkinaro, J., Insulander, C., MacLean, J., O'Maoileidigh, N.S. & Prusov, S. (1998) A method for estimating preliminary conservation limits for salmon stocks in the NASCO-NEAC area. International Council for the Exploration of the Sea, CM 1998/T:17.

Rago, P.J. (1993) Two randomisation tests for estimation of regional changes in fish abundance indices: application to North Atlantic salmon. International Council for the Exploration of the Sea, CM 1993/D:35.

Rago, P.J., Meerburg, D.J., Reddin, D.G., Chaput, G.J., Marshall, T.L., Dempson, B., Caron, T., Porter, T.R., Friedland, K.D. & Baum, E.T. (1993a) Estimation and analysis of pre-fishery abundance of the two-sea winter population of North American Atlantic salmon (*Salmo salar*), 1974–1991. International Council for the Exploration of the Sea CM 1993/M:24.

Rago, P.J., Reddin, D.G., Porter, T.R., Meerburg, D.J., Friedland, K.D. & Potter, E.C.E. (1993b) A continental run reconstruction model for the non-maturing component of North American Atlantic salmon: Analysis of fisheries in Greenland and Newfoundland-Labrador, 1974–1991. International Council for the Exploration of the Sea CM 1993/M:25.

Reddin, D.G., Friedland, K.D., Rago, P.J., Dunkley, D.A., Karlsson, L. & Meerburg, D.J. (1993) Forecasting the abundance of North American two-sea winter salmon stocks and the provision of catch advice for the West Greenland salmon fishery. International Council for the Exploration of the Sea CM 1993/M43.

Ricker, W.E. (1976) Review of the rate of growth and mortality of Pacific salmon in salt water, and non-catch mortality caused by fishing. *Journal of the Fisheries Research Board of Canada*, **33**, 1483–524.

Salmon Advisory Committee (1994) *Run Timing of Salmon.* MAFF Publications; London.

Shearer, W.M. (1984) The natural mortality at sea for North Esk salmon. International Council for the Exploration of the Sea CM 1984/M:23.

Summers, D.W. (1995) Long-term changes in the sea-age at maturity and seasonal time of return of salmon, *Salmo salar* L., to Scottish rivers. *Fisheries Management and Ecology*, **2**, 147–56.

Chapter 4
Description of Marine Growth Checks Observed on the Scales of Salmon Returning to Scottish Home Waters in 1997

J.C. MacLEAN, G.W. SMITH and B.D.M. WHYTE

Introduction

Atlantic salmon (*Salmo salar* L.) catch information has been collected in a systematic manner from salmon fisheries throughout Scotland since 1952. These statistics show a steady decrease in total catch from the late 1960s until the present day (Anon. 1997). Although some of this decline can be explained by reductions in net fishing effort, in response to a reduced real value of salmon (Smart 1986), and a number of buy-out initiatives, these factors do not fully explain the reductions observed. Changes in marine survival may also be implicated in the decline. Records from the North Esk, a monitored river on the east coast of Scotland, indicate that while smolt production has been well maintained, estimates of marine survival have fallen by a factor of two (Shearer 1992). Furthermore, these observations are not confined to Scotland; similar trends in both home water catches and in marine survival have been observed throughout the North Atlantic (Anon. 1998).

The marine survival of salmonids is thought to be heavily influenced by events during the first few months of their lives at sea by factors such as ocean productivity and growth related predation pressures (e.g. Holtby *et al.* (1990) in Pacific salmon and Salminen *et al.* (1995) in Baltic salmon). As a result, recent research has focused on the early marine phase of the salmon's life cycle. In Atlantic salmon, observations by Friedland & Reddin (1993) revealed a correlation between the north-east Atlantic salmon catch and an index of spring habitat thought to be preferred by post-smolts. In addition, Friedland *et al.* (1998) demonstrated that the marine survival of smolts from two stocks, one in Scotland (North Esk) and the other in Norway (River Figgio) entering the eastern Atlantic Ocean were both positively correlated with a spring index of preferred habitat as defined by sea surface temperatures.

Recent surveys (e.g. Shelton *et al.* 1997, Holm *et al.* 1996 and Holst *et al.* 1993) have described the trophic conditions of the immediate post-smolt phase by intercepting these fish at sea. Results suggest that post-smolts have ample feeding opportunities in the early weeks at sea. An alternative line of enquiry to such direct observations, which covers the whole of the marine phase, seeks to follow past marine feeding opportunities from the scale patterns of returning adult salmon. Thus, Friedland *et al.* (in press) showed that the size of the growth increment in the first year at sea was correlated positively with the marine survival rate of the North Esk smolts.

The Freshwater Fisheries Laboratory, Pitlochry, routinely examines scale samples from maturing salmon returning to Scottish home waters. Analysis of samples from fish caught in 1997 indicated a higher than previously recorded incidence of summer checks. Such summer checks are recognised as a number of successive narrowly-spaced circuli occurring within a period of otherwise more widely-spaced circuli. Patterns of circuli spacing may be interpreted as reflecting changes in the immediate environment of the salmon. Thus variation in growth opportunity may be inferred from an analysis of scale patterns (Bagenal & Tesch 1978, Casselman 1987). Summer checks therefore, may indicate periods of reduced growth opportunity which may in turn be associated with low marine survival (Friedland et al. 1998).

A description is given of these unusual growth patterns in the scales of salmon returning to river catchments within a wide geographic area across Scotland and a comparison is made of their frequency of occurrence with historical data. The relevance of such observations to marine survival is considered in general and with particular reference to marine survival indices from the North Esk.

Material and methods

During 1997, samples of scales were collected from salmon caught in a number of fisheries throughout Scotland (Figure 4.1). In addition, scales from adult recaptures in Scottish home waters of North Esk salmon tagged as emigrating smolts were also analysed.

Scales were impressed and age was determined according to the methods given in Anon. (1985). Figure 4.2 illustrates the growth patterns observed on 1SW salmon in the presence and absence of summer checks. Figure 4.3 illustrates the comparable growth patterns in 2SW salmon.

The incidence of summer checks in scales obtained in 1997 samples was determined among both 1SW and 2SW salmon. In addition, scale samples from the North Esk net and coble fishery in a number of previous years were re-examined to test whether the frequency of summer checks was significantly different from those observed previously. In order to test whether summer checks were associated with differences in marine survival, scale sample data from smolt years showing both a relatively high return rate (1973 and 1983) and a relatively low return rate (1969, 1980 and 1990), as indicated by catch (Anon. 1997), were chosen and compared with indices of marine survival for smolts emigrating from the North Esk for the period 1965–96 derived using the methods described in Friedland et al. (1998).

Results

Incidence of summer checks

The occurrence of a group of tightly-spaced circuli can have one of two interpretations. It may be interpreted as a summer growth check or, alternatively, it may be

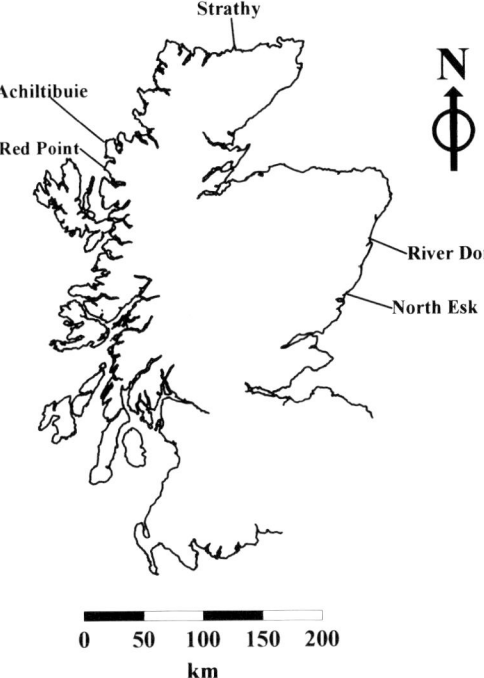

Fig. 4.1 Location of sites throughout Scotland from which samples of scales from returning adult salmon were collected.

considered to represent a winter annulus. The North Esk recapture data set provides scale samples from salmon tagged as emigrating smolts whose sea-age is therefore known and thus summer checks and annuli can be distinguished with confidence.

Substantial proportions of salmon returning in 1997 exhibited summer checks (Table 4.1). Evidence that checks were not misclassified winter annuli is provided by the observation that the incidence of validated summer checks from scale samples taken from the North Esk recapture data set (n = 50) was 26.0% in 1SW salmon, which was within the range observed in the fisheries examined throughout Scotland (20.0% (n = 1115) to 45.0% (n = 20)). Although the incidence of summer checks observed from 2SW North Esk recaptures (12.2% (n = 41)) was greater than that from the samples from the other sites (0% (n, range = 1–16) to 5.0% (n = 20)), this may be explained by the small sample sizes available.

All summer checks occurred during the first marine growing season in 1SW salmon or during the second marine growing season in 2SW salmon. Thus, the summer checks were laid down in the same calendar year (1996) in both sea-age groups examined. The incidence of summer checks in 2SW salmon was significantly less than in 1SW salmon (Wilcoxon test, n = 7, z = –2.366, p = 0.018) (Table 4.1).

Table 4.2 shows the incidence of summer checks in historical data sets from the North Esk net and coble fishery. For both 1SW and 2SW salmon, the incidence of

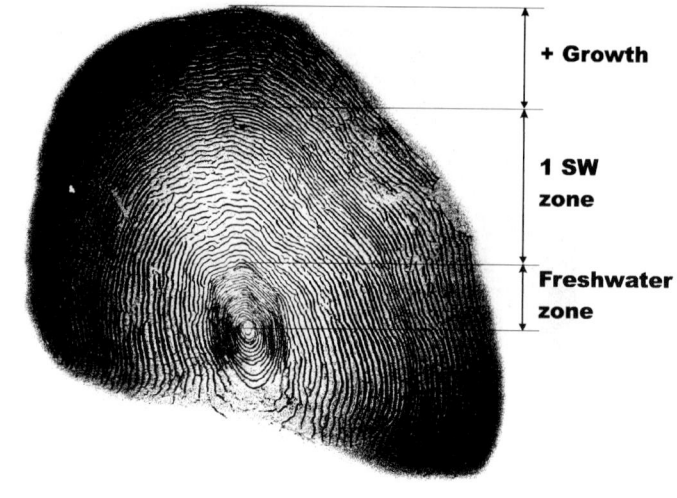

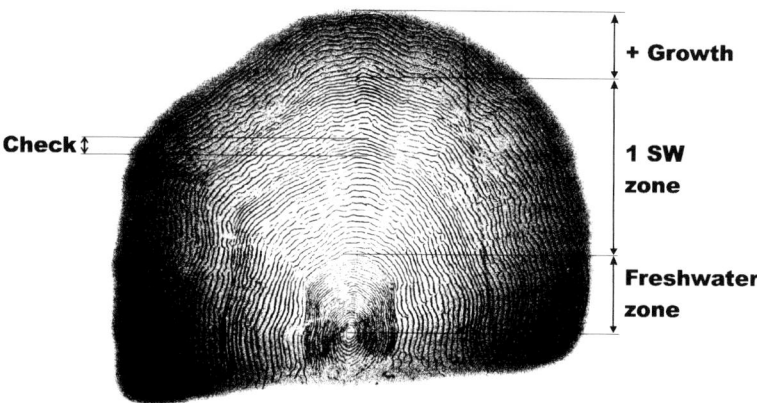

Fig. 4.2 Illustration of scale growth patterns observed in 1SW adult returns in the absence of (above) and in the presence of (below) a summer check.

summer checks in 1997 was outwith the 95% confidence limits for the historical data sets examined.

Incidence of summer checks in relation to month of capture

Table 4.3 shows the incidence of summer checks, by month, in 1SW and 2SW fish sampled from the largest data set, the North Esk net and coble fishery. In 1SW salmon, the incidence of summer checks varied significantly among months (G-test, $G = 25.76$, $df = 3$, $p < 0.001$), the incidence of summer checks rose through May and June, peaking in July (26.5%) before dropping again in August. For 2SW salmon

Description of Marine Growth Checks   41

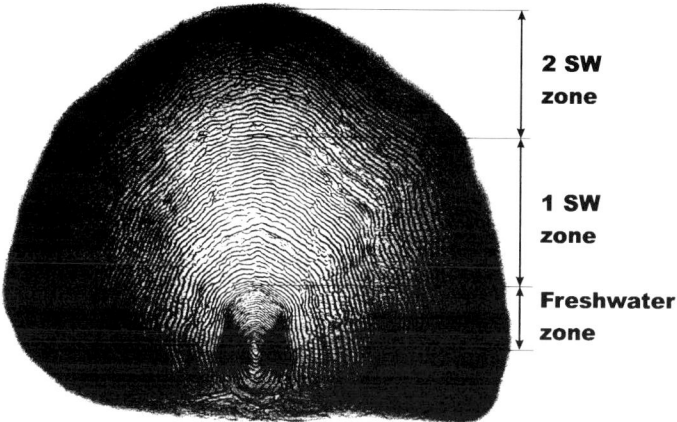

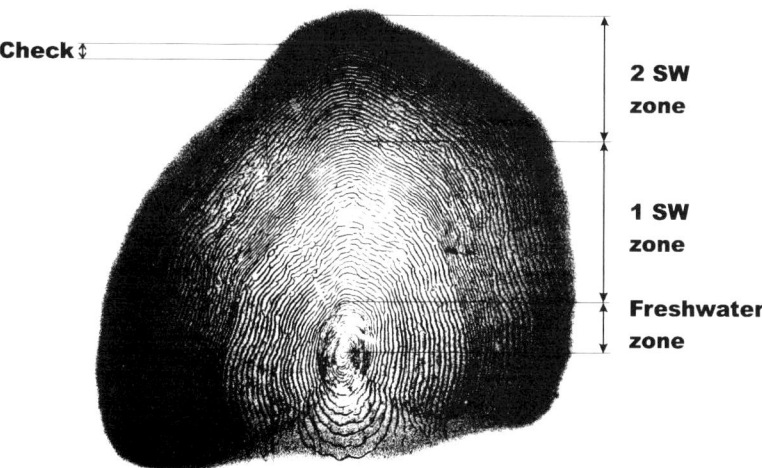

Fig. 4.3 Illustration of scale growth patterns observed in 2SW adult returns in the absence of (above) and in the presence of (below) a summer check.

there was no discernible temporal pattern in the incidence of summer checks (G-test, $G = 0.25$, $df = 2$, $p > 0.75$).

Effect of summer check on size at capture

Figure 4.4 compares the average fork length at capture of salmon with and without summer checks, by month, for both 1SW and 2SW salmon for the North Esk net and coble fishery data set. There were no significant differences in fork length between salmon with and without summer checks (Table 4.4).

Table 4.1 Incidence of summer checks in 1SW and 2SW salmon sampled from sites around Scotland in 1997.

Sample site	1SW salmon		2SW salmon	
	Number sampled	Number with summer checks (%)	Number sampled	Number with summer checks (%)
Recaptures of North Esk smolts, various sites	50	13 (26.0)	41	5 (12.2)
North Esk, net & coble	1115	223 (20.0)	528	14 (2.7)
North Esk, rod & line	52	15 (28.8)	185	3 (1.6)
Redpoint, fixed engine	32	13 (40.6)	1	0 (0)
Achiltibuie, fixed engine	20	9 (45.0)	1	0 (0)
Strathy, fixed engine	285	66 (23.2)	16	0 (0)
River Don, netting for broodstock	73	28 (38.4)	20	1 (5.0)

Table 4.2 Incidence of summer checks in annual samples of 1SW and 2SW salmon from the North Esk net and coble fishery.

Year	1SW salmon		2SW salmon	
	Number sampled	Number with summer checks (%)	Number sampled	Number with summer checks (%)
1969	1335	4 (0.3)	860	4 (0.5)
1973	1419	14 (1.0)	1045	14 (1.3)
1980	456	0 (0)	928	7 (0.8)
1983	226	0 (0)	403	3 (0.7)
1990	619	0 (0)	431	2 (0.5)
1997	1115	223 (20.0)	528	14 (2.7)

Table 4.3 Incidence of summer checks, by month, in 1SW and 2SW salmon sampled from the North Esk net and coble fishery in 1997.

Month	1SW salmon		2SW salmon	
	Number sampled	Number with summer checks (%)	Number sampled	Number with summer checks (%)
February			38	2 (5.3)
March			56	1 (1.8)
April	1	0 (0)	40	0 (0)
May	20	2 (10)	203	7 (3.4)
June	379	69 (18.2)	102	2 (2.0)
July	461	122 (26.5)	50	2 (4.0)
August	254	30 (11.8)	39	0 (0)

Location of summer checks with respect to growth attained in year of incidence

The position of the summer check was expressed as a proportion of the total scale growth attained in the year the summer check occurred for both 1SW and 2SW salmon. Figure 4.5 shows the distributions of these proportions where 0 represents the beginning of the relevant growth season on the scale and 1 indicates the end of that period. For 1SW salmon, the proportions are spread around a mean of 0.70 ± 0.004 and for 2SW salmon around a mean of 0.71 ± 0.015. These distributions were not significantly different (Mann-Whitney U test, n = 237, U = 1269.5, p = 0.238).

The relationship between summer checks and marine survival indices

Figure 4.6 presents the marine survival indices of North Esk smolts, 1965–96 both for 1SW and 2SW returns. In general, survival indices have been lower in the latter part of the time series for both 1SW and 2SW salmon. Although the marine survival indices for the 1SW returns from the 1996 smolt cohort and for the 2SW returns from the 1995 smolt cohort (both of which returned to home waters in 1997) are the lowest on record, there is no correlation between marine survival and the incidence of growth checks for the years examined (Spearman rank correlation, $r_s = 0.152$, n = 6, p > 0.5, for 1SW salmon and $r_s = 0.029$, n = 6, p > 0.5, for 2SW salmon).

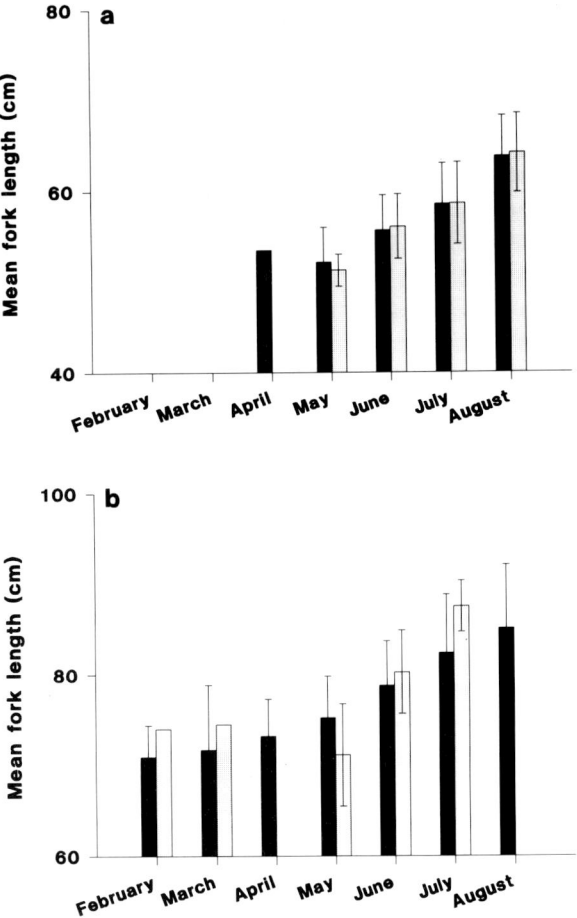

Fig. 4.4 Comparison of fork lengths at capture by month between adult returns with (white bars) and without (black bars) summer checks for (a) 1SW and (b) 2SW salmon. Error bars indicate 1 standard deviation of each mean value.

Discussion

Summer checks have been observed on scales taken during routine sampling from various locations throughout Scotland since 1963. Their incidence has, however, always been rare; in the data sets re-examined, less than 1.5% of 1SW and 2SW salmon showed the checks. The occurrence of summer checks on the scales of salmon returning to Scottish home waters in 1997 was significantly greater than these historical levels. Scale samples were taken from fisheries located in the east, north and west of Scotland, suggesting that the phenomenon affected salmon both originating from, and returning to, areas at least as large as Scotland.

The spacing between the circuli reflects the rate of growth (Bagenal & Tesch 1978, Casselman 1987). In general, widely-spaced circuli represent periods of rapid growth

Table 4.4 Results of a comparison (Mann Whitney U test) of the lengths of returning adult salmon between those with summer growth checks on their scales and those without.
§ indicates not enough data to carry out the analysis.

	1SW salmon			2SW salmon		
	N	U	P	N	U	P
February	§			38	12.0	0.116
March	§			58	11.5	0.309
April	§			§		
May	20	19	0.899	203	961.5	0.071
June	379	10020.5	0.412	102	79.5	0.620
July	461	20352.5	0.067	50	19.0	0.151
August	254	3128.5	0.540	§		

(e.g. in the summer) while narrowly spaced circuli represent slower winter growth. Thus, summer checks reflect periods of limited growth. However, the cause of the summer checks is unknown. One possibility is that there have been changes in the marine environment that limit the growth of salmon through lack of food. For example, Perry *et al.* (1996) demonstrated a link between dietary composition and growth in Pacific salmon. Furthermore, Brett (1979) has shown that, in salmonids in general, there is an association between water temperatures and growth. Alternatively, oceanographic and/or climatological changes may alter the environmental cues perceived by salmon resulting in atypical migration routes being pursued through less-suitable habitat.

Although the cause of the summer checks is unclear, their relative position in the scale may allow some inference to be made as to the timing and geographical scope of the phenomenon. The location of the summer checks in relation to the overall growth increment on the scale is fairly restricted (Fig. 4.5), indicating that the causal event is synchronised in time over all returns. While summer checks are present in scale samples taken from salmon returning to home waters throughout the sampling season (February to August) the extent to which they occur varies both with month and sea age at return (Table 4.3). A possible mechanism to explain this observation is that the causal effect is not uniform throughout the marine habitat and that run timing differences both between and within sea-age classes is related to different marine migration patterns. Thus, different groups of returning salmon may have been exposed to the causal event to differing degrees. This may also suggest that the temporal structuring of returning salmon populations is on a scale of a month, or possibly less.

The high incidence of summer checks reported in maturing salmon returning to the Scottish coast in 1997 is a new phenomenon. The position of the checks in the 1SW and 2SW salmon examined shows that the causal event occurred in 1996. Friedland *et al.* (in press) have demonstrated associations between changes in ocean climate with changes in growth and survival in Atlantic salmon. While the checks reported here

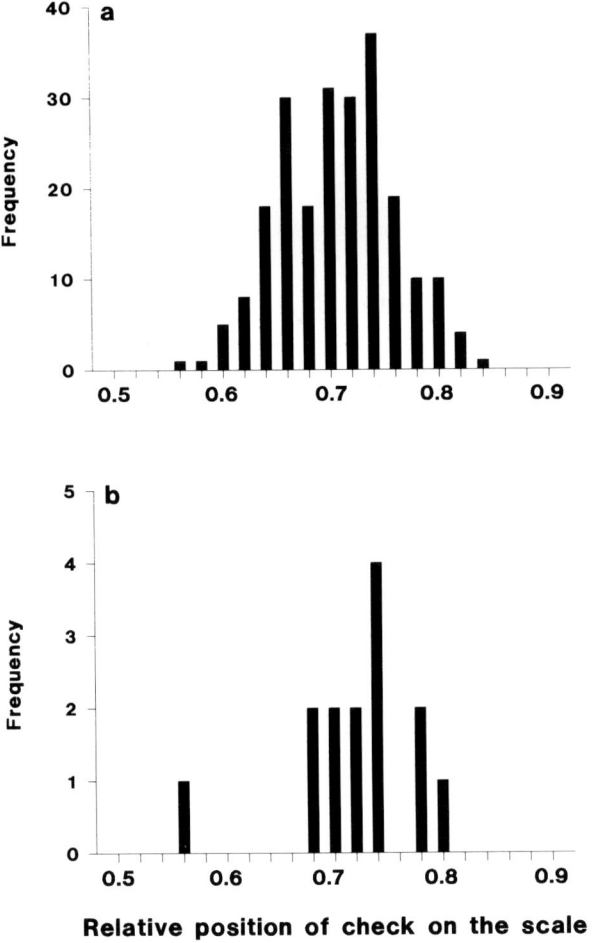

Fig. 4.5 Position of the summer check, expressed as a proportion of the total growth attained in the year in which the check occurred, for (a) 1SW and (b) 2SW salmon.

suggest that further, perhaps related, changes occurred in the ocean in the latter part of 1996, no link was shown with either growth or survival. In spite of this lack of apparent correlation, the observations described here further focus our attention on the marine phase of the salmon's life cycle and on changes in the marine environment that may have an impact upon growth and survival.

Acknowledgements

We thank David Hay and Alisdair MacDonald for producing Figs 4.2 and 4.3 and Alan Youngson, Dick Shelton and Tony Hawkins for reviewing the manuscript.

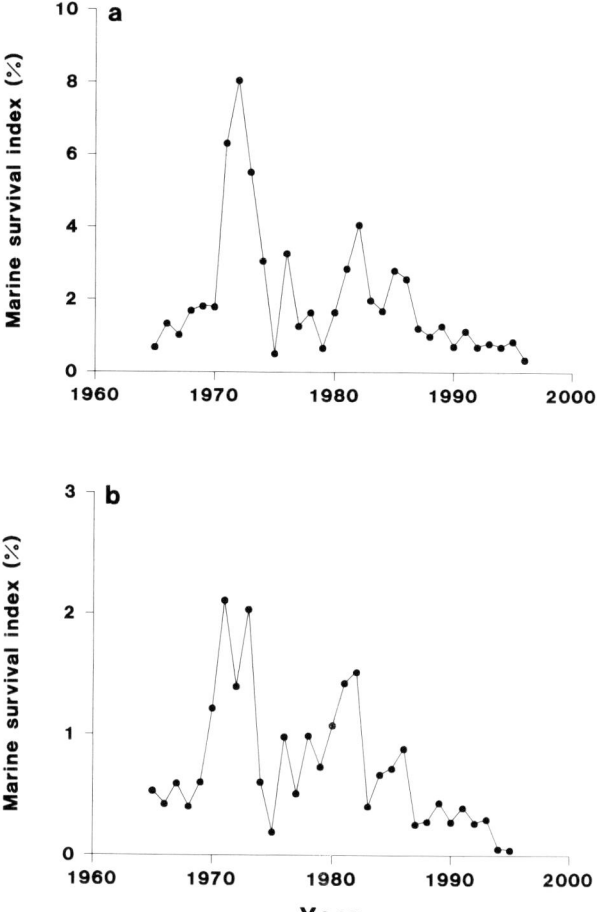

Fig. 4.6 Marine survival indices for North Esk smolts for (a) 1SW and (b) 2SW returns, 1965–96.

References

Anon. (1985) *Atlantic Salmon Scale Reading. Report of the Atlantic Salmon Scale Reading Workshop.* ICES, Copenhagen.
Anon. (1997) Scottish Salmon and Sea Trout Catches:1996. Statistical Bulletin No Fis/1997/1. Edinburgh, Scottish Office, October 1997.
Anon. (1998) Report of the ICES North Atlantic Salmon Working Group, CM 1998/ACFM 15. International Council for the Exploration of the Sea, Copenhagen, 293 pp.
Bagenal, T.B. & Tesch, F.W. (1978) Age and Growth. In: *Methods for Assessment of Fish Production in Fresh Waters*, 3rd edn (ed. T.B. Bagenal), pp. 101–36. Blackwell Scientific Publications, Oxford.
Brett, J.R. (1979) Environmental factors and growth. In: *Fish Physiology* (eds J.S. Hoar, D.J. Randall & J.R. Brett), vol. 8, pp. 599–675. Academic Press, London.
Casselman, J.M. (1987) Determination of Age and Growth. In: *The Biology of Fish Growth* (eds A.H. Weatherly & H.S. Gill), pp. 209–42. London, Academic Press.
Friedland, K.D. & Reddin, D.G. (1993) Marine survival of Atlantic salmon from indices of post-smolt

growth and sea temperature. Fourth International Atlantic Salmon Symposium, St Andrews, Canada. In: *Salmon in the Sea and New Enhancement Strategies* (ed. D. Mills), pp 119–38. Fishing News Books, Oxford.

Friedland, K.D., Hansen, L.P. & Dunkley, D.A. (1998) Marine temperatures experienced by postsmolts and the survival of Atlantic Salmon, *Salmo salar* L. in the North Sea area. *Fisheries Oceanography*, **7**(1), 22–34.

Friedland, K.D., Hansen, L.P., Dunkley, D.A. & MacLean, J.C. (in press) Linkage between ocean climate, post-smolt growth, and survival of Atlantic Salmon (*Salmo salar* L.) in the North Sea area. *ICES Journal of Marine Science*.

Holm, M., Holst, J.C. & Hansen, L.P. (1996) Salmon in the Norwegian Sea. Fishing experiments and recorded captures of salmon in the Norwegian Sea and adjacent areas, July 1991–August 1995. *Fisken og Havet*, **1**, 13.

Holst, J.C., Nilsen, F., Hodneland, K. & Nylund, A. (1993) Observations of the biology and parasites of postmolt Atlantic salmon, *Salmo salar*, from the Norwegian Sea. *Journal of Fish Biology*, **42**, 962–6.

Holtby, L.B., Andersen, B.C. & Kadowaki, R.K. (1990) Importance of smolt size and early ocean growth to interannual variability in marine survival of coho salmon (*Onchorhynchus kisutch*). *Canadian Journal of Fisheries and Aquatic Sciences*, **47**, 2181–94.

Perry, R.I., Hargroves, N.B., Waddell, B.J. & Mackas, D.L. (1996) Spatial variations in feeding and condition of juvenile pink and chum salmon off Vancouver Island, British Columbia. *Fisheries Oceanography*, **51**, 73–88.

Salminen, M., Kuikka, S. & Erkamo, E. (1995) Annual variability in survival of sea ranched Baltic salmon, *Salmo salar* L.: significance of smolt size and marine conditions. *Fisheries Management and Ecology*, **2**, 171–84.

Shearer, W.M. (1992) *The Atlantic Salmon, Natural History, Exploitation and Future Management.* Halsted Press, New York.

Shelton, R.G.J., Turrell, W.R, MacDonald, A., McLaren, I.S. & Nicoll, N.T. (1997) Records of post-smolt Atlantic salmon, *Salmo salar* L., in the Faroe-Shetland Channel in June 1996. *Fisheries Research*, **31**, 159–62.

Smart, G.N.J. (1986) Recent changes in fishing methods. In: *The Status of Atlantic Salmon in Scotland.* (eds D. Jenkins & W.M. Shearer) 50–54. Institute of Terrestrial Ecology, Huntingdon.

Chapter 5
Tracking Atlantic Salmon Post-smolts in the Sea

A. MOORE, G.L. LACROIX and J. STURLAUGSSON

Introduction

During the spring Atlantic salmon smolts emigrate from fresh water to the marine environment. The period is one of potentially high mortality when the timing of smolt movement into coastal waters may be important to their survival and subsequent return as adult fish (Hansen & Jonsson 1989). There is an indication that the strength of a particular sea year class is set early in the post-smolt stage (Anon. 1988, Eriksson 1994, Salminen *et al.* 1995). For example, Friedland *et al.* (1998) have suggested that when cool surface waters dominate the Norwegian coast and North Sea (Scotland) during May, survival of two Atlantic salmon stocks from the Rivers Figgjo and North Esk has been poor. Mortality may also be high during the subsequent migration to the feeding grounds, where a number of marine environmental factors may continue to act to regulate the size of populations (see reviews in Mills 1993).

However, we still have only a very limited understanding of the early marine phase of the Atlantic salmon and the environmental factors that may influence the behaviour and distribution of post-smolts in the sea. One method for increasing our knowledge of this particular aspect of salmon biology is the use of tracking and telemetry techniques to monitor the movements of post-smolts in the open sea. This approach has been successfully used to describe the environmental and biological factors influencing salmon smolt migration in fresh water and estuaries (Greenstreet 1992, Moore *et al.* 1995, 1996, 1998, Lacroix & McCurdy 1996).

The migrations of Atlantic salmon have long fascinated biologists, and there has been a resurgence of interest in the movements of post-smolts and adults in the sea due to dramatic declines in adult returns over the last decade. Atlantic salmon are distributed over large areas of the North Atlantic and may be found feeding in the Norwegian Sea, north of the Faroe Islands and between West Greenland and Labrador during their migration. There is, however, evidence that the fish are not evenly distributed and that the marine distribution of the Atlantic salmon is dependent upon ocean temperature (Reddin & Shearer 1987, see Chapter 7 of this volume).

In 1992, surveying and tracking salmon in the sea was the theme for a multinational workshop organised jointly by the Atlantic Salmon Trust and the Atlantic Salmon Federation (Potter & Moore 1992). More recently, a workshop in Seattle specifically addressed the application of acoustic and archival technology to assess the estuarine, nearshore and offshore habitat utilisation and movement of Pacific salmonids (Boehlert 1997). An ICES Study Group was also established to advise on the feasibility and design of experiments to tag salmon with data logging tags in the North

Atlantic area, and the Reports were presented to Council in 1997 and 1998 (Anon. 1997, 1998). All three forums described the methods and technology available at the time for studying post-smolt and adult movements in the open sea and outlined the requirements for further developments and collaborations. More generally, a workshop was recently convened in Boston to discuss the requirements for the development of the next generation of marine animal telemetry tags (Stone et al. 1998). The present paper reviews current electronic tagging methods used to examine the behaviour and movements of post-smolts in the sea, and outlines the problems and limitations involved in tagging and tracking salmon in coastal waters and open ocean. It then describes the most promising strategies for future open ocean tracking of post-smolts and the technological developments required to improve our ability to monitor and describe the distribution of these fish in the marine environment.

Definition of post-smolt

Mills (1989) separated the Atlantic salmon after they have left fresh water as smolts into three main groups, post-smolt, adult and kelt. The post-smolt phase can be defined either on a spatial scale (the stage between the mouth of the estuary and oceanic waters) or on a temporal scale (the stage between the fish emigrating from fresh water in spring and the first over-wintering stage). For the purpose of this paper the post-smolt stage will be considered as the migratory phase during which salmon move from the mouth of the estuary to the open ocean.

Established methods of tracking post-smolts and current knowledge

Our knowledge of post-smolt biology has been limited by our inability to readily find salmon in the sea, and there have been relatively few attempts to use telemetry techniques to track post-smolts on a large scale during their marine migration. Studies that have attempted to have involved: actively following individual fish tagged with acoustic tags (pingers); systematically searching geographical grids in coastal areas for tags using mobile receivers; and monitoring fixed sites for the passage of tagged fish using receivers that automatically record detected tags. These studies have all used acoustic pinger tags miniaturised specifically for use with smolts.

Active tracking

Active tracking requires that the tracking vessel remains in constant contact with the tagged fish as it migrates. The signals from the tag are detected by a hydrophone and the tracking vessel has to be kept within detection range. The position of the vessel is then fixed periodically using modern navigational systems (e.g. Global Positioning System (GPS) and Differential GPS systems) or, if actual resolution of the fish position is necessary, triangulation from multiple hydrophones can be used. Environmental data considered important in controlling orientation during the initial entry into coastal waters (e.g. current speeds, water temperature, salinity, etc.) can be collected in real time throughout each track using standard methods. Holm et al. (1982) individually released and tracked nine smolts (one wild and eight hatchery-reared) in a Norwegian fjord, using telemetry acoustic transmitters. During this study the swimming depth of the fish was estimated and the ambient temperature was also monitored. The post-smolts tended to move seawards in the direction of the prevailing currents.

Active tracking of post-smolts in coastal waters has also been carried out at three sites in England and Wales (Moore et al. 1995, 1996). In these studies wild smolts were trapped in fresh water, tagged with acoustic transmitters and allowed to migrate normally through the estuary before being tracked in the open sea. In the Rivers Conwy, Tawe and Avon, the smolts were monitored in the lower estuary using fixed acoustic 'sonar buoys' before being actively tracked from a small research vessel as the fish entered coastal waters (Moore et al. 1992, 1995, 1996). These studies have shown that the post-smolts migrated rapidly (ground speeds of approximately 1.8 km hour^{-1}), close to the surface, and there was a strong tidal current component to the speed and direction of movement.

More recently, Holm et al. (1998) have studied the movements of hatchery-reared post-smolts in Norwegian coastal waters, again using miniature acoustic transmitters and active tracking. Twenty-one post-smolts were tracked for distances up to 60 km and the results indicated that surface water currents and wind direction had a significant effect on their patterns of movement.

Automated receivers

Submersible automated receivers, fixed to the sea-bed, that continuously listen for and record detected tags, have been successfully used in a series of studies to track the coastal movements of post-smolts released in groups at the mouths of several rivers (Lacroix & McCurdy 1996, Voegeli et al. 1998). The movements of both hatchery-reared and wild post-smolts, tagged as smolts with miniature acoustic tags were monitored using submersible automated receivers that were moored to the sea-bed in areas of strong tidal currents. These studies also included extensive grid searches for tagged fish using active tracking to fill in details of fish movement between areas of fixed automated receiver deployment. During the study up to 96 fish were released at the same time. The migratory routes of wild and hatchery fish were similar and the post-smolts moved rapidly away from coastal waters. The tidal cycle had a strong influence on the behaviour and timing of migration through channels and passages with strong tidal currents, and the fish movements reflected the overall circulation patterns of surface waters. This tracking approach also made it possible to determine the survival of post-smolts from the time of release to locations where fixed automatic receivers were moored along the migration route (see Fig. 5.1 for an example of receiver deployment to determine survival to fixed points along a migration route).

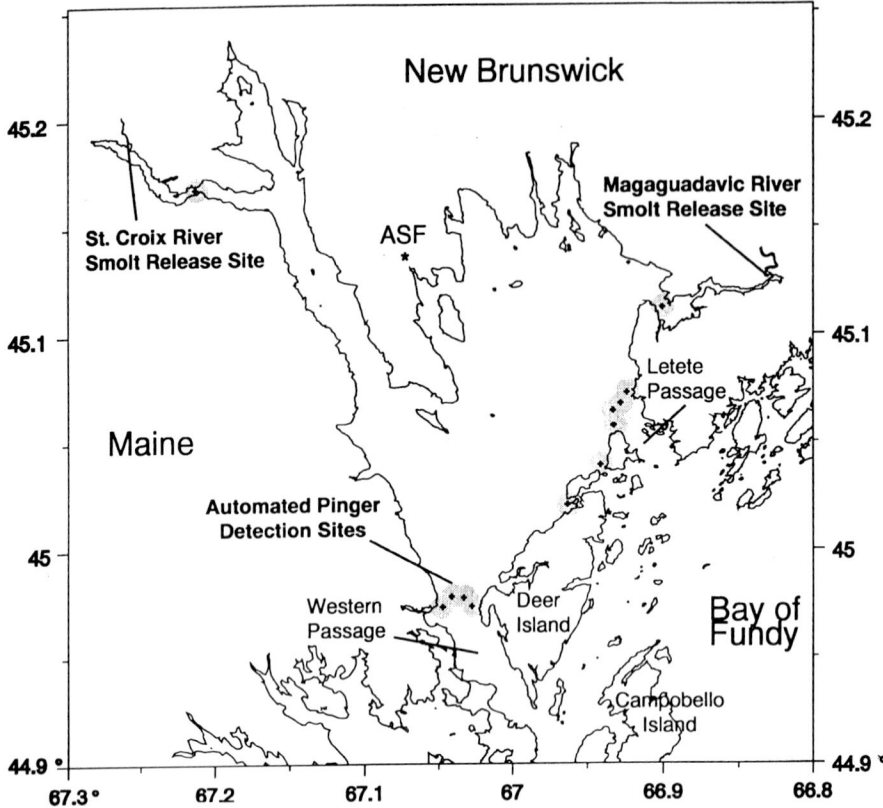

Fig. 5.1 Example of deployment of submersible automated receivers at fixed sites on the sea bed (+ with shaded circle showing detection range) to determine the survival of Atlantic salmon post-smolts to fixed points in the seaward migration. The map is of Passamaquoddy Bay on the east coast of Canada and the USA (reproduced from Voegeli et al. 1998).

All the above studies have demonstrated that tracking post-smolts in the marine environment is possible, but there are a number of limitations and difficulties with the available techniques that prohibit the study of individual fish throughout the post-smolt stage. Most of the tracks obtained so far have been of short duration (usually a matter of hours), and have only managed to describe the movements of the fish over relatively small distances (15–60 km). The principal problems have been the short life of the tags (15–90 days), limited detection range in coastal waters (200–400 m) and the difficulties in remaining in contact with the fish during adverse weather conditions. In addition, there are limitations with the use of multiple listening stations, principally the high number of stations required, the high cost of the equipment, and a knowledge of the migration routes to enable the correct siting of the equipment.

Tag recapture surveys

An alternative method for describing the movements of post-smolts in the sea that does not rely on telemetry techniques is the use of drift-nets to sample post-smolts in the sea that have been microtagged and sequentially released throughout the smolt season (Sturlaugsson 1994, Sturlaugsson & Thorisson 1995a,b). This technique was used in the coastal waters around Iceland and involved releases of large numbers of smolts (up to 1 million a day). The results of these studies indicated that the smolts migrated rapidly through the coastal zone at ground speeds of up to 2.4 km hour^{-1}. The smolts migrated close to the surface either singly or in small groups. There was evidence of some limited feeding during the movement towards oceanic waters.

Potential strategies for tracking post-smolt salmon in the sea

Ideally, tracking post-smolts in the sea would involve tagging emigrating smolts in fresh water and then describing their marine distribution and behaviour, the migration routes they follow to the oceanic feeding grounds, and the environmental conditions they experience during their migrations. In some cases, the strategy used could allow survival to be determined between specific points along the migration route. The application of tracking and telemetry technology therefore promises to be one of the principal methods for increasing our understanding of this stage of the salmon life cycle.

Specific hypotheses and questions concerning post-smolt behaviour, and in some cases survival in the marine environment, that could be tested using tracking technology are numerous, and the following are but a few examples:

- Does the timing of smolt migration from fresh water affect the initial survival of post-smolts in coastal areas?
- What is the extent of post-smolt mortality in estuaries and coastal areas and what are the potential causes?
- Do origin or age/size of smolts influence the timing and route of post-smolt migration?
- Do post-smolts use tidal streams and shelf edge currents to migrate to feeding grounds?
- Do post-smolt migration routes follow preferred temperature and salinity fronts?
- Do extremes in the range of environmental conditions interrupt migration?
- Is the distribution of post-smolts at sea correlated with prey distribution and abundance?
- What is the behaviour of post-smolts in relation to that of potential predators such as seals, and vice versa?

However, it is clearly not possible at present to address adequately all these questions with the available tracking technology and methods. In order to increase our knowledge of the marine distribution and behaviour of post-smolts in the sea or to determine their survival, different strategies and methods need to be appraised and considered. Those that are likely to be successful for tracking post-smolts at sea are described here and their advantages and limitations are examined. The future technological developments that will be required to improve the systems for use on post-smolts in the sea are then described in the next section.

Active tracking and grid searching

Active tracking of individual fish is best suited to studying the detailed movements and behaviour of post-smolts as they emigrate from estuaries and first enter the marine environment. Grid searching for tagged fish can then be used to map their distribution at sea and is best used in conjunction with other tracking methods. Although a number of miniature acoustic tags (pingers) small enough for attaching to wild smolts are currently available, telemetry tags, which provide information on the physiological or environmental conditions, are still too large to attach or implant on wild smolts. However, the additional information that telemetry tags could provide would complement active tracking studies of post-smolts in the sea.

The following procedure has been devised for the active tracking of post-smolts in the sea and is based upon the experiences derived from several recent behaviour studies (Moore *et al.* 1992, 1995, Moore 1995, Lacroix & McCurdy 1996, Voegeli *et al.* 1998). The procedure ensures that the effects of handling and tag attachment on the fish are minimised, which is especially important when obtaining detailed behavioural data on salmon migration. Smolts should be trapped and tagged in fresh water and then allowed to migrate normally through the estuary before being tracked in the open sea. Tags should be less than 10% of body length (to allow insertion) and less than 5% of fish weight (a goal of 1–3% is preferable) to minimise effects on behaviour and survival. Capsule-shaped tags and special coatings should be used to maximise tag retention. However, it is recommended that detailed behavioural and physiological studies to assess the effects of each model of tag and method of attachment be carried out in the laboratory prior to large-scale field-based studies. After tagging, the fish should be retained only long enough for them to recover from the effects of anaesthesia before being released back into fresh water with other untagged smolts trapped at the same time. Retaining wild migrating fish in recovery tanks for longer periods can result in increased stress and mortality. Active tracking or grid searches may begin after fish are detected passing fixed listening stations in lower estuaries. It is recommended that a variety of environmental data (e.g. current speed and direction, water depth, temperature and salinity) be collected along a track to assist in interpretation of fish behaviour and movements.

These methods have been successfully used to look at behaviour and patterns of movement of post-smolts on a small scale in open waters (Moore 1995, 1996, Lacroix & McCurdy 1996, Voegeli *et al.* 1998). However, their use for studying the large-scale marine migrations of post-smolts may be limited by a number of factors. These techniques are logistically difficult, time-consuming and have to date provided only relatively short tracks (approximately 40 miles). In addition, only a single fish can be actively tracked at a time. Modern GPS provide considerable precision in ship position when following or locating a single animal, but the cost per datum in the open ocean is very high. Noise from turbulent flows associated with the moving ship and towing hydrophone systems during a track can be problematic, and coupled with the short tag detection range, may make it difficult to remain in contact with the tagged fish for any length of time. Specialised hydrophone systems for active tracking also remain expensive and require an expensive platform.

Active tracking can, however, provide detailed information on the migration routes and behaviour of post-smolts as they first enter the marine environment. Such an approach when used with telemetry tags (when sufficiently miniaturised) may also provide physiological and environmental data relative to the fish. Active grid searches can be valuable when used to complement large scale studies using fixed listening stations to fill in gaps in plotting migration routes, and to find fish lost during active tracking in order to resume the track.

Automated listening stations and hydrophone arrays

One strategy for future tagging studies in coastal waters involves deploying arrays of automated listening stations that continuously record the presence of tags. These may be deployed at specific locations, along known or expected migration routes, or by saturating a particular area in the sea. The listening stations will detect and record the time of passage of each tagged fish at a known position, and the data can be used to examine behaviour and determine survival. The same miniature acoustic tags and procedure described for active tracking also apply when using this approach. A sufficiently large number of tagged fish must be released to ensure some margin of detection success when used on a large scale; the larger the area, the greater the number of tags required. The method could also be used in the future to record telemetry data at the time of detection (real time), or to query and obtain stored information (e.g. historic tag depth and temperature records) from Communicating History Acoustic Transponders tags (CHAT tags). However, telemetry and CHAT tags are at present too large for use on post-smolts, but current telemetry tags could be attached to or inserted in kelts to obtain environmental information that could be related to post-smolt migration routes.

The use of fixed listening stations has been previously used to track smolts in areas where the migration routes of the fish are known. For instance, 'sonar buoys' have been used in estuaries (Moore *et al.* 1995, 1996), and underwater automated receivers have been used in estuaries and large bays (Lacroix & McCurdy 1996, Voegeli *et al.* 1998). However, effective deployment of listening stations requires some prior knowledge of post-smolt movements and migration routes. Active grid searches can help in finding potential migration routes and in the design of such studies.

With the recent availability of relatively cheap (approximately US$1000) and simple submersible automated receivers, it is possible to greatly increase the number of listening stations deployed (Voegeli et al. 1998). Long lines (10–20 nautical miles) of 50–100 receivers can be deployed as effective listening barriers to record the passage of fish and to monitor increasingly large areas of the sea. This approach could be used in situations where post-smolts are known to migrate along a narrow area at sea, such as in the Faroe-Shetland channel (see Chapter 6). Other situations where post-smolts either remain within specific areas or must migrate out of coastal waters through specific channels (e.g. the Bay of Fundy on the east coast of Canada) are also amenable to this approach. Initial cost is still high if used on a large scale, but the cost per datum could be lower if there is homing behaviour to a known area (where fewer listening stations would need to be deployed).

Many fish can be released and detected, especially if the recently tested uniquely coded tags are used (Voegeli et al. 1998). Data recovery from receiver or hydrophone arrays and regular redeployment of these units are still logistically difficult and a major expense of such an approach. International collaboration could make it feasible to saturate an environment with such receivers (e.g. specific sectors of known feeding grounds at sea or migration corridors) and have simultaneous releases of large numbers of tagged smolts, each with a unique code, from both sides of the Atlantic. The system would have a number of advantages once the migration routes of post-smolts have been determined and verified using other methods. Fixed listening stations sited at strategic positions could then be used to examine changes in the timing of smolt runs annually.

Underwater noise pollution, resulting in false detection or preventing detection of tagged fish, can also be a problem when using automated receivers. Such a problem was identified in recent studies in an area where seal scaring devices that transmit low frequency noise at high volume under water are used at a marine cage site for salmon aquaculture (Voegeli et al. 1998). Therefore protection of the acoustic band width used for tagging is a major issue, and there is a need for developing and encouraging standard codes.

Archival tags

Another strategy for obtaining information about the movements and environmental conditions experienced by salmon at sea includes the potential use of archival or data storage tags (DST – Fig. 5.2). These are microprocessor-controlled data-logging tags that record information about their surroundings on internal memory and operate independently of any external recording devices. DSTs are not actively monitored but rely on the tag being found and returned in order to retrieve the stored information. DSTs can provide a means of collecting information about the geographical position and behaviour of salmon in the sea as well as the marine environmental conditions they experience (Anon. 1997, 1998, Stone et al. 1998). The present generation of tags can collect environmental data such as temperature, depth, light, salinity, compass direction etc. DSTs when equipped with a geo-locating system (GLS) will provide data on the geographic position of the fish during its migration. This information can

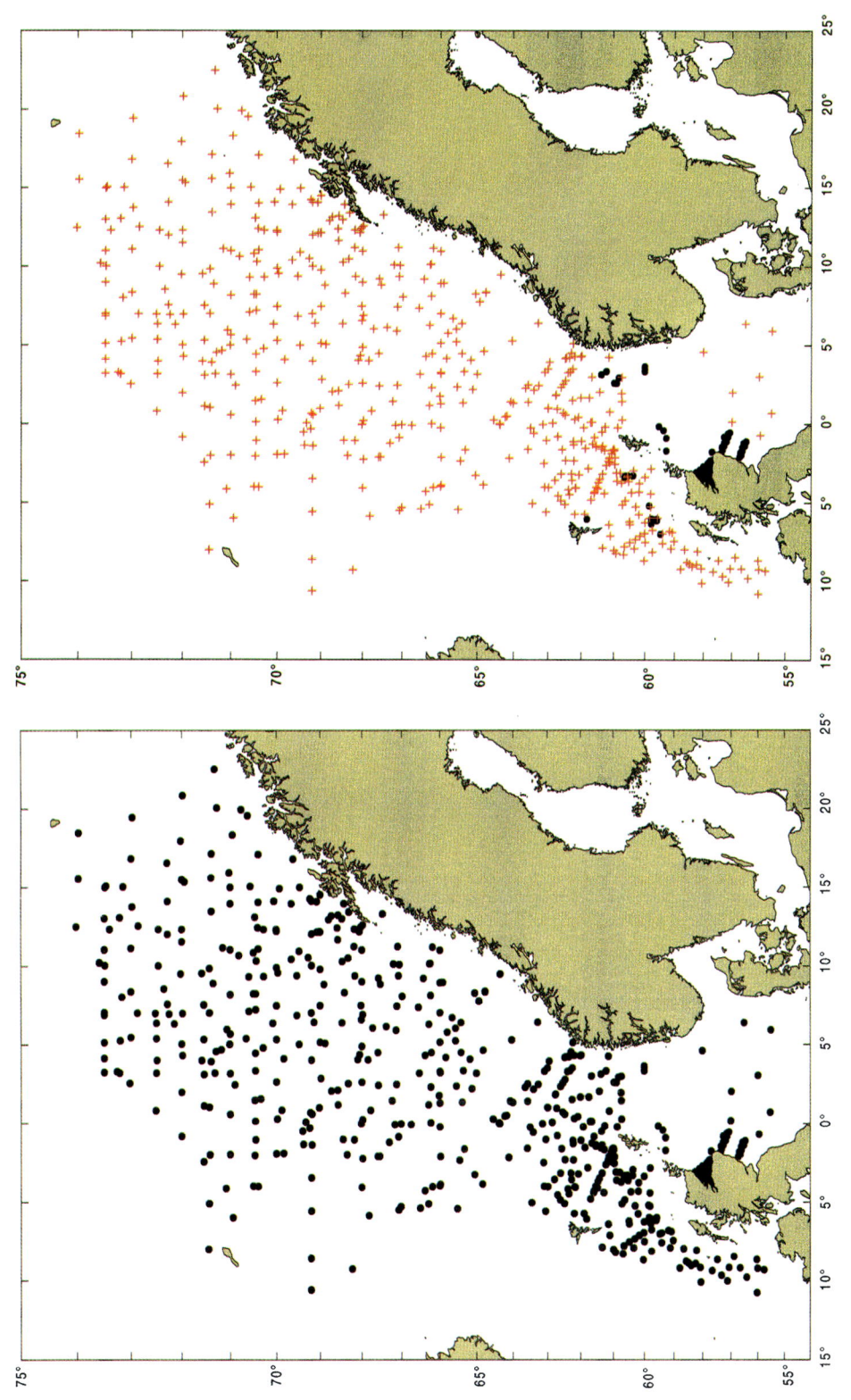

Plate 6.1 Trawling positions 1991–98.

Plate 6.2 Scottish (black circles) and Norwegian (red crosses) hauls.

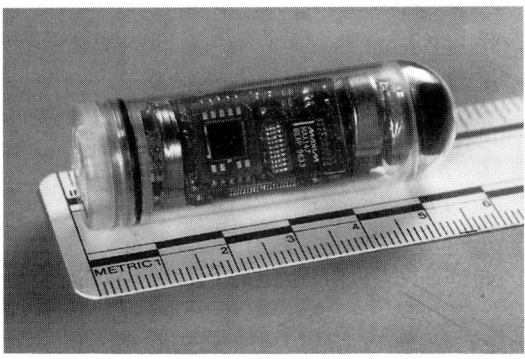

Fig. 5.2 The Lotek Marine Technologies Ltd 100 Archival Tag. The tag, developed at the CEFAS Laboratory, Lowestoft, incorporates pressure (depth), temperature and light sensors. The light sensor allows an estimation of day length and time of local noon, which can be used to determine geographic location.

then be coupled to environmental data such as sea surface temperature and productivity information derived from earth observation satellites to further describe their distribution in the sea.

The use of DSTs on post-smolts would involve implanting the tags into smolts according to the procedures described above for active tracking. Planning a study using this method would require a recovery mechanism for the fish, usually via a fishery (existing or instituted as part of the study), and the release of a sufficient number of tags in relation to the expected rate of recovery. However, most DSTs that currently have a geo-locating system are too large relative to the size of both wild smolts and post-smolts to implant or attach externally. Large post-smolts (>35 cm fork length) caught later at sea or kelts ('surrogate post-smolts') leaving the rivers could be tagged using the smaller of the existing DSTs. In addition, smaller DSTs that collect information on just a single environmental factor (e.g. temperature) may be suitable to attach to larger post-smolts and subsequently examine their thermal range in the ocean (Sturlaugsson & Thorisson 1997, Anon. 1998, Reddin et al. Chapter 8).

Information already obtained from applications of DSTs on adult salmon in coastal waters clearly shows the potential importance of these tags in studies of salmon in the sea (Sturlaugsson 1995ab, Karlsson et al. 1996, Sturlaugsson & Thorisson 1997). Large post-smolt salmon have been caught in sufficient numbers at several locations in the north-west Atlantic (Reddin & Short 1991), and the north-east Atlantic (Holst et al. 1996, Shelton et al. 1997, Holst et al. Chapter 6), to encourage the planning of studies using DSTs to obtain detailed records of fish movements and the environment over part of the migration. However, it is unlikely that the use of these fish would provide information on the behaviour and distribution in coastal waters or their migration routes to the feeding grounds in the North Atlantic.

A major limitation to the use of DSTs in general is that imposed by the absence of an effective mechanism or fishery to recover tags and retrieve the stored data. Returns to rivers are generally less than 2–5% and returns from a target fishery (not a popular option) could be as low. This would mandate a huge release of DSTs that are still relatively expensive (> US$1000 each) in order to recover a few tags. There are also other issues that need to be resolved for effective use of these tags. Methods of external tag attachment or of implanting the tag with the GLS light stalk trailing externally need to be developed and tested. Algorithms for position correction need to be developed and tested to obtain better resolution of geographical position in order to interpret the movements of salmon in relation to environmental conditions.

Position fixing DSTs have been successfully used with Atlantic bluefin tuna where the migrations are on the scale of oceans and the present GLS precision is considered adequate. The large size of these fish also allows for effective tag implantation and, most importantly, there is a fishery ensuring recovery of some tags to make studies logistically and economically feasible (Block et al. 1998).

Archival tags have already provided valuable information on salmon migration, and with GLS options they will make significant advances in describing the behaviour and distribution of salmon in the sea in relation to their environment. Further miniaturisation of DSTs with position fixing capabilities is now required and an effective, non-lethal recovery mechanism for the tags. However, until the developments have been satisfactorily carried out, the use of DSTs for tracking post-smolt movements during the early part of their marine life is limited.

Future development required to implement tracking strategies

Active tracking and acoustic tags

Acoustic tags (pingers) are at present small enough (8 mm diameter × 18 mm length, and less than 2 g) for implanting in wild smolts (13–18 cm) without adverse effects on behaviour and survival (Moore et al. 1992). Lighter tags would be preferable, but size is now a function of the commercially available batteries that can power the tag.

Power consumption of tags is one of the major constraints to successful tracking of post-smolts over long distances in the marine environment. Batteries are the heaviest component of most acoustic tags and there is always a trade-off between power (affecting signal range and duration) and weight/size when designing a tag to limit adverse effects on the fish. Power consumption can be addressed with improved battery power densities (although there are physically defined limits to this), decreasing transmission frequency using 'smart pingers' (e.g. tags that turn on and off at pre-set times), and increasing the number of listening stations and their sensitivity (Voegeli et al. 1998).

Currently, miniature pingers usable in smolts have a detectable signal range of about 200–400 m and will allow tracking for up to 90 days. Tracking at sea requires that these limits be improved while maintaining current miniature tag size and weight. Because of the limitations of active tracking described in the previous section (e.g. only a single fish can be tracked at a time), there is a requirement for methods by which many tagged fish can be released simultaneously and individually identified. This is being addressed with the current development and testing of miniature coded tags using single-microchip technology. This may permit the production of tags with thousands of unique identification codes and should allow large-scale collaborative releases of tagged smolts. This new technology emphasises the need to develop and encourage the use of standard codes, so that each tagged fish could then be recognised and reported from anywhere that acoustic receivers were in use.

The recent availability of acoustic towed body systems with multiple hydrophones (with increased sensitivity) and receivers with directional analysis and display features (F.A. Voegeli, pers. comm.) should increase the ability to extend the range and duration of active tracking. However, these systems are still very expensive and will benefit from the increased availability of new technology.

Telemetry tags

A wide variety of environmental and physiological factors may be telemetered using specific transducers that convert them into electrical signals (see list in Wolcott 1995). Some form of signal processing is usually necessary to compress data for transmission; most acoustically telemetered field data are 'interval' encoded (i.e. the interval between pings is proportional to the measured environmental factor). For parameters that change rapidly (e.g. physiological variables such as electromyogram) power consumption of the tag is large and limits the amount of tracking that can be achieved. High power requirements and limited bandwidth for acoustic signals are also limitations of these tags. The present size of currently-available telemetry tags also limits their use on fish the size of post-smolts. As for most other tags, there is a need for miniaturisation, but this may also lead to problems associated with decreased range, similar to those of miniature acoustic tags.

The type of measurement that can be made is limited only by the ability to turn that measurement into a change in electrical potential. The variables presently include physiological measurements and environmental parameters that would help provide an understanding of factors affecting salmon at sea. Miniaturisation of telemetry tags that measure depth (to assess habitat utilisation), ambient temperature (to assess thermal preferences) and an activity related to feeding (to assess foraging activity) is probably the most immediate requirement.

Automated listening stations

Recent developments in electronic automated listening stations have increased their potential use in the study of tagged marine fishes (Klimley *et al.* 1998). However, further developments are required to increase the area of coverage of listening stations if they are to be used to describe the distribution of post-smolts in the sea. Increasing the range of detection can be achieved by increasing the sensitivity of the stations, the use of sophisticated hydrophones, amplification of signal strength, filtering noise, and improving codes and detection algorithms. The disadvantage with this approach is that the developments will significantly increase the cost of the receivers with only a small gain in reception range. This is largely due to the fact that it is the limited output of the tags themselves that ultimately limits any gain. Another approach being pursued to increase coverage at sea is to reduce the cost and complexity of the receivers, while maintaining range of detection. This would permit more stations to be deployed at a greater density and increase the chances of detecting tags (Voegeli et al. 1998, O'Dor & Webber 1998). The development of an effective strategy for the use and deployment of long lines of listening stations would improve the potential contribution of listening stations for tracking post-smolts in the marine environment. There is also a need to establish a mechanism to utilise existing marine acoustic systems developed and used by the military, e.g. hydrophone arrays and ultra-acoustic signal detection cables, for marine animal telemetry work.

Archival tags

The size and weight of current DSTs need to be significantly reduced if they are to be used on smolts or small post-smolts (<35 cm fork length). A number of currently available DSTs record only one or two environmental factors such as ambient temperature and depth and these may be used on post-smolts >30 cm in length (Anon. 1998). However, there is an increasing need to incorporate a GLS (e.g. light sensor) to obtain the position of post-smolts in the marine environment, and this has been addressed by a number of manufacturers (Fig. 5.2). This is critical to understanding the migration routes and distribution of post-smolts in relation to environmental conditions (e.g. to interpret their movements in relation to thermal and other oceanic features or to examine feeding behaviour in relation to availability of prey items). Improvement in the accuracy and resolution of the GLS will require the development of algorithms for position correction, e.g. depth correction for light changes. Light sensors used in conjunction with sensors measuring the angle of magnetic dip could improve information on both the latitude and longitude of the fish and hence fix its position. The next generation of DSTs now being developed at CEFAS, Lowestoft, that have position fixing capabilities, should be of a size suitable for use on larger smolts and post-smolts (T. Storeton-West pers. comm.). However, before positioning fixing DSTs are deployed in large numbers, validation studies on their accuracy are required, particularly in study areas such as the North Atlantic.

Where large amounts of information are collected by DSTs during a single project, e.g. via frequent sampling, multiple sensors and long term studies, there may be a need to compress data and employ on-board signal processing. However, many biologists prefer to obtain the raw data for post-processing once the tag has been retrieved. Close collaboration between the manufacturers and biologists may therefore be required in the design of DSTs to provide a tool that will adequately address the biological question being addressed.

There may also be a requirement to develop a mechanism for transferring the stored data from a DST on a fish back to the researcher without the need for capturing and killing the fish. Fishery-independent tags such as CHAT tags are being developed by two groups to attempt to address this issue (Klimley *et al.* 1998). Unlike other data-storage archival tags, these can be reprogrammed whilst attached to a fish, and multiple memory fills and downloads are possible during the life of the tag. The process requires the fish to swim within the vicinity of a listening station to offload the data which can include the same information as on DSTs. Listening stations 'wake-up' tags that then transmit processed and compressed data using an ultra-acoustic signal. As for most other tags, CHAT tags currently available are usable only on large fish such as sharks and would need considerable miniaturisation for eventual use on post-smolts. In addition, the range of the tag would need to be significantly increased and some previous knowledge on the movements and migration routes of the salmon would be needed to site the listening stations correctly.

'Pop-up' tags

The low tag projected return rate during a number of DST studies on tuna has fuelled the development of a 'pop-up' archival tag that transmits stored data to satellites via radio telemetry. The recovery rate of DSTs could be greatly improved by the associated development and testing of a small 'pop-up' mechanism and the incorporation of radio technology for transmitting stored data to an aircraft, and eventually to satellites. The recent launch of the Iridium satellite system may improve the feasibility of such a tracking system. Such a tag could detach from the fish after specified time intervals or with changes in water conductivity, e.g. upon return of the fish to fresh water. DSTs currently need to be recovered, usually via a fishery, to retrieve stored information, limiting their usefulness even if they are to be suitably reduced in size for use in post-smolts. Alternatively, the number of retrievals could also be increased by the manufacture of a buoyant tag that would allow an additional small number to be returned after stranding on the shore (Anon. 1998).

In addition, the development of a small release and float mechanism to bring currently available miniature radio tags to the water surface for transmission of position could lead to the use of a new, unexplored approach to track the migration of post-smolts. Searches using standard radio telemetry technology from aircraft or ships could then be used to detect signals and determine position. By releasing a number of fish with tags having different 'pop-up' intervals, the timing and routes of post-smolts migration could be determined. This approach would probably involve attaching the tags externally and would require development and testing of procedures to ensure minimal effect on the fish. The development of suitable attachment techniques would also be required to ensure tag retention until release, particularly during long-term studies. Radio tags are currently the smallest (5 mm × 5 mm × 12 mm) and lightest (1.0 g) of the commercially available electronic tags and could be readily carried by wild smolts. However, the technique would require the development of a miniature attachment, release and float mechanism, and in addition the range of the tags would need to be in the order of kilometres to allow for adequate detection from low flying aircraft and 'homing in' by marine vessels.

Tag design and attachment

Electronic tags should have a minimal effect on the behaviour and survival of the fish, whether they are implanted internally or externally attached. There is a need for more extensive testing of the effects of tags on fish physiology and behaviour if the results of tracking studies are to be interpreted accurately. Further developments in tag shape, casing materials and attachment methods are necessary to increase and ensure retention when implanted for long-term studies. This is also particularly important where there is considerable growth in the fish and for energetic considerations. Internal implant of acoustic tags has been successful, but specific methods need to be developed and tested for implanting archival geolocation tags that need to trail a light stalk or other sensors and for attaching 'pop-up' radio tags. New methods of tag insertion should be explored, e.g. oviduct insertion (Peake *et al.* 1997). A major disadvantage of internally implanting DSTs is the need for the fish to be later recaptured and killed to retrieve the tag and obtain the stored information. Attachment methods that would allow the tag to be recovered and the fish to be released alive need to be developed, particularly when studying fish from small or endangered populations.

Potential collaborations

Tracking post-smolts in the sea can be an expensive and logistically difficult enterprise. Large-scale studies of salmon in the sea and the development of appropriate technology should involve international collaboration and careful cost-benefit analyses of methods. The following collaborations are recommended for future developments and tracking of salmon at sea:

- Establish an international working group to be involved in validation and standardisation of data formats and identification codes. Data standards would provide more uniform products from manufacturers and make data sharing among researchers feasible. It would reduce the increasing probability of one study interfering with another, with potentially disastrous results. Code sharing would allow the information to be recovered from an increased number of sources.
- Establish a working group to determine and compare the costs and benefits of the different tracking approaches outlined above and to identify practicable options.
- Explore and develop links with the military, i.e. those involved in naval research, and other agencies with existing ocean monitoring experiments in order to utilise existing marine acoustic networks to study salmon in the sea.
- Explore the possibility of conducting studies involving simultaneous releases of tagged smolts or post-smolts from east and west sides of the North Atlantic with

large scale collaborative efforts at tracking post-smolts at sea with either acoustic tags or DSTs.
- Establish a working group that includes researchers involved in tagging marine species other than salmon to develop collaborative projects where potential predators and prey species for salmon are tagged and tracked simultaneously in coastal waters and at sea.
- Extend studies using DSTs on adult salmon at sea, kelts and post-smolts and smolts when archival tags have been miniaturised and suitable recovery mechanisms have been developed.

References

Anon. (1988) Report of the Working Group on North Atlantic Salmon, Copenhagen, International Council for the Exploration of the Sea CM 1988/Assess:16.

Anon. (1997) Report of the study group on ocean tagging experiments with data logging tags. International Council for the Exploration of the Sea CM 1997/M:3.

Anon. (1998) Report of the study group on ocean tagging experiments with data logging tags. International Council for the Exploration of the Sea CM 1998/G:17.

Block, B.A., Dewar, H., Williams, T., Prince, E.D., Farwell, C. & Fudge, D. (1998) Archival tagging of Atlantic bluefin tuna (*Thunnus thynnus thynnus*). *Marine Technology Society Journal*, **32**, 37–46.

Boehlert, G.W. (1997) Application of acoustic and archival tags to assess estuarine, nearshore and offshore habitat utilization and movements by salmonids. NOAA Technical Memorandum NMFS, NOAA-TM-NMFS-SWFSC-236, Boston.

Eriksson, T. (1994) Mortality risks of Baltic salmon during downstream migration and early sea-phase: effects of body size and season. *Nordic Journal of Freshwater Research*, **69**, 100.

Friedland, K.D., Hansen, L.P. & Dunkley, D.A. (1998) Marine temperatures experienced by post-smolts and the survival of Atlantic salmon, *Salmo salar* L. in the North Sea area. *Fisheries Oceanography*, **7**, 22–34.

Greenstreet, S.P.R. (1992) Migration of hatchery reared juvenile Atlantic salmon, (*Salmo salar*) smolts down a release ladder. 1: Environmental effects on migratory behaviour. *Journal of Fish Biology* **40**, 683–94.

Hansen, L.P. & Jonsson, B. (1989) Salmon ranching experiments in the River Imsa; Effect of timing of Atlantic (*Salmo salar*) smolt migration on the survival to adults. *Aquaculture*, **82**, 367–73.

Holm, M., Axelsen, B.E., Sturlaugsson, J., Hvidsten, N.A., Ikonen, E. & Johnsen, B.O. (1998) Behaviour of acoustically tagged post-smolts in the Trondheim fjord – Influence of hydrographical and meteorological conditions on migration. International Council for the Exploration of the Sea CM 1998/N:17.

Holm, M., Huse, I., Waatevik, R., Doving, K.B. & Aure, J. (1982) Behaviour of Atlantic salmon smolts during seaward migration. 1. Preliminary report on ultra-acoustic tracking in a Norwegian fjord system. International Council for the Exploration of the Sea CM 1982/M:7.

Holst, J.C., Hansen, L.P. & Holm, M. (1996) Observations of abundance, stock composition, body size and food of post-smolts of Atlantic salmon caught in the North East Atlantic during summer. International Council for the Exploration of the Sea CM 1996/M:4.

Karlsson, L., Ikonen, E., Westerberg, H. & Sturlaugsson, J. (1996) Use of data storage tags to study the spawning migration of Baltic salmon (*Salmo salar* L.) in the Gulf of Bothnia. International Council for the Exploration of the Sea CM 1996/M:9.

Klimley, A.P., Voegeli, F., Beavers, S.C. & Le Boeuf, B.J. (1998) Automated listening stations for tagged marine fishes. *Marine Technology Society Journal*, **32**, 94–101.

Lacroix, G.L. & McCurdy, P. (1996) Migratory behaviour of post-smolt Atlantic salmon during initial stages of seaward migration. *Journal of Fish Biology*, **49**, 1086–101.

Mills, D. (1989) *Ecology and Management of Atlantic Salmon*. Chapman and Hall, London.

Mills, D. (ed.) (1993) *Salmon in the Sea and New Enhancement Strategies*. Fishing News Books, Oxford.

Moore, A. (1995) The movements of Atlantic salmon (*Salmo salar* L.) and sea trout (*Salmo trutta* L.) smolts through the river and lower estuary of the River Conwy, North Wales. *NRA Report*, 312.

Moore, A., Potter, E.C.E. & Buckley. A.A. (1992) Estuarine behaviour of migrating Atlantic salmon smolts. In: *Wildlife Telemetry* (eds I.M. Priede & S.M. Swift), pp. 390–9. Ellis Horwood Limited, Chichester.

Moore, A., Potter, E.C.E., Milner, N.J. & Bamber, S. (1995) The migratory behaviour of wild Atlantic salmon smolts in the estuary of the River Conwy, North Wales. *Canadian Journal of Fisheries and Aquatic Sciences*, **52**, 1923–35.

Moore, A., Stonehewer, R., Kell, L.T., Challiss, M.J., Ives, M., Russell, I.C., Riley, W.D. & Mee, D.M. (1996) The movements of emigrating salmonid smolts in relation to the Tawe barrage, Swansea. In: *Barrages: Engineering Design & Environmental Impacts* (eds N. Burt & J. Watts), pp. 409–17. H.R. Wallingford Ltd, John Wiley & Sons Ltd, Chichester.

Moore, A., Ives, S., Mead, T.A. & Talks, L. (1998) The migratory behaviour of wild Atlantic salmon (*Salmo salar* L.) smolts in the River Test and Southampton Water. *Hydrobiologia*, **371/372**, 295–304.

O'Dor, R.K. & Webber, D.M. (1998) A brief history of marine fish and invertebrate tagging. In: *Marine Animal Telemetry Tags* (eds G.S. Stone, H.C. Tausig & J.R. Schubel) pp. 17–27. New England Aquarium, Aquatic Forum Series, Report 98-3, Boston.

Peake, S.R., McKinley, R.S. & Beddow, T.A. (1997) New procedure for radio transmitter attachment: oviduct insertion. *North American Journal of Fisheries Management*, **17**, 757–62.

Potter, E.C.E. & Moore, A. (1992) *Surveying and Tracking Salmon in the Sea*. Atlantic Salmon Trust, Pitlochry.

Reddin, D.G. & Shearer, W.M. (1987) Sea surface temperature and distribution of Atlantic salmon (*Salmo salar* L.) in the north-west Atlantic. *American Fisheries Society Symposium*, **1**, 262–75.

Reddin, D.G. & Short, P.B. (1991) Post-smolt Atlantic salmon in the Labrador Sea. *Canadian Journal of Fisheries and Aquatic Sciences*, **48**, 2–6.

Salminen, M., Kuikka, S. and Erkamo, E. (1995) Annual variability in survival of sea ranched Baltic salmon, *Salmo salar* L.: significance of smolt size and marine conditions. *Fisheries Management and Ecology*, **2**, 171–84.

Shelton, R.G.J., Turrell, W.R., Macdonald, A., McLaren, I.S. & Nicoll, N.T. (1997) Records of post-smolt Atlantic salmon *Salmo salar* L. in the Faroe-Shetland Channel in June 1996. *Fisheries Research*, **31**, 159–62.

Stone, G.S., Tausig, H.C. & Schubel, J.R. (eds) (1998) *Marine Animal Telemetry Tags*. New England Aquarium, Aquatic Forum Series, Report 98-3, Boston.

Sturlaugsson, J. (1994) The food of ranched Atlantic salmon postsmolts (*Salmo salar* L.) in coastal waters, W-Iceland. *Nordic Journal of Freshwater Research*, **69**, 43–7.

Sturlaugsson, J. (1995a) The possible usage of data storage to study the migration and growth of salmon in the sea – in Iceland). Institute of Freshwater Fisheries Research, Research Report VMST-R/95007, Reykjavik.

Sturlaugsson, J. (1995b) Migration study on homing of Atlantic salmon (*Salmo salar* L.) in coastal waters in west Iceland – depth movements and sea temperature recorded at migration routes by data storage tags. International Council for the Exploration of the Sea CM 1995/M:17.

Sturlaugsson, J. & Thorisson, K. (1995a) Postsmolts of ranched Atlantic salmon (*Salmo salar* L.) in Iceland: II. The first days of the sea migration. International Council for the Exploration of the Sea CM. 1995/M:15.

Sturlaugsson, J. & Thorisson, K. (1995b) Postsmolts of ranched Atlantic salmon (*Salmo salar* L.) in Iceland: III. The first food of sea origin. International Council for the Exploration of the Sea CM.1995/M:16.

Sturlaugsson, J. & Thorisson, K. (1997) Migratory pattern of homing Atlantic salmon (*Salmo salar* L.) in coastal waters of West Iceland, recorded by data storage tags. International Council for the Exploration of the Sea CM. 1997/CC:09.

Voegeli, F.A., Lacroix, G.L. & Anderson, J.M. (1998) Development of miniature pingers for tracking Atlantic salmon smolts at sea. *Hydrobiologia*, **371/372**, 35–46.

Wolcott, T.G. (1995) New options in physiological and behavioural ecology through multi-channel telemetry. *Journal of Experimental Marine Biology and Ecology*, **193**, 239–56.

Chapter 6
Distribution and Possible Migration Routes of Post-smolt Atlantic Salmon in the North-east Atlantic

JENS CHRISTIAN HOLST, RICHARD SHELTON, MARIANNE HOLM and LARS PETTER HANSEN

Introduction

Until recently, records of the occurrence of Atlantic salmon in the north-east Atlantic were predominantly confined to the sites of directed high seas and coastal fisheries and isolated records of salmon caught incidentally in fisheries for other species. In 1991, the senior author noticed that samples of herring, *Clupea harengus* L., taken near the surface of the Norwegian Sea during a pair-trawl survey, sometimes also contained young salmon (Holst *et al*. 1993). Following this observation, the Norwegian Institute of Marine Research (IMR) in Bergen put in hand a programme of opportunistic but systematic sampling of young salmon in the surface waters of the north-east Atlantic (Holst *et al*. 1996, Holm *et al*. 1998). The best results were achieved by towing pelagic trawls rigged to fish at the surface at some 2.5–4.0 knots (Holm *et al*. 1996). From 1996 onwards, the Norwegian programme was supplemented by additional observations made in the early summer from research vessels operated jointly by staff from the Freshwater Fisheries Laboratory, Pitlochry, and the Marine Laboratory, Aberdeen. Between them, these two laboratories comprise The Scottish Office Agency, Fisheries Research Services (FRS) (Shelton *et al*. 1996, 1997 and in press).

The IMR and FRS laboratories have worked closely together on the marine sampling of young salmon since 1996, co-ordinating research cruises wherever possible. The present paper is the first compilation of the total data set collected by the two organisations and aims to sum up their joint knowledge concerning post-smolt distribution and migrations during the first 4 months at sea in the north-east Atlantic. The paper also briefly considers pelagic fisheries for other species with the potential to reduce sea-year class strength in European salmon stocks.

Methods

The data presented in this study were collected during the years 1991–98 by the IMR (1991–98) and the FRS Laboratories (1996–98). The surveys were conducted during May–June in the Ireland/Scotland/Shetland area and during July and August in the Norwegian Sea. The Scottish surveys in 1997 and 1998 were planned as salmon fishing cruises, although restraints existed on the geographical range of sampling. During the Norwegian surveys and the Scottish 1996 survey, the sampling of salmon was done opportunistically, and so the sampling strategy could not be optimised with regard to the distribution of salmon.

The post-smolts were caught using pelagic trawls fitted with additional flotation on the sweeps and upper front panel to achieve sampling from the surface downwards to 15–30 metres according to the individual net used (Valdemarsen & Misund 1995, Holm *et al.* 1996, Shelton *et al.* 1997). All Norwegian hauls lasted for 30 minutes while the duration of the Scottish hauls varied from about 40 to 120 minutes. The trawling speed varied from 2.5 to 4 knots. The results of 582 trawl hauls distributed over a large area are included in this study (Plate 6.1 in Colour Section between pp. 56 & 57). Of these hauls 84 hauls were taken by the Scottish vessels while 497 were taken by the Norwegian vessels (Plate 6.2).

Trawling was undertaken in late May–early June in the areas north of Ireland, west of Scotland, in the Faroe-Shetland Channel, in the Cromarty and Moray Firths, in the North Sea out to 50 miles between Montrose and Cruden Bay and on the Viking Bank from about 60°N to 61°N. Sampling was carried out in the Norwegian Sea in July–August from approximately 62°N and northwards to 74°N.

Results

Geographical distribution of catches

Post-smolt salmon were caught over a wide geographical range throughout the sampled area (Plates 6.3 & 6.4). The young salmon were caught in late May–early June in a narrow area stretching northwards from north of Ireland, west of Scotland and into the Faroe-Shetland Channel (Fig. 6.1). Post-smolts were also caught over the same period off the eastern coast of Scotland out to 50 miles, in the Cromarty and Moray Firths, in the northern North Sea and off the west coast of Norway. Yearly herring surveys covering most of the Norwegian Sea and using the described trawling methods in May 1995–98 yielded no young salmon, indicating that the post-smolts

Fig. 6.1 Three post-smolt Atlantic salmon taken in the surface waters of the Faroe–Shetland Channel in June 1997 by FRS Scotia.

had not yet reached this area in May. Catches of post-smolts were made in July–August in the Norwegian Sea, from 62°N to 73°N.

Post-smolt distribution and hydrography

Over much of the sampled area, catches were closely linked to the main surface currents. Thus, in the areas north of Ireland and west of Scotland and in the Faroe-Shetland Channel, the post-smolts were caught in a narrow band close to and running parallel with the 200 m depth contour (Plates 6.3 & 6.4). The strong Slope Current follows this depth contour along the continental shelf north-eastwards from the Porcupine Bank west of Ireland, off western Scotland, the Outer Hebrides and Shetland. It is the major surface hydrographic feature of the area (Huthnance & Gould 1989). The main filament of transport is some 20–40 km in width. Trawling to either side of it only rarely yielded catches of post-smolts.

The pattern observed off Scotland continued into the Norwegian Sea, where the catches were generally linked to the main surface currents extending northwards via the Slope Current along the western edge of the Vøring plateau approximately parallel with the 2000 m contour northwards to about 64°N (Poulain *et al.* 1996). Above this latitude, the post-smolts were spread more diffusely over a wider area. This observation is consistent with the less pronounced current systems at these latitudes in the Norwegian Sea (Poulain *et al.* 1996). Less exploratory fishing has been undertaken in the North Sea. Results to date suggest that post-smolt salmon from east coast Scottish rivers move rapidly offshore in a northerly and easterly direction only loosely related to surface currents.

Geographical origins of the post-smolts

Broad clues to the likely geographical origins of the post-smolt samples come from their river-age compositions. Generally speaking, post-smolts caught in the Faroe-Shetland Channel and the western sector of the Norwegian Sea had lower river-ages than post-smolts caught in the eastern Norwegian Sea. On this evidence, fish from southern Europe (principally the British Isles, France and the Iberian Peninsula) dominate the western samples while post-smolts from northern Europe (principally Norway) are better represented in the eastern samples.

Corroborative evidence in favour of the above interpretation comes from recoveries of micro-tags applied to hatchery smolts in their rivers of origin in Spain, England, Wales and Ireland.

One Spanish tag (River Asón), one Welsh tag (River Ogmore) and 26 Irish tags (Rivers Erne, Shannon, Burrishoole, Bundorragha and Lee) were taken in the Faroe-Shetland Channel during the Scottish surveys (Tables 6.1 and 6.2), while one Welsh (unknown river origin) and one Irish (River Ballynahinch) tag was taken during the Norwegian survey in the same area. One English tag (River Test) and one Irish tag

Table 6.1 Summary details of fish tagged as smolts caught by FRV Scotia close to the south-eastern end of the Wyville-Thomson Ridge (59°47′N, 6°10′W) in June 1996.

Recapture			Release			Journey		
Haul no.	Date	No of fish	Location	Position	Date	No. of days	Distance travelled (km)	Minimum speed (cm/sec)
212	06.06.1996	4	Shannon	52°40′N 09°00′W	25.04.1996	42	874	24
215	08.06.1996	2	Bundorragha (Delphi)	53°35′N 09°51′W	01.05.1996	38	713	22
		2	Burrishoole	53°51′N 09°41′W	18.04.1996	51	713	16
		1	Burrishoole	53°51′N 09°41′W	24.04.1996	45	713	18

An additional fish tagged as a 1-year-old parr in the River Ogmore (Wales) in June 1995 was taken in haul 212.

Distribution and Possible Migration Routes of Post-smolt Atlantic Salmon

Table 6.2 Summary details of fish tagged as smolts caught close to the south-eastern end of the Wyville-Thomson Ridge in June 1997.

Haul no.	Recapture Position	Recapture Date	No. of fish	Location	Release Position	Release Date	No. of days	Journey Distance travelled (km)	Journey Minimum speed (cm/s)
212	59°47'N 06°22'W	03.06.1997	1	Erne	54°30'N 08°12'W	10.03.1997	85	740	10.1
	"	"	2	Shannon	52°40'N 06°22'W	24.04.1997	40	1000	28.9
	"	"	1	Burrishoolee	53°53'N 09°35'W	25.04.1997	39	750	22.3
214	59°47'N 06°17'W	03.06.1997	1	Delphi	53°36'N 09°47'W	23.04.1997	41	820	23.1
	"	"	2	Erne	54°20'N 08°12'W	10.03.1997	85	730	9.9
	"	03.06.1997	1	Bundorragha (Delphi)	53°37'N 09°40'W	16.01.1997	138	800	6.7
216	59°47'N 06°19'W	06.06.1997	2	Shannon	52°40'N 06°22'W	24.04.1997	43	1010	27.2
217	59°44'N 06°09'W	06.06.1997	3	Shannon	52°40'N 06°22'W	24.04.1997	43	1010	27.2
	"	06.06.1997	1	Shannon	52°40'N 06°22'W	29.04.1997	38	1010	30.8
	"	06.06.1997	1	Burrishoole	53°53'N 09°35'W	25.04.1997	42	820	22.6
	"	06.06.1997	1	Burrishoole	53°53'N 09°35'W	10.03.1997	88	820	10.8
	"	06.06.1997	1	Lee	52°15'N 09°43'W	11.04.1997	56	1050	21.7

An additional fish tagged as a 1 year old parr in the river Asón (Spain) in October 1996 was taken in haul 212.

were taken in the western Norwegian Sea at about 71°N during the Norwegian surveys.

Post-smolt distribution and intercepting fisheries

Given the evidence of the Norwegian and Scottish surveys that post-smolt salmon are surface-swimming shoaling fishes, we suggest that only two major fisheries have the potential to intercept the post-smolt migration and catch substantial numbers of young salmon as a by-catch. These fisheries are by purse-seine and pelagic trawls, rigged to fish close to the surface. Purse-seine fisheries seem to be of secondary concern since this gear encloses only a very limited area of sea surface when the seine is set. However, in the most densely populated salmon feeding areas, post-smolts could be caught in unacceptable numbers if the level of fishing effort were high enough. We have adequate data for assessing the situation only for the area north of 62°N. No large purse-seine fishery operates at present in this area in a manner that would conflict with the observed distribution of post-smolts. In the area south of 62°N, the data available to us are too limited for an exhaustive assessment. However, to our knowledge, no large-scale purse-seine fisheries are concentrated in the observed migration paths of the post-smolts.

On current evidence, we identify only one trawl fishery that has the potential to catch large numbers of post-smolts, namely the pelagic fishery for mackerel in the international zone in the Norwegian Sea. This is the only major surface trawl fishery known to us that takes place in the area where we have found post-smolts. According to available statistics, this fishery has taken place since 1989, yielding catches up to approximately 50 000 t per year (Belikov *et al.* 1998, Anon. 1998). Data obtained from the Norwegian Coast Guard indicate that the fishery is prosecuted by a fleet of up to 30 vessels, mainly of Russian nationality. The main fishery takes place in the international zone in the Norwegian Sea, close to the Norwegian Economic Exclusion Zone (EEZ) (Plate 6.5). It is apparent from the distribution of the catches of post-smolts and the reported distribution of the fishery, that they overlap markedly in both space and time.

Discussion

The observed distribution of post-smolts considered in relation to the prevailing hydrographic regime suggests a close correlation between strong northerly or northeasterly surface currents and post-smolt migrations in the north-east Atlantic. This conclusion is in accordance with the suggestions made by Hansen *et al.* (1993), who suggested a relationship between currents and post-smolt migrations based on post-smolt catches made in coastal areas off western Norway. From the available catches and current knowledge of the Slope Current running north-east in the areas west of Scotland, the general pattern appears to be that most of the post-smolts are caught close to the core of the current. Staying in the core of the current gives the post-smolts a maximum northerly transportation vector (an along-slope poleward velocity of 30–40 cm/s) that may enable them to reach their northern feeding areas with the least expenditure of energy in migratory swimming.

However, comparison of dates and places of release with those for recapture in the Faroe-Shetland Channel (Tables 6.1, 6.2), shows that the calculated migration speed may well fall below that of the Slope Current (Shelton *et al.* 1997). Swimming in local pursuit of food or entrainment in eddies may form part of the explanation. Certainly the evidence of scale analysis suggests that, for wild smolts, growth as a post-smolt is rapid from the beginning of sea entry. Thus the migration north is better regarded as a mobile banquet than as a headlong rush from the comparative poverty of the river to the relative richness of the sub-arctic.

Given the patchiness of our data, our conclusions about the main routes followed are inevitably tentative. They are summarised below by areas of origin.

Ireland and Wales

On the evidence of catches and tag recaptures, it seems evident that many of the post-smolts originating from Ireland and some from Wales enter the Slope Current and are transported first north and later north-eastwards past the Wyville-Thomson Ridge, through the Faroe-Shetland Channel and into the Norwegian Sea (Plate 6.6). The high catches consistently achieved at the south-eastern end of the Wyville-Thomson Ridge suggest that this small area may be an important intermediate feeding zone for post-smolts following this pathway north.

Scottish west coast

In the absence of direct evidence, we consider it most probable that post-smolts originating from south-west Scottish rivers enter the Slope Current as described above and follow it northwards via the Faroe-Shetland Channel into the Norwegian Sea (Plate 6.6). It is likely that post-smolts from more northerly west coast rivers enter the Slope Current via the Northern Minch.

Scottish east coast

The data available are too sparse for us to draw firm conclusions about the migration pattern of post-smolts originating from Scottish east-coast rivers. It is possible that they follow the Dooley current, which runs across the North Sea approximately from Aberdeen in a north-easterly direction at about 58°N (Plate 6.6). This current would lead the post-smolts towards the coast of Norway, where they could enter the coastal current that runs northwards along the Norwegian coast. Alternative routes might be to pass through the Fair Isle Channel between the Orkney and Shetland Islands or to swim northwards to pass east of Shetland. Both of the two last routes would involve counter current swimming for relatively long distances and so appear less likely.

Norwegian coast south of 62°N

Unfortunately, very few trawl hauls have been carried out along the Norwegian coast south of 62°N and we have no data to support a given migratory pattern of post-smolts from these areas. For the time being, we rely on the general hypothesis that the post-smolts will follow the main northerly surface currents (Plate 6.6). The Norwegian coastal current, originating in the Baltic, runs northwards along the entire coast. We believe that the post-smolts from these areas follow the Norwegian coastal current to about 62°N where two things may occur; either the fish pass directly northwards into the Norwegian Sea, where they mix with the Irish, Welsh and western Scottish fish entering from the Faroe-Shetland Channel, or they turn north-east and follow the currents onwards along the Norwegian coast. Only further investigations would enable us to choose which interpretation is correct.

Norwegian coast 62°N–68°N

The total absence of catches made along the Norwegian coast in the area north of 62°N during July–August was surprising. The area has been sampled (Plate 6.1) but no post-smolt was taken close to the coast. Based on the described relationship between the Slope Current and the post-smolt migrations off Scotland, we suggest a similar mechanism for fish originating in this area. After leaving the fjordic system, the fish are rapidly flushed north-eastwards along the Norwegian coast within either the Coastal Current or the Slope Current farther offshore (Plate 6.6). This means that the fish reach more northerly positions than those sampled along the Norwegian coast in July–August and explains why no fish have been caught here. A suitable period to catch fish in these areas appears to be from early June onwards. The post-smolts most probably stay in the current until it moves off the coast temporarily at about 68°N and they then disperse in a more west-north-westerly direction. Some post-smolts may follow the northerly currents even farther and not turn westwards until they pass Bear Island or maybe Spitsbergen. It is suggested that this mechanism applies for most of the fish leaving the Norwegian coast between 62°N and 71°N.

Norwegian coast north of 70°N

Post-smolts leaving the Norwegian coast north of 70°N and east of approx 19°E may, given the main hydrographic features of the Barents Sea, take an anticlockwise turn more or less north-east into the Barents Sea before swimming westwards in more northerly parts of the Barents Sea. It is possible that many of these fish will either leave the Barents Sea close to Bear Island or feed in the supposedly attractive Polar Front area close to Bear Island during their first summer.

Future priorities

The present account may be regarded as marking the end of the initial descriptive phase of exploratory fishing for post-smolt salmon in north-east Atlantic waters. Much remains to be done. Future priorities include the following items.

The geographical coverage achieved is far from adequate, even for the earliest weeks of the post-smolt phase. Future studies should pay special attention to inadequately-sampled areas within the inferred migration pathways. Coverage of the northern North Sea is especially poor and there is also scope for more work off the western coasts of both the British Isles and Norway. In planning this new work there may well be scope for employing computer simulation techniques, using predictions based on known current speeds and rates of migration, to define the sampling grids.

Much of the sampling reported here has been undertaken during the long daylight hours of the northern summer. Records of post-smolts taken in darkness are rather rare. The possibility that post-smolt and later life stages of salmon may leave the immediate surface at night (Shelton *et al.* 1997) is worthy of further study.

Separator panels have been used in conjunction with live fish cod ends in attempts to obtain fish in externally and neurologically intact condition. On occasion, the presence of jellyfish has led to unavoidable damage to catches. Nevertheless, both Norwegian and Scottish experience has shown that fish in good condition are potentially obtainable using existing methods, perhaps in conjunction with net-mounted closed circuit television (CCTV). This work should be pursued vigorously. Potential outputs include a better understanding of shoaling behaviour and a critical account of the geographical distribution of infestations by external parasites, especially the sea louse *Lepeophtheirus salmonis*.

Preliminary analyses of wild post-smolt samples show that growth in the early weeks at sea is rapid, at least in the southern part of their range. The presence of late summer growth checks on some returning adults (Chapter 4) suggests that marine growth and survival opportunities may be more critical at this time than in the first weeks after sea entry. Sampling salmon at the end of their first sea summer is now a matter of priority.

Our understanding of the extent to which young salmon may be caught inadvertently in fisheries for other species is still very incomplete. The systematic sampling of human consumption and industrial fisheries for salmon by-catch should be considered. On the evidence currently available, the highest priority should be accorded to the screening of pelagic catches obtained using gears rigged to fish close to the surface.

References

Anon. (1996) Report of the working group on North Atlantic salmon. International Council for the Exploration of the Sea CM 1996/Assess: 16.

Belikov, S.V., Jakupstovu, S.H., Shamrai, E. and Thomsen, B. (1998) Migration of mackerel during summer in the Norwegian Sea. International Council for the Exploration of the Sea CM 1998/AA:8.

areas in the ocean during late spring and summer (Thorpe 1988, Mills 1989). When moving down rivers (Fried et al. 1978) and entering marine waters (Carlin 1969, McCleave 1978, LaBar et al. 1978, Holm et al. 1982, Jonsson et al. 1993), their movements seem to depend on the speed and direction of surface currents. Most of the records available suggest that post-smolts move relatively quickly into the ocean. Further evidence of rapid migration comes from the fact that very few post-smolts are recorded in fjords and coastal waters during summer and autumn, although they are already present in oceanic areas in the east Atlantic (Holst et al. 1993, 1996, Holm et al. 1996, Shelton et al. 1997) and west Atlantic (Reddin & Short 1991) at this time of the year. However, Dutil & Coutu (1988) caught many post-smolts in a nearshore zone of the northern Gulf of St Lawrence, suggesting that this trait may vary among populations or areas.

Based on the distribution of catches north of Scotland, the fish appeared to move northwards with the shelf edge current (Shelton et al. 1997). Farther north in the Norwegian Sea post-smolts were caught beyond 70°N in July. Analysis of growth and smolt age distribution strongly suggested that a relatively high proportion of the post-smolts captured in the northern Norwegian Sea originated from rivers in southern Europe (Holst et al. 1996). This was supported by the recapture of a salmon that had been tagged in April 1995 in southern England, and recovered about 2000 km farther north three months later, demonstrating the capacity for rapid travel by post-smolts (Holst et al. 1996).

In the Labrador Sea, Reddin & Short (1991) caught many post-smolts in the early autumn of 1987 and 1988, using surface drift-nets. The highest catch rates occurred between 56 and 58°N, and they also caught many older salmon in the area. Based on age analysis, they concluded that post-smolts originating from Maine to Labrador were present. This was supported by recoveries of two post-smolts that had been tagged in the Penobscot River in Maine and Western Arm Brook in Newfoundland. In addition, four older fish in the Labrador Sea had tags, three from the Penobscot River and one from the Middle River, Nova Scotia.

It is easier to document the marine distribution of Atlantic salmon when the fish have reached catchable size. Many countries have been tagging smolts and adults, and some of these fish have been recaptured in high seas fisheries for salmon. It is difficult to know the true distribution in time and space of salmon at sea, as recoveries depend on the distribution of the fishery. The waters off the western coast of Greenland contain many immature salmon that have spent one winter at sea (one-sea-winter or 1SW) or more than one winter at sea (multi-sea-winter or MSW). Here, fish from both North America and Europe feed during June–September and are exploited.

Several studies have been carried out to assess the composition of the stocks present in the West Greenland area and establish the origin of these fish (Parrish & Horsted 1980, Reddin 1988). Swain (1980) analysed a time series of smolt tagging in European rivers in relation to recaptures off West Greenland, as did Ruggles & Ritter (1980) for North American smolt tagging. Møller Jensen (1980a) used tag recoveries from West Greenland in 1972 to assess the distribution along the West Greenland coast of salmon origination from North America and Europe. These investigations demonstrated that Atlantic salmon from a large number of different rivers in North America and Europe were present in the area. Furthermore, based on much more comprehensive material, Reddin (1988) and Reddin *et al.* (1988) used a discriminant analysis of scale characteristics and concluded that catches of salmon off West Greenland are split fairly evenly between the two continents. It is more difficult to determine the country of origin, but Canada apparently accounts for most of the North American component and Scotland accounts for most of the European fish (Møller Jensen 1980a). All in all, salmon tagged as smolts in home waters and recaptured at West Greenland have originated from Canada, USA, Scotland, England, France, Norway, Sweden, Iceland and Ireland, and some Russian salmon may also be present in the area.

From 1965 to the beginning of the 1970s, tagging experiments took place off West Greenland (Møller Jensen 1980b). In total 4657 salmon were tagged and 93 fish were recaptured outside Greenland, 28 in North America and 65 in Europe. The tags were reported from United Kingdom (44), Canada (28), Ireland (16), Spain (3) and France (2). Taking into account the production of smolts in these countries, this suggests that larger proportions of salmon from more southern areas of Europe are present in this area than from farther north.

In the East Atlantic, salmon are found in large areas in the Norwegian Sea. In the 1970s there was an important commercial long-line fishery far north in the Norwegian Sea in February–May. Recoveries of fish in this fishery that had been tagged as smolts, and recaptures in coastal and freshwater fisheries of salmon tagged in the Norwegian Sea suggested that Norwegian salmon were most abundant, although fish from the United Kingdom, Sweden and Russia were also present. Most of the fish were recaptured in home waters the same year as when they were tagged, suggesting that they were maturing (Rosseland 1971). Towards the end of the 1970s fishing for salmon in the northern Norwegian Sea was banned, and fishing was limited to the area within the Faroese Economic Exclusion Zone (EEZ).

The abundance of salmon within the Faroese Economic Exclusion Zone (EEZ) has been assessed from sampling the fishery for a number of years. Jákupsstovu (1988) reported on a tagging programme at sea from 1969 to 1976 in which 1946 salmon caught on long lines were tagged and released. The fish were tagged in more southerly areas of the Faroes, and 1SW fish were probably highly over-represented. In total, 90 fish were recovered, 33 in Scotland, 31 in Norway, 15 in Ireland, eight in other European countries, and three at West Greenland. The great majority of the tags were reported in home waters the same year as they were tagged. However, it is interesting to note that some fish in the area may have been on their way westwards, as they were reported from West Greenland later that year.

In recent years a tagging programme has been carried out in the major fishing grounds north of the Faroes (Fig. 7.1). From November to March in the 1992–93, 1993–94 and 1994–95 fishing seasons, 5448 salmon (3811 wild and 1637 farmed) were caught by long-line, tagged and released (Hansen & Jacobsen 1997, and in press). The

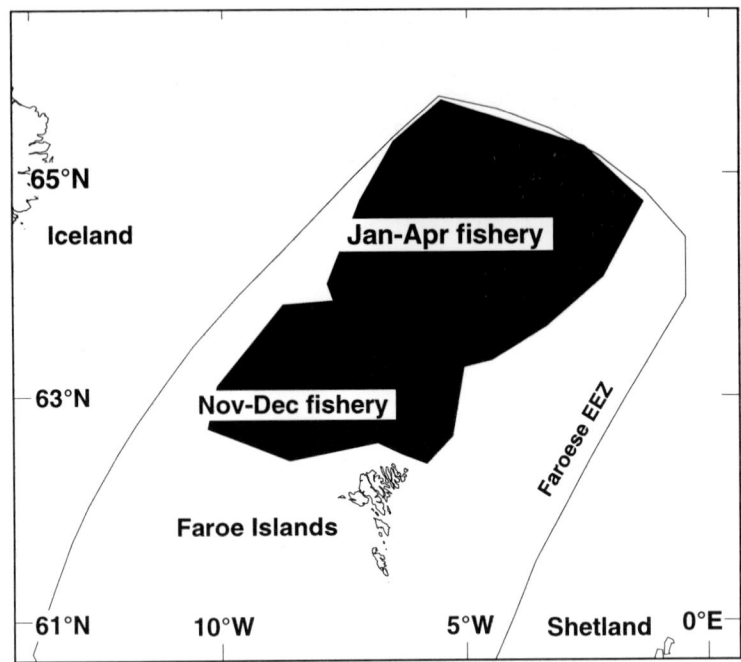

Fig. 7.1 Areas of tagging north of the Faroes. The autumn fishery is located closer to the isles and as the season progresses the fishery moves in a north-easterly direction farther into the Norwegian Sea.

overall recapture rate of the number of salmon tagged was small, up to November 1998 106 fish had been recovered, which is 1.9% of the number of salmon tagged. Of wild fish 87 individuals (2.3% of the number tagged) were recaptured, whereas 19 (1.2%) of the fish identified as fish farm escapees were recovered (Hansen & Jacobsen in press). The recapture rate of fish of farmed origin is significantly lower than for wild fish. For both wild and farmed salmon, the recapture rates of fish tagged in the autumn were lower than those of fish tagged in winter. Tags were reported from large areas in the North Atlantic, both from marine fisheries and in fresh water. No tags were recovered from the research fishery at Faroes or from West Greenland; the latter could be due to the fact that there has been a marked decrease in the level of exploitation in that fishery in recent years.

In wild fish the majority of the tags were reported from Norway, but there were relatively large numbers from Scotland and Ireland (Table 7.1, see Hansen & Jacobsen in press). Additional recaptures were reported from Ireland, Iceland, Spain, Sweden, Denmark, England, and even Canada. The geographic distribution of recaptures of fish tagged in the three respective fishing seasons (pooled data) is shown in Fig. 7.2 (Hansen & Jacobsen in press). Tag returns were scattered over large areas of Norway and Scotland, suggesting that fish from large areas of these countries were found together at the Faroes. It is interesting to note, however, that four recaptures were reported from Canada, one tagged in March 1993 and recaptured in Miramichi

Distribution and Migration of Atlantic Salmon in the Sea

Table 7.1 Recaptures in number of wild salmon in different countries tagged at Faroes during the 1992–93, 1993–94 and 1994–95 fishing seasons. From Hansen & Jacobsen (in press).

Country	Tagged Rec.93	1992–93 Rec.94	Tagged Rec.94	1993–94 Rec.95	Tagged Rec.95	1994–95 Rec.96	Total No	%
Norway	22	3	2		17	3	47	54.0
Scotland	8		1		3		12	13.8
Ireland	3		2		4		9	10.3
Sweden	2	1			1		4	4.6
Russia	1	1	3		1		6	6.9
Canada	1				3		4	4.6
Denmark	2						2	2.3
England	1						1	1.1
Iceland	1						1	1.1
Spain	1						1	1.1
Total	42	5	8	0	29	3	87	99.8

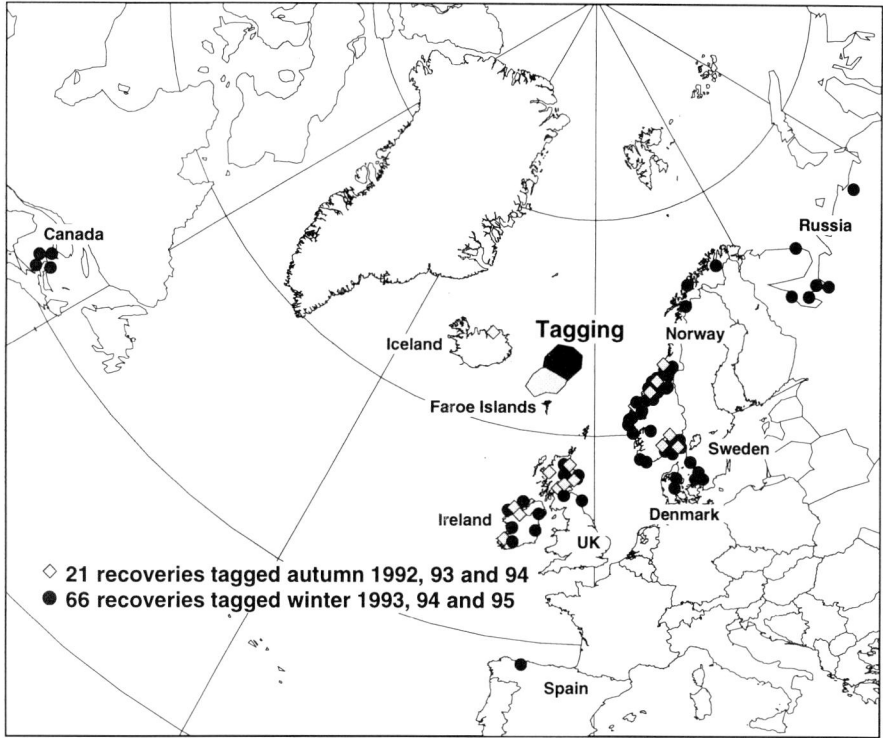

Fig. 7.2 Recoveries of wild Atlantic salmon tagged during the 1992–93 to 1994–95 fishing seasons. Fish were tagged north of the Faroes during autumn (November–December: light shading) and winter (February–March: dark shading). Recoveries from each tagging period are shown as light diamonds (autumn tagging) and dark circles (winter tagging). From Hansen & Jacobsen (in press).

River in September 1993, and three tagged in February–March 1995, two of which were subsequently recaptured in the Miramichi in September 1995 and one in Kouchibouguac River (close to Miramichi) in October 1995. It is also evident that salmon from distant areas, such as north-east Russia and Spain, were present at Faroes. Of fish recaptured relatively close to the tagging site there was no apparent difference in distribution between salmon tagged in the autumn and those tagged in the winter. This does not seem to be the case for fish that were recaptured relatively far away from the tagging sites. They were all tagged in the winter. This may suggest that salmon of distant origin are not present in the Faroese area in the autumn.

The overall estimates of the proportion of wild salmon originating from different countries in the research fishery during these three fishing seasons are presented in Table 7.2 (Hansen & Jacobsen in press) together with assumptions and approximations made. It is not surprising that Norway accounts for the major proportion (41%), whereas the mean estimated proportion of salmon from Scotland and Russia is close to 20%. For the other countries there are only a relatively small number of fish in the area.

Table 7.2 Results of 'at risk' simulation to estimate proportion (%) of fish tagged at Faroes and returning to different countries. Confidence limits (95%) were applied based on 1000 simulations. Recoveries were adjusted for homewater exploitation rates and tag reporting rates as provided by the North Atlantic Salmon Working Group members, 1997. From Hansen & Jacobsen (in press).

Country	No. recaptured	Tag reporting rate		Exploitation rate		Estimated recaptured	Simulation		
		min	max	min	max		'–5%	Mean (%)	'+95%
Norway	47	0.40	0.60	0.50	0.80	144.6	28.6	41.1	55.3
Scotland	12	0.80	1.00	0.10	0.30	66.7	9.1	19.9	34.3
Russia	6	0.60	0.80	0.10	0.15	57.1	8.2	18.9	31.0
Ireland	9	0.60	0.80	0.50	0.75	20.6	2.7	5.8	9.7
Denmark	2	0.40	0.60	0.14	0.34	16.7	0.0	4.9	12.3
Canada	4	0.65	0.85	0.35	0.55	11.9	0.8	3.4	6.8
England	1	0.40	0.60	0.15	0.35	8.0	0.0	2.4	7.2
Sweden	4	0.55	0.75	0.55	0.90	8.5	0.6	2.4	4.9
Spain	1	0.60	0.80	0.55	0.85	2.0	0.0	0.6	1.7
Iceland	1	0.80	1.00	0.40	0.60	2.2	0.0	0.6	1.9
Total	87					338.3		100.0	

All in all, the geographical distribution of recaptured wild fish in home waters strongly suggests that fish from most areas of the distribution range of Atlantic salmon are at some life stage present in the area in the Norwegian Sea north of the Faroe Islands. This is supported by the fact that tagged wild fish have been recovered in North America, in Spain and the eastern part of European Russia, as well as in all major salmon producing countries in Europe. This does not imply that all stocks are always abundant in the area, but some of them may pass through occasionally, or components of stocks sometimes use the area for feeding.

Of the 19 fish farm escapees recaptured, 18 were recovered from different areas of Norway, and one from the west coast of Sweden, at Ugglarp (Fig. 7.3, see Hansen & Jacobsen in press). Analysing scale samples from the Faroes fishery between the 1981–82 and the 1995–96 fishing seasons showed that the proportion of farmed fish was relatively low from the 1981–82 to the 1986–87 fishing season, and increased considerably thereafter, and reached a peak in the 1989–90 and 1990–91 fishing seasons, when more than 40% of the samples were farmed salmon. Then the proportion declined, and in the 1995–96, 1996–97 and 1997–98 fishing seasons the proportion of farmed fish was estimated to be around 20%. The estimated proportion of farmed salmon in the Faroese fishery was significantly correlated with the total production of farmed salmon in the north-east Atlantic. Taking into account the fact that the great majority of tagged farmed salmon were recovered in Norway and that Norway is the major producer of farmed salmon strongly suggests that most of the farmed salmon in the areas have escaped from Norwegian farms. At West Greenland the incidence of farmed fish appeared to be very small (Hansen et al. 1997).

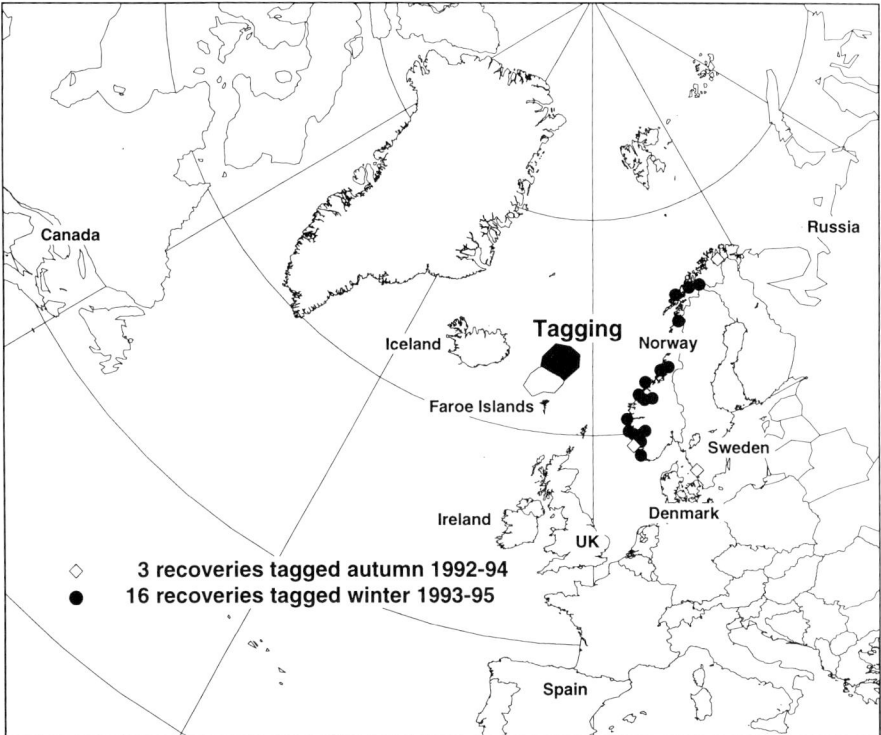

Fig. 7.3 Recoveries of escaped farmed Atlantic salmon tagged in the 1992/93–1994/95 fishing seasons north of the Faroes. Fish were tagged north of the Faroes during autumn (November–December: light shading) and winter (February–March: dark shading). Recoveries from each tagging period are shown as light diamonds (autumn tagging) and dark circles (winter tagging). From Hansen & Jacobsen (in press).

The distribution of Atlantic salmon in the ocean is still not well known but the limited information indicates that salmon are not evenly distributed. Salmon from North America seem to stay in the West Atlantic area, although some fish can move into the north-east Atlantic. It is also evident that a relatively large proportion of the European MSW salmon moves far into the west Atlantic to feed. Salmon from many populations differ in how long the fish stay at sea, and hence different sea-age classes from the same populations may be present in different areas. For example, MSW salmon may move farther away from home than grilse. A better map of Atlantic salmon distributions at sea in time and space would help us to understand the fluctuations in their survival and life history. The distribution of Atlantic salmon in the sea seems to reflect environmental factors like surface temperature and surface currents (Reddin & Friedland 1993), and probably also the availability of suitable food organisms (Chapter 13), as growth and survival are important fitness characters as well.

Homing migration

Atlantic salmon leave the ocean feeding grounds after 1–4 years. The factors initiating the homeward migration are unknown, but salmon have circannual rhythms of reproductive hormones, synchronised by photoperiod (Bromage et al. 1993). The usual pattern is that the older individuals return earlier in the season than younger ones (Dunkley 1986, Jonsson et al. 1990, Shearer 1992). However, the internal rhythms and responses to photoperiod are probably population-specific. For example, salmon ascend several Scottish rivers during all months of the year (Mills 1989), whereas in Norway salmon ascend rivers only from May to October. There is direct evidence for a genetic component in the seasonal return migration of Atlantic salmon (Hansen & Jonsson 1991).

From extensive data on tagged salmon, Hansen et al. (1993) concluded that the homing migration from the feeding areas in the north Norwegian Sea to natal rivers in Norway consists of two phases, an initial phase with orientation from the feeding areas towards the Norwegian coast and a second phase in coastal and estuarine waters with more precise orientation towards the home river. Hansen et al. (1993) estimated that salmon travelled 50–100 km per day in the Norwegian coastal current.

Little is known about the mechanisms of homing by salmon on the open ocean. Behavioural experiments with juvenile Pacific salmon indicate that they are sensitive to the magnetic field of the earth (Quinn 1980, Quinn & Brannon 1982) and that magnetite particles may be a component of the transduction mechanisms (Walker et al. 1988, Moore et al. 1990). Such a magnetic field detection system could aid in navigation but no conclusive experiments have been conducted on this subject.

Alternatively, the salmon may possess a compass orientation ability and head in a homeward direction without regard to their location at sea. This hypothesis is consistent with interannual variation in timing and landfall of Fraser River sockeye salmon (Blackbourn 1987, Groot & Quinn 1987), and with experimental evidence that juvenile salmon can orientate using the sun's position (Groot 1965) and magnetic field (Quinn 1980, Quinn & Brannon 1982).

After salmon accomplish the migration from distant, oceanic feeding grounds to the nearshore environment, the coastal waters and estuaries present another set of challenges to their orientation systems. The homing migration of Atlantic salmon does not seem to be passive nor current-guided, as suggested by Taguchi (1957), Mathisen & Gudjonsson (1978) and Isaksson (1980). When homing to Norwegian rivers the fish approached the fjords from both the north and the south, and may have followed the coastline for long distances (Hansen et al. 1993). A similar pattern has also been observed along the east coast of Canada (Reddin and Lear 1990). Salmon tracked by Smith et al. (1980) off the east coast of Scotland showed the ability to swim in fixed directions at energetically efficient speeds but were also affected by tidal currents.

Hansen et al. (1993) reported that Atlantic salmon travel rates slowed in fjords, perhaps because they require time to locate their natal rivers. Westerberg et al. (1982a) and Døving et al. (1985) reported that telemetered salmon moved up and down in the water column in association with fine-scale hydrographic stratification, and they concluded that salmon search for vertical gradients of odours from the home river rather than horizontal gradients.

Alternatively, the apparently random movements of many salmon in coastal waters and estuaries may not reflect orientation mechanisms but the tendency of many populations to wait for suitable conditions for upstream migration (Jonsson et al. 1990) or to remain in the estuary as they undergo the physiological processes of maturation and osmoregulatory adaptation to fresh water. Atlantic salmon tracked in estuaries of the Baltic Sea (Westerberg 1982b), Nova Scotia (Brawn 1982) and England (Potter 1988, Priede et al. 1988) tended to move back and forth with tidal currents for some time before ascending their home river.

The juvenile salmon probably learn their way (imprint) sequentially during seaward migration and use that information when they return as adults (Hasler 1966, Harden Jones 1968, Hasler & Scholz 1983, Hansen et al. 1989, Dittman & Quinn 1996). They can even locate the small tributary where they grew up (Youngson et al. 1994). Olfaction is probably the most important cue guiding the fish on the final route to their natal stream. It has also been hypothesised that population-specific odours (pheromones) from conspecifics living in the river or migrating to sea guide maturing salmon (Nordeng 1977, 1989). Most evidence, however, supports the sequential learning hypothesis. Atlantic salmon released as smolts in a small stream devoid of salmon returned there to spawn at maturity (Hansen et al. 1989). Pacific salmon also returned to their release site rather than to a site containing members of their population (Donaldson & Allen 1957, Brannon & Quinn 1990), and can home to rivers scented with artificial odours on which they imprinted, even in the absence of pheromones (Hasler & Scholz 1983).

Experimental transportation and release studies indicate that salmon may not merely imprint once in fresh water but rather that odours are learned in a more complex sequence. Atlantic salmon released as smolts in the River Imsa returned there as adults with high precision (Hansen et al. 1989). However, when smolts imprinted to River Imsa water were transported by boat about 45 km towards the ocean, and thus deprived of a part of the migration route, the great majority of the adults failed to return to the Imsa, suggesting that the salmon needed to experience the outward migration in order to find their home river at maturity. Atlantic salmon transported and released as post-smolts on the feeding grounds north of the Faroe Islands failed to home as adults to their home river; in fact, no fish were reported caught in fresh water (Hansen et al. 1993). This may indicate a reduced preference for these fish to enter rivers and spawn. All adult recaptures, however, were made in Norwegian home waters eastward of the site of release. No maturing salmon were reported from elsewhere despite heavy salmon fishing in the UK, Ireland and North America, and in Icelandic rivers. Natural mortality of post-smolts in the sea is high (Mills 1989), and the recaptures are few and scattered over large areas. However, the fact that all adult recaptures of the released post-smolts were reported from Norwegian home waters supports the idea that the salmon may have an inherited crude compass sense of direction (Walker et al. 1988, Moore et al. 1990), and that sequential learning may not be mandatory to sense the general direction back to Norway.

Hansen & Jonsson (1994) provided evidence to support the hypothesis that the learning process during migration to the sea becomes fixed, and is not overridden by a new learning process at the post-spawning stage. Kelts transplanted from their home river to several distant rivers returned as second time spawners to the river they left as smolts, not to the river from which they entered the ocean as adults.

References

Blackbourn, D.J. (1987) Sea surface temperature and the prediction of age-at-return in stocks of chum salmon from the Fraser River, British Columbia, and other areas. *Proceedings of the 1987 Northeast Pacific Pink and Chum Salmon Workshop, Juneau, Alaska*, Alaska Department of Fish and Game, Seattle.

Brannon, E.L. & Quinn, T.P. (1990) A field test of the pheromone hypothesis for homing by Pacific salmon. *Journal of Chemical Ecology* **16**, 603–9.

Brawn, V.M. (1982) Behavior of Atlantic salmon (*Salmo salar*) during suspended migration in an estuary, Sheet Harbour, Nova Scotia, observed visually and by ultrasonic tracking. *Canadian Journal of Fisheries and Aquatic Sciences*, **39**, 248–56.

Bromage, N., Randall, C., Duston, J., Thrush. M. & Jones, J. (1993) Environmental control of reproduction in salmonids. In: *Recent Advances in Aquaculture*, Vol. 4. (eds J.F. Muir & R.J. Roberts), pp. 55–65. Blackwell Scientific, London.

Carlin, B. (1969) Migration of salmon. *Lectures Series, Atlantic Salmon Association Special Publication*, Montreal, Canada, 14–22.

Dittman, A.H. & Quinn, T.P. (1996) Homing in Pacific salmon: mechanisms and ecological basis. *Journal of Experimental Biology*, **199**, 83–91.

Donaldson, L. & Allen, G. (1957) Return of silver salmon, *Oncorhynchus kisutch* (Walbaum) to point of release. *Transactions of the American Fisheries Society*, **87**, 13–22.

Døving, K.B., Westerberg, H. & Johnsen, P.B. (1985) Role of olfaction in the behavioral neuronal responses of Atlantic salmon, *Salmo salar*, to hydrographic stratification. *Canadian Journal of Fisheries and Aquatic Sciences*, **42**, 1658–67.

Dunkley, D. (1986) Changes in the timing and biology of salmon runs. In: *The Status of Atlantic Salmon in Scotland.* (eds D. Jenkins & W.M. Shearer), *Institute of Terristrial Ecology Symposium*, **15**, 20–7. ITE, Abbot's Ripton.

Dutil, J.-D. & Coutu, J.-M. (1988) Early marine life of Atlantic salmon, *Salmo salar*, postmolts in the northern Gulf of St Lawrence. *Fisheries Bulletin*, **86**, 197–212.

Fried, S.M., McCleave, J.D. & LaBar, G.W. (1978) Seaward migration of hatchery-reared Atlantic salmon, *Salmo salar*, smolts in the Penobscot River estuary, Maine: riverine movements. *Journal of the Fisheries Research Board of Canada*, **35**, 76–87.

Groot, C. (1965) On the orientation of young sockeye salmon (*Oncorhynchus nerka*) during their seaward migration out of lakes. *Behaviour., Supplement*, **14**, 1–198.

Groot, C. & Quinn, T.P. (1987) The homing migration of sockeye salmon to the Fraser River. *Fisheries Bulletin*, **85**, 455–69.

Hansen, L.P. & Jacobsen, J.A. (1997) Migration, growth and origin of wild and escaped farmed Atlantic salmon, *Salmo salar* L., tagged and released north of the Faroe Islands. CM 1997/AA:05.

Hansen, L.P. & Jonsson, B. (1991) Evidence of a genetic component in the seasonal return pattern of Atlantic salmon, *Salmo salar*. L. *Journal of Fish Biology*, **38**, 251–8.

Hansen, L.P. & Jonsson, B. (1994) Homing of Atlantic salmon: effects of juvenile learning on transplanted post-spawners. *Animal Behaviour*, **47**, 220–2.

Hansen, L.P., Jonsson, B. & Andersen, R. (1989) Salmon ranching experiments in the River Imsa: is homing dependent on sequential imprinting of the smolts? In: *Proceedings of the Salmonid Migration and Distribution Symposium* (eds E. Brannon & B. Jonsson), *NINA, Trondheim, Norway*, pp. 19–29. School of Fisheries, University of Washington, Seattle, USA.

Hansen, L.P. & Jacobsen, J.A. (in press) Origin and movements of Atlantic salmon tagged and released at the Faroe Islands.

Hansen, L.P., Jonsson, N. & Jonsson, B. (1993) Oceanic migration of homing Atlantic salmon, *Salmo salar*. *Animal Behaviour*, **45**, 927–41.

Hansen, L.P., Reddin, D.J. & Lund, R.A. (1997) The incidence of reared Atlantic salmon in the commercial fishery at West-Greenland. *ICES Journal of Marine Science*, **54**, 152–5.

Harden Jones, F.R. (1968) *Fish Migration*. Edward Arnold, London.

Hasler, A.D. (1966) *Underwater guideposts: homing of salmon*. University of Wisconsin Press, Madison, Wisconsin.

Hasler, A.D. & Scholz, A.T. (1983) *Olfactory Imprinting and Homing in Salmon*. Springer-Verlag, Berlin.

Holm, M., Holst, J.C. & Hansen, L.P. (1996) Sampling Atlantic salmon in the NE Atlantic during summer: methods of capture and distribution of catches. International Council for the Exploration of the Sea CM 1996/M:12.

Holm, M., Huse, I.., Waatevik, E., Døving, K.B. & Aure, J. (1982) Behaviour of Atlantic salmon smolts during seaward migration. I. Preliminary report on ultrasonic tracking in a Norwegian fjord system. International Council for the Exploration of the Sea CM 1982/M:7.

Holst, J.C., Hansen, L.P. & Holm, M. (1996) Preliminary observations of abundance, stock composition, body size and food of postsmolts of Atlantic salmon caught with pelagic trawls in the NE Atlantic in the summers 1991 and 1995. International Council for the Exploration of the Sea CM 1996/M:4.

Holst, J.C., Nilsen, F., Hodneland, K. & Nylund, A. (1993) Observations of the biology and parasites of postsmolt Atlantic salmon, *Salmo salar*, from the Norwegian Sea. *Journal of Fish Biology*, **42**, 962–6.

Isaksson, A. (1980) Salmon ranching in Iceland. In: *Salmon Ranching* (ed. J.E. Thorpe), pp. 131–56. Academic Press, London.

Jákupsstovu, H.I. (1988) Exploitation and migration of salmon in Faroese waters. In: *Atlantic Salmon: Planning for the Future* (eds D.H. Mills & D.J. Piggins), pp. 458–82. Croom Helm, London & Sydney, Timber Press, Portland, Oregon.

Jonsson, N., Jonsson, B. & Hansen, L.P. (1990) Partial segregation in timing of migration of different aged Atlantic salmon. *Animal Behaviour*, **40**, 313–21.

Jonnsson, N., Hansen, L.P. & Jonsson, B. (1993) Migratory behaviour and growth of hatchery-reared post-smolt Atlantic salmon *Salmo salar*. *Journal of Fish Biology*, **42**, 435–43.

LaBar, G.W., McCleave, J.D. & Fried, S.M. (1978) Seaward migration of hatchery-reared Altantic salmon

(*Salmo salar*) smolts in the Penobscot River estuary. Maine: open-water movements. *Journal du Conseil International pour Exploration de la Mer*, **38**, 257–69.

McCleave, J.D. (1978) Rhythmic aspects of estuarine migration of hatchery-reared Atlantic salmon (*Salmo salar*) smolts. *Journal of Fish Biology*, **12**, 559–70.

Malloch, P.D.H. (1910) *Life History of Salmon, Trout and Other Freshwater Fish*. A. and C. Black, London.

Mathisen, O.A. & Gudjonsson, T. (1978) Salmon management and ocean ranching in Iceland. *Journal of Agriculture Research, Iceland*, **10**, 156–74.

Mills, D. (1989) *Ecology and Management of Atlantic Salmon*. Chapman and Hall Ltd, London.

Mills, D. (1993) (ed.) *Salmon in the Sea and New Enhancement Strategies*. Fishing News Books, Blackwell Scientific Publications Ltd, Oxford.

Møller Jensen, J. (1980a) Recaptures of salmon at West Greenland tagged as smolts outside Greenland waters. In: *ICES/ICNAF Joint Investigation on North Atlantic Salmon* (eds B.B. Parrish & Sv. Aa. Horsted). *Rapports et Procès-Verbaux de la Rèunion, Conseil International pour l'Exploration de la Mer*, **176**, 114–21.

Møller Jensen, J. (1980b) Recaptures from international tagging experiments at West Greenland. In: *ICES/ICNAF Joint Investigation on North Atlantic Salmon* (eds B.B. Parrish & Sv. Aa. Horsted) *Rapports et Procès-Verbaux de la Réunions, Conseil International pour l'Exploration de la Mer*, **176**, 122–35.

Moore, A., Freake, S.M. & Thomas, I.M. (1990) Magnetic particles in the lateral line of the Atlantic salmon (*Salmo salar* L.). *Philosophical Transactions of the Royal Society of London*, **B. 329**, 11–15.

Nordeng, H. (1977) A pheromone hypothesis for homeward migration in anadromous salmonids. *Oikos*, **28**, 155-9.

Nordeng, H. (1989) Salmonid migration; hypotheses and principles. In: *Proceedings of the Salmonid Migration and Distribution Symposium* (eds E. Brannon & B. Jonsson), pp. 19–29. School of Fisheries, University of Washington. Seattle.

Parrish, B.B. & Horsted, Sv. Aa. (eds). (1980) *ICES/ICNAF Joint Investigation on North Atlantic Salmon. Rapports et Procès-verbaux de la Réunion, Conseil International pour l'Exploration de la Mer*, **176**: 1–146.

Potter, E.C.E. (1988) Movements of Atlantic salmon, *Salmo salar* L., in an estuary in south-west England. *Journal of Fish Biology*, **33** (Supplement A), 153–9.

Potter, E.C.E. & Russell, I.E. (1994) Comparison of the distribution and homing of hatchery-reared and wild Atlantic salmon, *Salmo salar* L., from north-east England. *Aquaculture and Fisheries Management*, **25** (Supplement 2), 31–44.

Priede, I.G., Solbé, J.F. de L.G., Nott, J.E., O'Grady, K.T. & Cragg-Hine, D. (1988) Behaviour of adult Atlantic salmon, *Salmo salar* L., in the estuary of the River Ribble in relation to variations in dissolved oxygen and tidal flow. *Journal of Fish Biology*, **33** (Supplement A), 133–9.

Quinn, T.P. (1980) Evidence for celestial and magnetic compass orientation in lake migrating sockeye salmon fry. *Journal of Comparative Physiology*, **137**, 243–8.

Quinn, T.P. (1990) Current controversies in the study of salmon homing. *Ethology, Ecology and Evolution*, **2**, 49–63.

Quinn, T.P. & Brannon, E.L. (1982) The use of celestial and magnetic cues by orienting sockeye salmon smolts. *Journal of Comparative Physiology*, **147**, 547–52.

Reddin, D.G. (1988) Ocean life of Atlantic salmon (*Salmo salar* L.) in the northwest Atlantic. In: *Atlantic Salmon: Planning for the Future* (eds D. Mills & D. Piggins), pp. 483–511. Croom Helm, London and Sydney.

Reddin, D.J. & Friedland, K.D. (1993) Marine environment factors influencing the movement and survival of Atlantic salmon. In: *Salmon in the Sea and New Enhancement Strategies* (ed. D. Mills), pp. 79–103. Fishing News Books, Blackwell Scientific Publications Ltd, Oxford.

Reddin, D.G. & Lear, W.H. (1990) Summary of marine taggings studies of Atlantic salmon (*Salmo salar* L.) in the Northwest Atlantic area. *Canadian Technical Report of Fisheries and Aquatic Scinces*, **1737**: 1–115.

Reddin, D.G. & Shearer, W.M. (1987) Sea-surface temperature and distribution of Atlantic salmon in the Northwest Atlantic Ocean. *American Fisheries Society Symposium*, **1**, 262–75.

Reddin, D.G. & Short, P.B. (1991) Post-smolt Atlantic salmon (*Salmo salar*) in the Labrador Sea. *Canadian Journal of Fisheries and Aquatic Sciences*, **48**, 2–6.

Reddin, D.J., Stansbury, D.E. & Short, P.B. (1988) Continent of origin of Atlantic salmon (*Salmo salar* L.) caught at west Greenland. *Journal du Conseil International pour l'Exploration de la Mer*, **44**, 180–8.

Rosseland, L. (1971) Fiske av atlantisk laks i internasjonalt farvann. *Jakt-Fiske-Friluftsiv*, **100**, 190–195, 238–242. (In Norwegian).

Ruggles, C.P. & Ritter, J.A. (1980) Review of North American salmon to assess the Atlantic salmon fishery off West Greenland. *Rapports et Procès-verbaux de la Réunion, Conseil International pour l'Exploration de la Mer*, **176**, 82–92.

Shearer, W.M. (1992) *The Atlantic Salmon. Natural History, Exploitation and Future Management*. Fishing News Books, Blackwell Scientific Publications Ltd, Oxford.

Shelton, R.G.J., Turrell, W.R., Macdonald, A., McLaren, I.S. & Nicoll, N.T. (1997) Records of post-smolt Atlantic salmon, *Salmo salar* L., in the Faroe-Shetland Channel in June 1996. *Fisheries Research*, **31**, 159–62.

Smith, G.W., Hawkins, A.D., Urquhart, G.G. & Shearer, W.M. (1980) The offshore movements of returning Atlantic salmon. *The Salmon Net*, **13**, 28–32.

Swain, A. (1980) Tagging of salmon smolts in European rivers with special reference to recaptures off West Greenland in 1972 and earlier years. *Rapports et Procès-verbaux de la Réunion, Conseil International pour l'Exploration de la Mer*, **176**, 93–113.

Swain, A. (1982) The migrations of salmon (*Salmo salar* L.) from three rivers entering the Severn estuary. *Journal du Conseil International pour l'Exploration de la Mer*, **40**, 76–80.

Taguchi, K. (1957) The seasonal variation of the good fishing area of salmon and movements of the water masses in the waters north Pacific. II. The distribution and migration of salmon populations in offshore waters. *Bulletin of the Japanese Society for Fisheries Science*, **22**, 515–21.

Thorpe, J.E. (1988) Salmon migration. *Science Progress Oxford*, **72**, 345–70.

Walker, M.M., Quinn, T.P., Kirschvink, J.L. & Groot, C. (1988) Production of single-doman magnetite throughout life by sockeye salmon, *Oncorhynchus nerka*. *Journal of Experimental Biology*, **140**, 51–63.

Westerberg, H. (1982a) Ultrasonic tracking of Atlantic salmon (*Salmo salar* L.), II. Swimming depth and temperature stratification. *Report of the Institute of Freshwater Research, Drottningholm*, **60**, 102–20.

Westerberg, H. (1982b) Ultrasonic tracking of Atlantic salmon (*Salmo salar* L.) I. Movements in coastal regions. *Report of the Institute of Freshwater Research, Drottningholm*, **60**, 81–101.

Youngson, A.F., Jordan, W.C. & Hay, D.W. (1994) Homing of adult Atlantic salmon (*Salmo salar* L.) to a tibutary stream in a major river catchment. *Aquaculture*, **121**, 259–67.

Chapter 8
Survival of Atlantic Salmon (*Salmo salar* L.) Related to Marine Climate

D.G. REDDIN, J. HELBIG, A. THOMAS, B.G. WHITEHOUSE and K.D. FRIEDLAND

Recruitment variability observed for anadromous salmonids can be divided into that arising from their life in fresh water and that arising from their life in the sea. Survival of salmon (*Salmo salar* L.) in the sea can be easily measured from counts of smolts as they leave fresh water compared with the number of returning adults 1 or 2 years later. Furthermore, these rates can be adjusted for marine exploitation to provide an estimate of natural mortality/survival. Recent survival rates for salmon stocks of Western Arm Brook and Conne River, Newfoundland, and the North Esk, Scotland, demonstrated at least a 50 per cent decline in survival rates in recent years compared with rates earlier in the 1970s and 1980s (Plate 8.1 in Colour Section between pp. 88 & 89). The source of this increased mortality is unknown but seems to be occurring while the fish are in the sea; although causes that arise in fresh water but are manifested in the sea have not yet been completely eliminated.

Total returns to the river for six stock complexes in North America (Labrador, Newfoundland, Quebec, Gulf of St Lawrence, Scotia-Fundy and USA) were analysed for coherence. For 1-sea-winter (1SW) and 2-sea-winter (2SW) maturing salmon, the northern stocks from Labrador and Newfoundland were correlated with each other; whereas the more southerly origin stocks were correlated with each other but not with Newfoundland and Labrador. The strong similarity of the pattern of returns suggests that common events are influencing the production of North American salmon. These common events influence production and abundance for stocks that originate over a large area despite variable fishing effort, different gear, and different management practices applied to the various stock components. The apparent widespread nature of these events suggests climate as a possible cause.

The life of salmon in the sea has been frequently compared to a 'black box' into which salmon enter but in which relatively few survive. Factors related to the marine environment (climate) have been blamed without knowing the specific source of mortality. Analysis of satellite sea surface temperature (SST) data provides a way of opening the black box, especially if coupled with information from research vessels fishing at sea. The factors affecting survival of Atlantic salmon were investigated by an exploratory analysis of inshore and offshore SSTs. The SSTs used are from the global weekly composite Multi-Channel Sea Surface Temperature (MCSST) data set, which is derived from the Advanced Very High Resolution Radiometer (AVHRR) instrument on board the NOAA polar orbiting satellites. The data set for the time

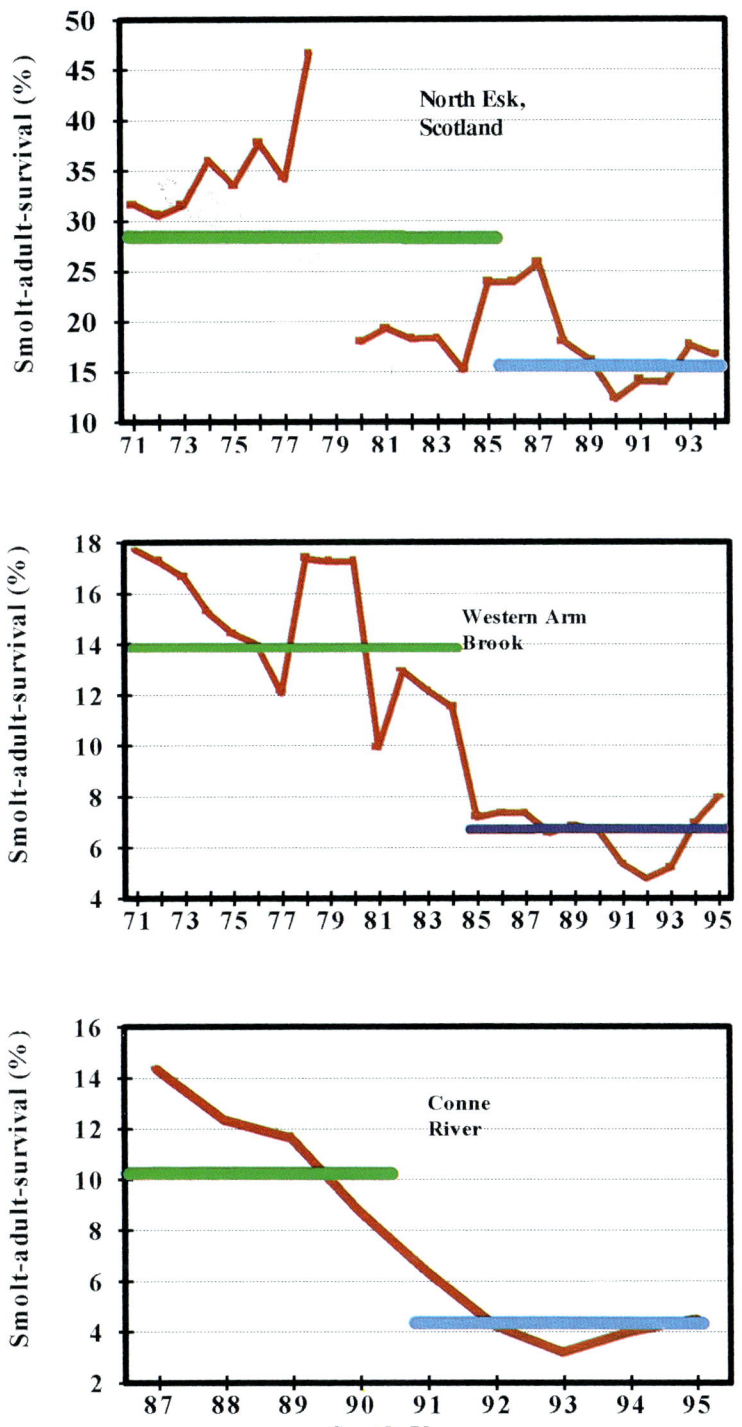

Plate 8.1 Sea survival rates corrected for commercial exploitation for North Esk, Scotland, Western Arm Brook and Conne River, Newfoundland. Values are 3-year moving averages and means are represented by bars.

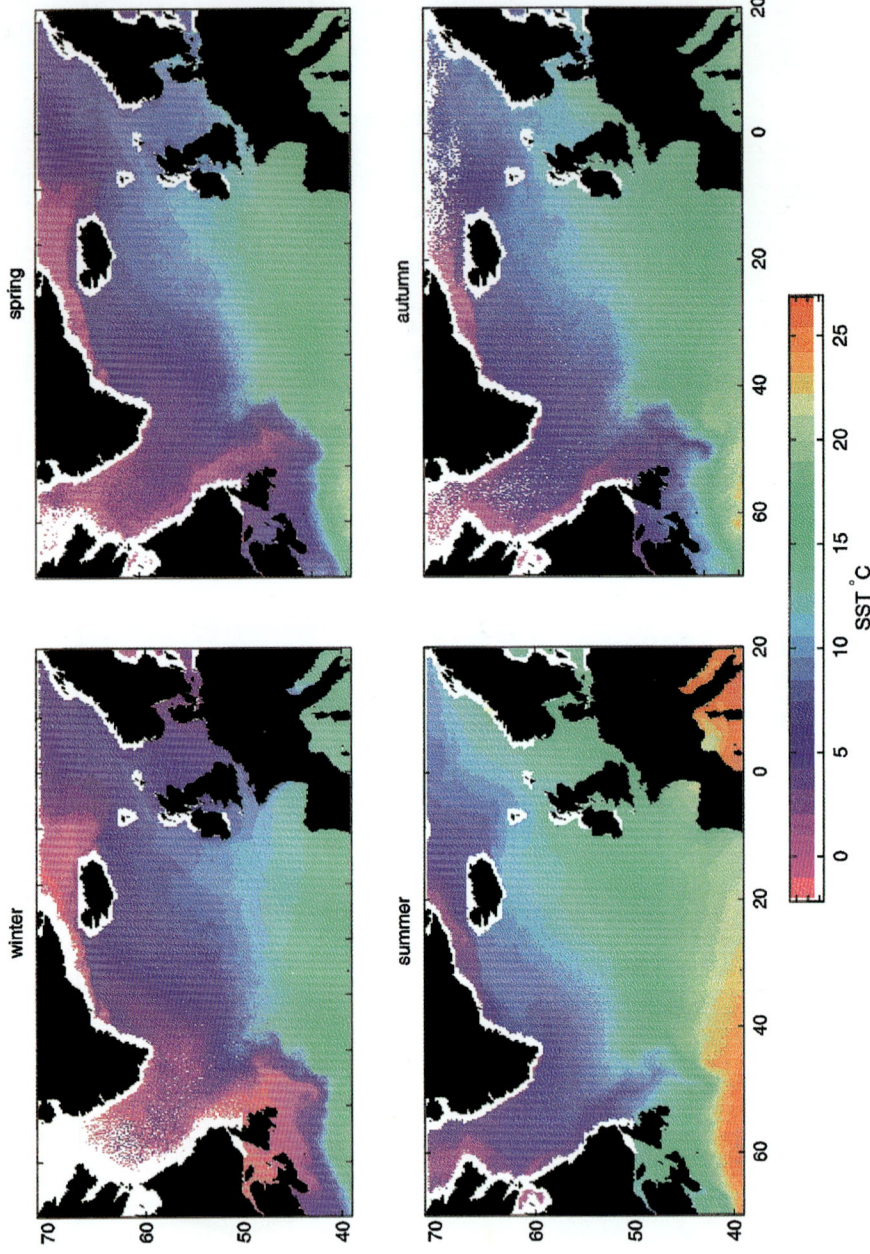

Plate 8.2 Seasonal climatologies of MCSST data, 1982–96.

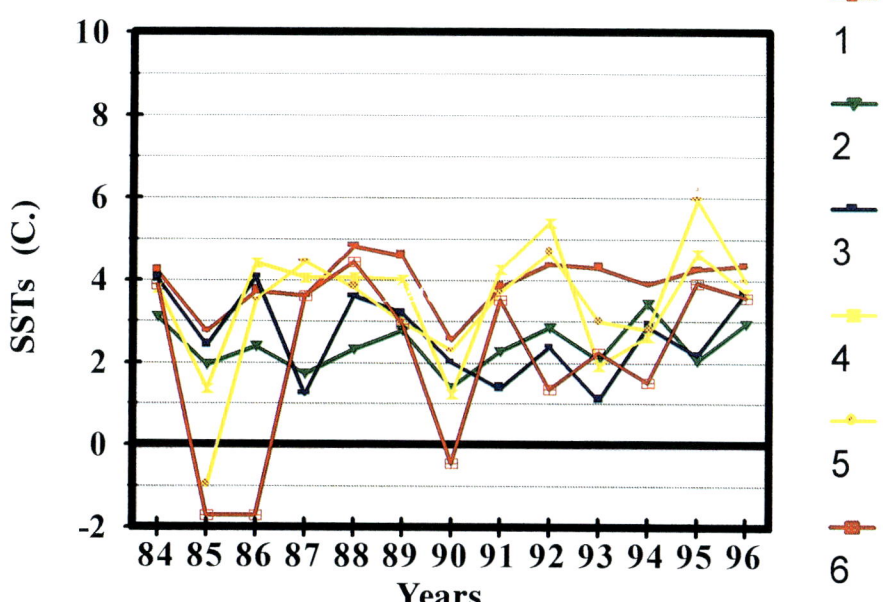

Plate 8.3 Inshore Newfoundland sea surface temperatures from MCSST data. Locations are (1) Bonavista Bay, (2) Trepassey, (3) Bay d'Espoir, (4) Bay St. George, (5) Humber Arm, and (6) Western Arm Brook, 1984–86.

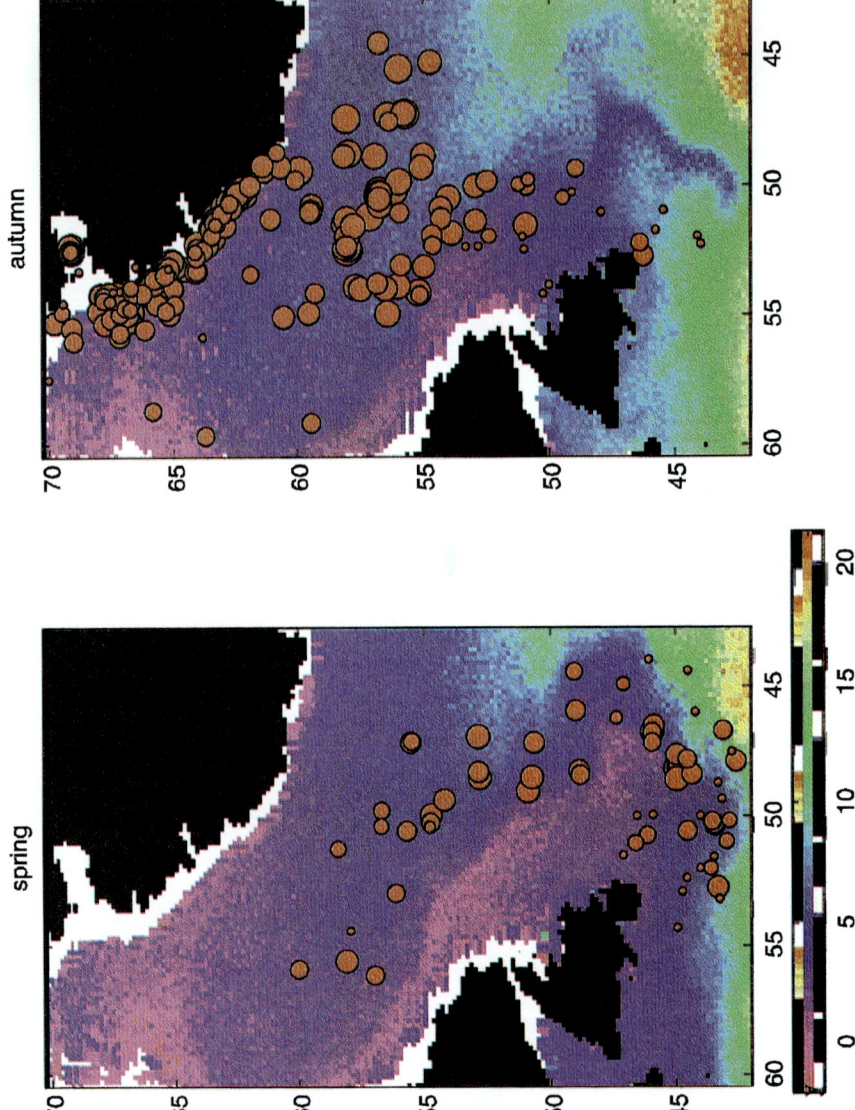

Plate 8.4 Research vessel catch rates represented by circles as the log of salmon caught per mile-hour gear fished for Atlantic salmon in the north-west Atlantic, and SSTs.

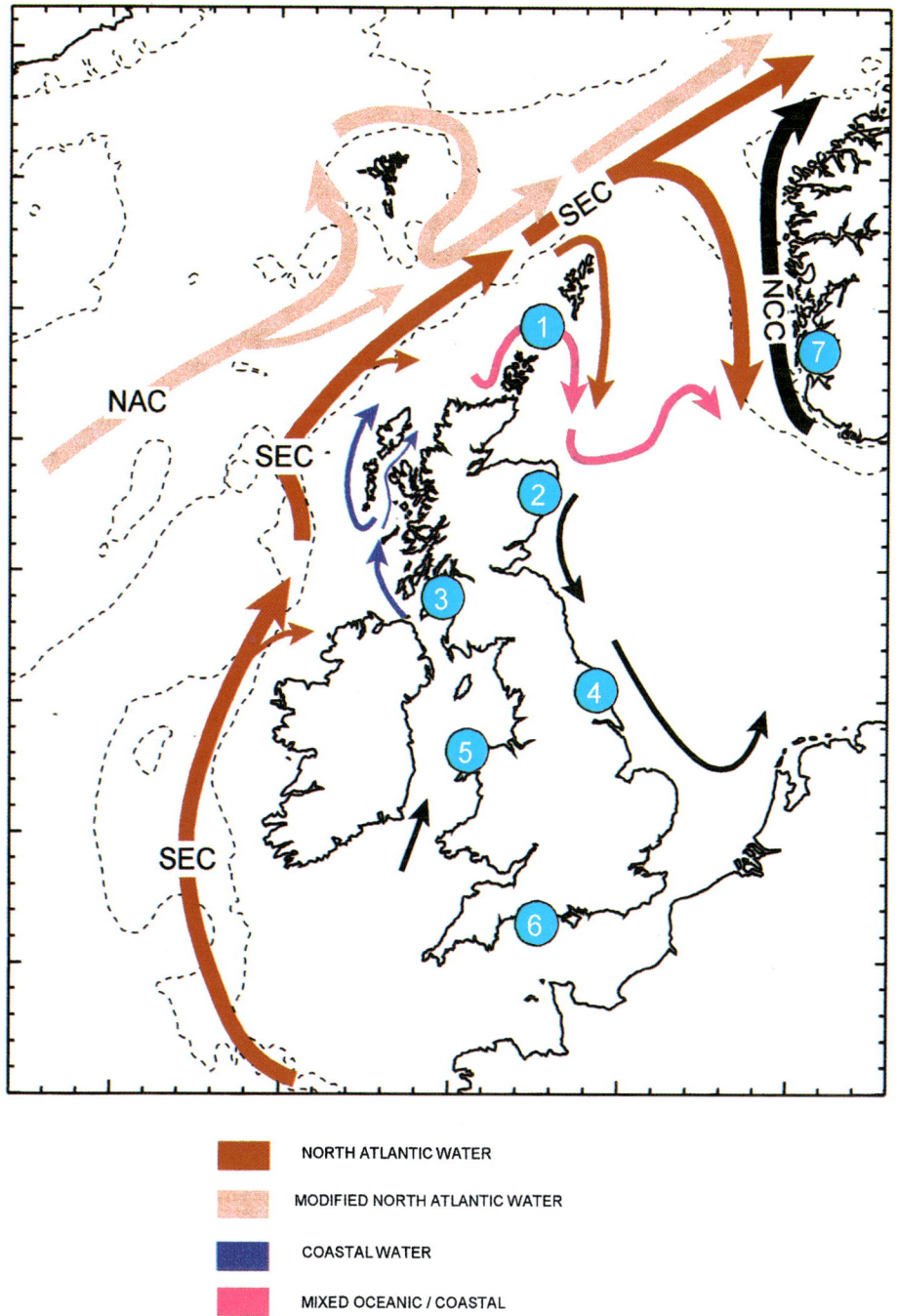

Plate 9.1 Map showing the principal currents flowing around the north-west European continental shelf, and the location of the sources of coastal temperatures. Numbered sites are referenced in the legend of Fig. 9.4 (text p. 100). The main currents referred to are the North Atlantic Current (NAC), Shelf Edge Current (SEC), and the Norwegian Coastal Current (NCC).

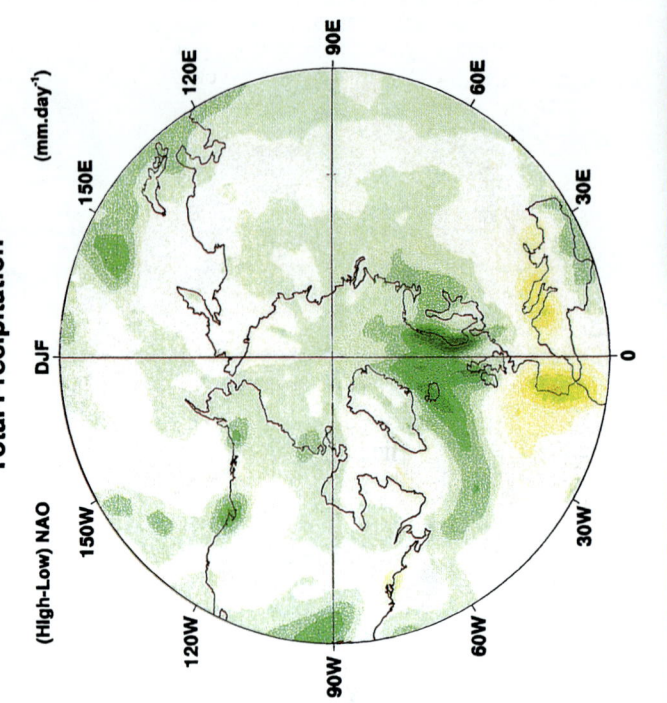

Plate 9.3 Change in the hemispheric winter precipitation between 'low-index NAO' and 'high-index NAO' composites of winter months. [Reconfigured from Xie and Arkin, 1996.]

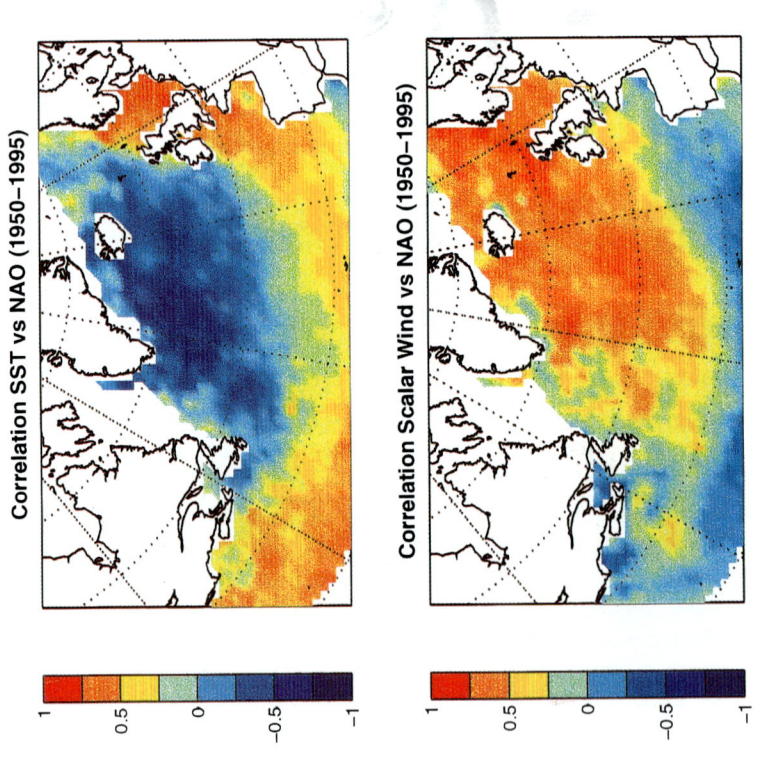

Plate 9.2 Spatial distribution of the Pearson correlation coefficient between (top) local winter sea surface temperature (°C) and (bottom) local winter scalar wind speed (m/s) and the NAO index for the period 1950–95. Winter is taken as the months January to March.

period 1982–96 represents the longest continuous time series of satellite data available for large-scale ocean measurements (Fig. 8.1). The seasonal cycle in SST over the North Atlantic is shown in terms of average patterns for four seasons: Winter (January–March), Spring (April–June), Summer (July–September) and Autumn (October–December) (Plate 8.2). These images dramatically illustrate the seasonal latitudinal shift of surface temperatures coincident with solar heating, especially on the Grand Banks, in the Greenland Sea north of Iceland and in the Labrador Sea. Particularly prominent are the Labrador, East Greenland, and West Greenland currents, which transport cold water of subarctic origin along the Greenland and Labrador coasts and around the seaward edge of the Grand Banks. The Gulf Stream/ North Atlantic Drift is prominent in summer, although its year-round influence is seen in the much weaker latitudinal thermal gradient in the eastern North Atlantic, where warmer temperatures extend to higher latitudes.

Of greater practical significance is the seasonal variability of the number of SST estimates that contribute to each weekly MCSST value (Fig. 8.1). The number of

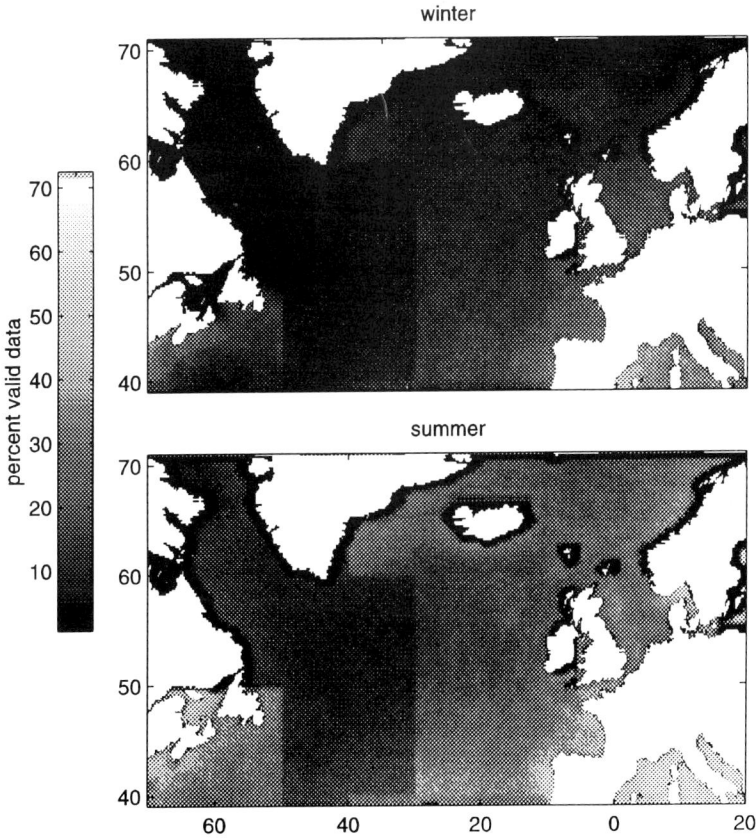

Fig. 8.1 Percentage of valid data obtained from MCSST series for the north Atlantic, 1982–96.

estimates is largest in summer, when the weather is fairest and least cloudy. The MCSST cloud detection scheme is very conservative and relatively few valid retrievals remain in the winter, especially at higher latitudes. However, this figure also displays a systematic bias in the number of estimates, and is due to the level of effort assigned to various geographic regions (J. Kelly pers. comm.). This bias does not exist in the new Pathfinder data set, which unfortunately is not yet completed. This figure also demonstrates that the MCSST processing sequence routinely blanks a number of nearshore zones, especially at higher latitudes.

In order to examine the hypothesis that survival is influenced by the nearshore environment through which post-smolt and adult salmon pass on their exit and return from rivers to the open sea, we correlated adult return numbers of nine individual salmon stocks to their respective rivers and seven stock complexes in Newfoundland with nearby inshore SSTs (Fig. 8.2 & Plate 8.3). At the post-smolt stage, four out of nine of the individual stocks were significantly correlated with annual SSTs when spawners were included in the relationship. For the seven stock complexes tested (1SW & 2SW returns) nine out of 14 were significant at less than the 10% level of significance. This suggests that annual SSTs inshore around the coast of Newfoundland may have a relationship to post-smolt stage survival and ultimately the number of salmon produced. For salmon returning to inshore waters as adults, five out of nine individual stocks were significantly correlated with inshore SSTs when spawners were included in the relationship. For the stock complex analyses, four out of 14 were significantly correlated at the 10% level of significance or less. This suggests that at the time of adult migration SSTs inshore can also influence production.

Salmon have been caught on research cruises in the Labrador Sea, coastal west Greenland, and near or on the Grand Banks in SSTs varying between 0.5 and 9.6°C. Catch rate (not shown) decreases rapidly as the SST increases above 8°C, but it is only

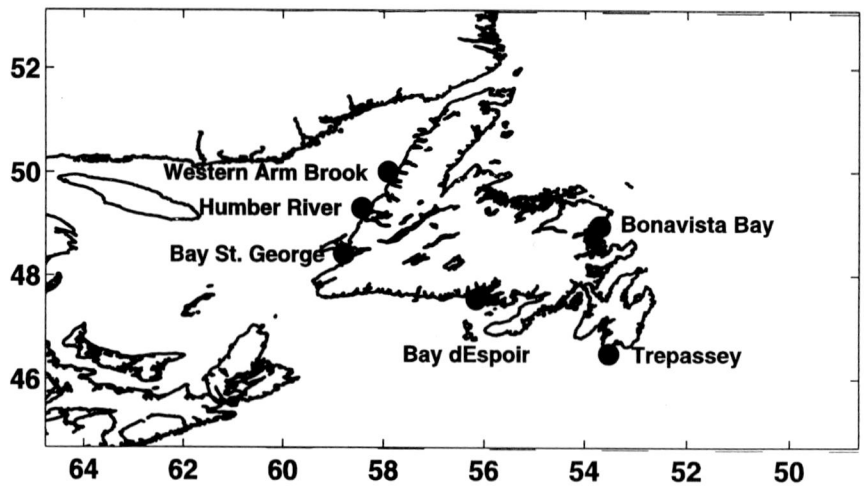

Fig. 8.2 Sites selected for inshore SSTs in Newfoundland.

weakly correlated with SST at the fishing site (0.25 in spring and 0.14 in autumn). On the other hand, a plot of catch rate displayed on mean spring and autumn SST patterns suggests that spatial patterns exist (Plate 8.4). The highest catch rates occur in three areas: the relatively warm waters of the central Labrador Sea (5–8°C), just seaward of the West Greenland Current in waters of about 4–7°C, and just seaward of the Labrador Current off the Labrador continental slope. Only a few large catches occurred in waters colder than 4°C. Conversely, the majority of zero catches, as indicated by the smallest circles in Plate 8.4, occurred in waters with SSTs less than 4°C. Most of these were on the Grand Banks proper or on the continental slope in the slightly warmer Slope Water separating the Labrador Current from the Gulf Stream. Other zero catches occurred on the continental shelf off the north-east Newfoundland coast. The proximity of both large and small catch rates to frontal zones that separate cold currents (West Greenland and Labrador) from warmer, offshore water is intriguing, and suggests that frontal zones may play some role in salmon life history.

It is concluded from this information and other studies in the literature relating growth, maturity, and survival of salmon to climatic factors suggest that the increased natural mortality being experienced in recent years may in part be climate related.

Acknowledgements

This chapter is an Abstract from a paper previously published in 'Wild Atlantic Salmon: New Challenges – New Techniques', Proceedings of the 5th International Atlantic Salmon Symposium, Galway, Ireland, 1997, by the Atlantic Salmon Federation, Atlantic Salmon Trust.

The project was partially funded by the Atlantic Salmon Trust, Atlantic Salmon Federation, the Molson Foundation, and the Ministry of Agriculture, Fisheries and Food, UK.

Chapter 9
The NAO: the Dominant Atmospheric Process Affecting Oceanic Variability in Home, Middle and Distant Waters of European Atlantic Salmon

R.R. DICKSON and W.R. TURRELL

Introduction

An oceanographer's fish

A large body of literature now exists relating oceanic climate, mainly using sea surface temperature (SST) or properties derived from SST, to marine growth, survival and maturation of Atlantic salmon, *Salmo salar* L. On making the change from fresh water to salt water, the relatively small, pelagic post-smolt enters an environment very different from the one in which it has spent its first years of life. The temperature and productivity of the coastal waters that post-smolts enter will vary from year to year and may be critical in determining their ultimate survival. Post-smolts rapidly migrate through coastal waters, and commence large scale oceanic migrations, reaching the extremes of their fairly restricted thermal range (e.g. optimal range 4–8°C, Saunders 1986) during which they will make some 90% of their somatic growth, reach sexual maturity and make the return migration home to their native rivers.

During their lives, depending on their location of origin and subsequent ocean migration, they will cross many oceanic regimes, and experience the physical forces associated with basin scale gyre circulations, oceanic eddies, frontal systems, small scale turbulence, waves and mid-ocean storms. All of these processes may affect, both directly and indirectly, their migration timing, routes and speeds, as well as their growth, maturation and survival. The salmon is indeed an oceanographer's fish, and would be the ultimate Autonomous Underwater Vehicle (AUV), if only it could reveal details of its ocean life. Developments in diverse disciplines such as otolith micro-chemistry and electronic data tags may one day reveal some of those details, but until then we must make educated guesses as to which ocean climate processes may be causing long-term changes in salmon populations.

Past correlations with SST

To try to determine which processes may be affecting population change, previous studies have focused mainly on stock-SST correlations. Statistics of change in stocks have been derived from fishery catch statistics and index rivers, all of which have some degree of error associated with them. However, the recent declines in European stocks most likely exceed the variations due to such errors. SST has been selected as the oceanographic variable for a number of reasons, not least because it is one of the few ocean properties for which there are time-series of equal length to those of salmon catch data, and also from large areas of the ocean range of salmon, derived from ships of opportunity and now from satellite observations. Fortunately, there are physiological and behavioural reasons why temperature should be a relevant property, and recent observations have shown that post-smolts and adults spend much of their lives in the surface layers (possibly in the upper 10 m) for which SST is representative.

One of the first quantitative correlations between SST and changes in a Scottish salmon stock was performed by Martin & Mitchell (1985). They examined sea-age at maturity using catch statistics by calculating the percentage of multi-sea-winter (MSW) fish in the total catch of the River Dee. They tried various marine temperature time series, but found that the only significant correlation was between catch data and water temperatures from the sub-arctic, and hence used a data set from the north of Iceland. They found that warmer temperatures north of Iceland produced more MSW fish in the catch.

The hypothesis they put forward to explain these results implied changes in the size of the marine habitat of European salmon. Warmer sub-arctic temperatures resulted in the northern boundary of salmon's ocean habitat moving north. Fish spent their first year in the sea farther north, away from the productive areas of the frontal regions at the southern boundary of the sub-arctic domain. In addition, the longer migration distances consumed more energy. Thus these fish would not mature after their first sea winter, and hence became MSW fish. Conversely, colder sub-arctic temperatures resulted in the southward shift of the northern limit of the salmon's ocean habitat, reducing energetic losses in migration and keeping fish within areas where food was more plentiful, thus increasing the numbers of one-sea-winter (1SW) fish.

Martin & Mitchell's analysis started a series of such quantitative statistical studies. For example, Dempson *et al.* (1986) conducted an intensive study of North American rivers, but could find no similar statistically significant evidence linking SST with sea age at maturity for several North American stocks of Atlantic salmon. However, Reddin & Shearer (1987) showed that the abundance of salmon west of Greenland could be directly related to the area enclosed by the 4°C and 10°C isotherms in the north-west Atlantic, and this correlation was used to provide advice to managers of Atlantic salmon stocks. These 'thermal habitat' ideas were further developed for the Labrador Sea and north-west Atlantic by Reddin & Friedland (1993a,b) who used research vessel catches to weight certain temperature bands in relation to their relative importance to the overall catch of salmon at sea. This weighted thermal habitat index was, and still is in a modified form (Anon. 1998), used to advise on the size of the Greenland high seas fishery, with the biological objective of maintaining a sustainable spawning stock size of fish returning to home waters (Reddin *et al.* 1993).

Quantitative thermal habitat concepts were applied to European salmon for the first time by Friedland *et al.* (1993a,b) and Friedland & Reddin (1993). They found a positive correlation between changes in the ocean area enclosed by the 7°C and 13°C isotherms, between 20°W and 0°W (a width of 1200 km at 57°N), in March, April and May, and the abundance of European salmon estimated from catch data. As temperature increased, both seasonally from March to May and interannually, habitat area decreased in size, as did catches.

The biological explanation for the link between habitat area and stock abundance relied on concepts such as inter-specific and intra-specific competition, food availability and predation. The period 1970–75 was a period of large habitat area and catch, which rapidly declined to low values in 1985–90. The change was attributed to a cooling North Sea (so that the northern habitat boundary moved south) and a warming along the southern boundary in the north-east Atlantic.

Turrell and Shelton (1993) examined the trajectory of UK salmon in coastal and middle distance waters, and demonstrated that post-smolts from this area could enter the sea and migrate north to feeding areas while remaining within water between 8°C and 10°C. Turrell (1994) examined the spring thermal habitat ideas for European salmon, and showed that habitat area was inversely proportional to the spring mean SST of the north-east Atlantic. He suggested that, as the then limited evidence describing the ocean life of salmon indicated that they remained associated with oceanic features such as fronts, gyres and boundary currents, the biological relevance of changes within large habitat areas might not be great, as salmon would not be evenly distributed through these areas. Significant correlations may indicate secondary processes linking changes in SST with stock abundance, rather than habitat area itself. In this way habitat area was simply a proxy for SST.

Actual observations of salmon at sea then began to become more frequent. Welch *et al.* (1995) demonstrated clear evidence, from direct observations of salmon distributions, that at least three species of Pacific salmon were clearly restricted in their ocean migrations by the location of particular temperatures. In the Pacific sub-arctic, salmon is the dominant fish species, and concepts such as thermal habitat areas are applicable there, as they also appear to be in the Labrador Sea and north-west Atlantic. However, the increasing numbers of observations of post-smolt migrations in the north-east Atlantic (Holm *et al.* 1996, Shelton *et al.* 1997) showed that post-smolts stayed within very narrow oceanic corridors, such as along the Slope Current at the edge of the European shelf edge (< 50 km wide), along the offshore boundary of the Norwegian Coastal Current and within the Norwegian Atlantic Current, which is fed by both the Slope Current and Norwegian Coastal Current.

Thermal habitat ideas in the north-east Atlantic were then refined by Friedland *et al.* (1998a), who recognised that the earlier concepts depended on large spatial and temporal averaging. A Scottish (North Esk) and a Norwegian (Figgio) index river were used in the analysis, where marine survival was estimated from tag returns. Monthly thermal habitat areas were calculated, although these were still fairly large and based on 22° wide longitudinal bands (approximately 1330 km at 57°N) . The most significant correlation was found between the return rate of one-sea-winter (1SW) fish and habitat area defined by the 8°C and 10°C isotherms, and centred on approximately 0°W in May. The main difference between the May marine habitats of good survival years (1971–74) and poor survival years (1985–88) were colder temperatures ($<8°C$) along the Norwegian coast in the North Sea. The biological mechanisms responsible were suggested to be mainly changes in growth-mediated effects on survival, food availability and migration cues.

Identifying the cause of environmental change

We attempt in this chapter to identify the cause of observed change in SST patterns in the north-east Atlantic, and to identify other environmental changes that may have been of relevance in the life history of European salmon. In selecting the arguments of greatest relevance, we have made several assumptions: first, that the maturing 1SW fish from countries bordering the north-east Atlantic are likely to range only as far as Faroe-Jan Mayen and the southern Norwegian Sea, but that the MSW fish can range throughout the Nordic (Norwegian-Greenland-Iceland) Seas to West Greenland (Turrell & Shelton 1993); second, that as the North Atlantic Oscillation (NAO) is the dominant recurrent mode of atmospheric forcing in the Atlantic sector, especially in winter but robustly present at other times of the year) the recent amplification of the NAO beyond the range of past experience in a century-long record (Fig. 9.1) is likely to be the partial or dominant cause of environmental changes experienced in recent years by salmon; and third, that the environmental features likely to be of greatest relevance to salmon are those producing significant changes in the ocean current system and/or in the input and distribution of heat and fresh water in the near-surface layers throughout their range.

Any of these assumptions may be challenged since we have little enough factual basis for ascribing importance to any particular set of environmental properties as controls on the salmon. However, if the apparent decrease in their marine survival means that salmon are experiencing some marked and sustained trend of change in their environment on time scales of years to decades, this focus will at least cover the greatest and most sustained trends of change known to us in recent years.

Characteristics of the NAO

The NAO is a large-scale alternation of atmospheric mass with centres of action near the Icelandic Low and the Azores High. It is the dominant recurrent mode of atmospheric behaviour in the North Atlantic sector throughout the year, accounting for more than one-third of the total variance in sea-level pressure. The NAO alternates between a 'high-index' pattern, characterised by an intense Iceland Low with a strong Azores Ridge to its south, and a 'low-index' pattern in which the signs of these anomaly cells are approximately reversed (Fig. 9.1b). The pressure difference between these two main cells is the conventional index of NAO activity.

The NAO index has exhibited considerable long-term variability. Cook *et al.* (1998) and Hurrell & van Loon (1997) describe concentrations of spectral power around

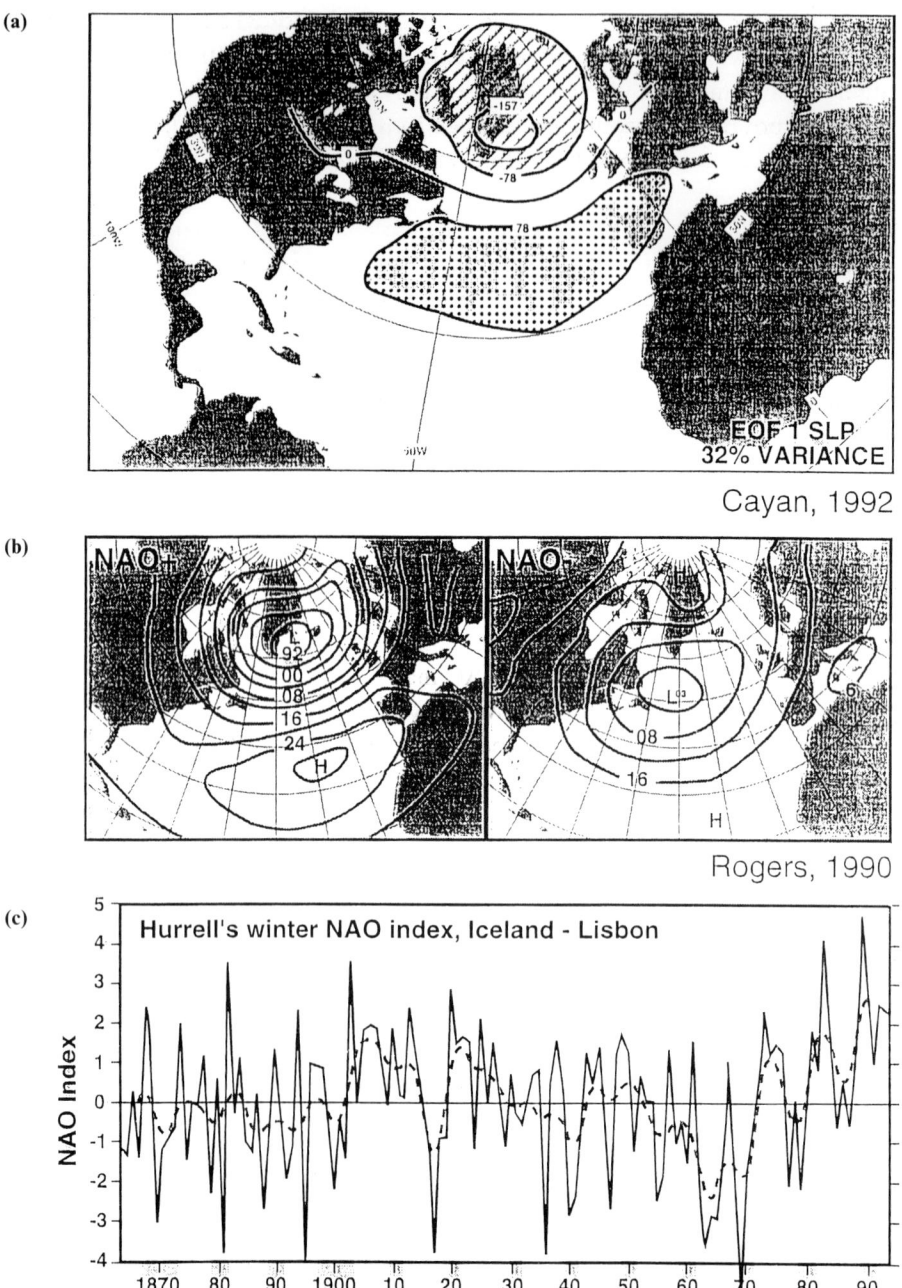

Fig. 9.1 The North Atlantic Oscillation (NAO) – its pattern and variability. (a) The first mode of variability of sea level atmospheric pressure. (b) Average air pressure distributions for high and low index years. (c) The winter NAO index showing its annual values (solid line) and smoothed values (broken line).

periods of 24, 8 and 2.1 years, but also identify a multi-decadal signal which appears to be amplifying with time. Thus the 1960s exhibited the most protracted and extreme negative phase of the Index, the late 1980s–early 1990s experienced its most prolonged and extreme positive phase, and the change from the low-index 1960s to the high index 1990s became the largest low-frequency change on record (Fig. 9.1c, from Hurrell 1995a).

These changes have a wide range of physical and biological responses in the North Atlantic. These include effects on wind speed, latent and sensible heat flux (Cayan 1992a,b,c), evaporation–precipitation (Cayan & Reverdin 1994, Hurrell 1995a), sea surface temperature (Cayan 1992c), the ocean circulation (Myers *et al.* 1989, McCartney *et al.* 1996, 1997), the distribution, prevalence and intensity of Atlantic storms (Rogers 1990, Hurrell 1995b, WASA Group 1997), the wave climate (Bacon & Carter 1993, Kushnir *et al.* 1997, Cotton *et al.* 1997), the iceberg flux past Newfoundland (Drinkwater, in Rhines 1994), and the intensity of deep convection at the main Atlantic sites (Greenland Sea, Labrador Sea and Sargasso) (Dickson *et al.* 1996, Dickson 1997). Effects on the marine ecosystem include changes in the production of zooplankton (Fromentin & Planque 1996) and the recruitment of cod (B. Planque & C.J. Fox pers. comm.).

In the following sections we follow a speculative trajectory of a northern European salmon, and examine in what ways the NAO may have modified the ocean conditions experienced by the fish between the low-index 1960s, when European catches were high, to the high-index 1990s, when stocks have dramatically declined.

Home waters (UK)

Freshwater life

Before following the ocean life of a salmon, it is worth noting that the first influence the NAO may have upon its life history is within its freshwater environment. Since NAO variability will dominate and determine the weather experienced across the catchment areas of many river systems, we envisage a wide range of possible effects. In Fig. 9.2, for example, we illustrate this for a range of variables in and around Lerwick, Shetland, including air temperature, rainfall and a variety of wind conditions. However, we must make the associated point also, that the effect of the NAO on the river systems of Britain is likely to vary in space as well as time. We illustrate this in Fig. 9.3 using one of the most relevant 'weather' factors at the highest resolution available to us, the percentage change in winter rainfall in different regions of the UK between negative and positive extremes of the NAO (i.e. NAO < –1 SD to NAO > +1 SD; reconfigured from Wilby *et al.* 1997). With the precipitation balance across the UK changing from +31% in north-east Scotland to –27% in south-east England as the NAO amplifies, it seems likely that the freshwater life of young salmon will have been influenced by just such a change in recent decades. The less obvious point which that figure also makes is that the changes which young fish encounter in

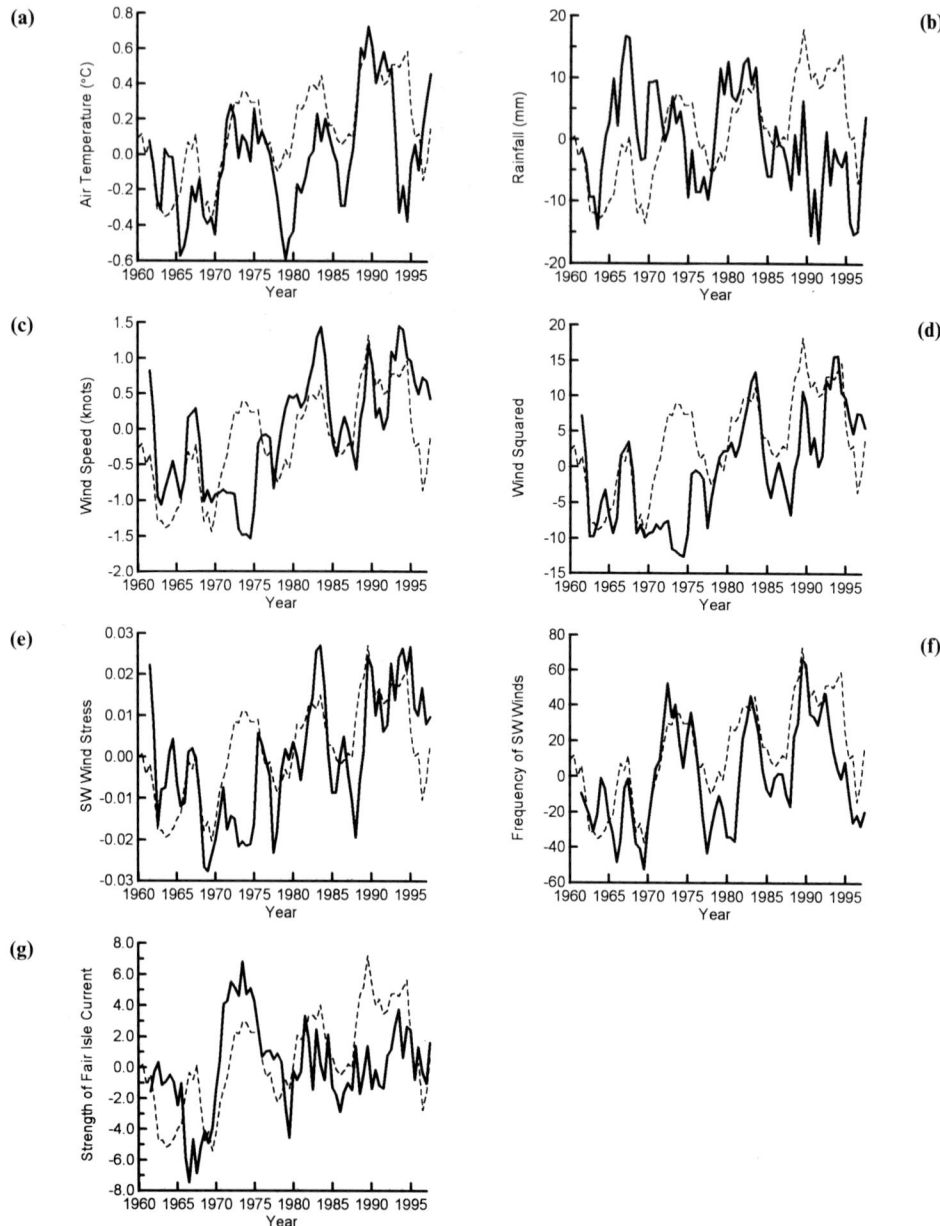

Fig. 9.2 The effect of the NAO (broken line) on weather conditions at a station close to the northern North Sea (Lerwick, Shetland). (a) Air temperature. (b) Rainfall. (c) Wind Speed. (d) Wind speed squared (indicating changes in total wind stress on the sea surface). (e) South-westerly wind stress. (f) Frequency of south-westerly winds. (g) Strength of the wind-driven inflow to the North Sea past Fair Isle. All values have been de-seasoned and smoothed using a 2-year running mean. The NAO index is an update of Hurrell's (1995) winter index, smoothed using a 2-year running mean.

The NAO   99

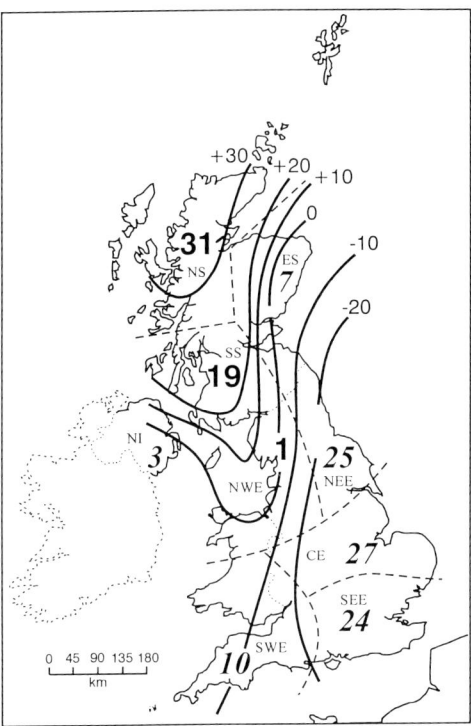

Fig. 9.3 Regional change in UK winter rainfall (%) associated with the change from negative to positive extremes of the NAO, 1865–1996. Data from Wilby et al. (1997).

their freshwater life are likely to be more regionally variable than the larger-scale changes that we assume to affect their oceanic environment. Though we may not yet be able to describe the full range of such effects in adequate detail, it seems clear enough that the NAO will have influenced key aspects of the freshwater lives of salmon during the last three decades, and this should be borne in mind when stock/environment correlations are attempted.

First entry to sea

On first entry to the sea, salmon from United Kingdom rivers enter a variety of sea areas and conditions. Five representative sites have been selected around the UK (Plate 9.1 in Colour Section between pp. 88 & 89), characterising the north coast of Scotland (Fair Isle), the east and west coasts of Scotland (Peterhead and Millport), the east and west coasts of England and Wales (Scarborough and Amlwch) and the English south coast (Weymouth). The interannual variability of SST at these sites, compared with the variations in the NAO, may be seen in Fig. 9.4. Clearly the NAO has influenced changes in the temperature of coastal waters around the UK in the last few decades. The increasingly positive NAO since 1970 is mirrored by the upward

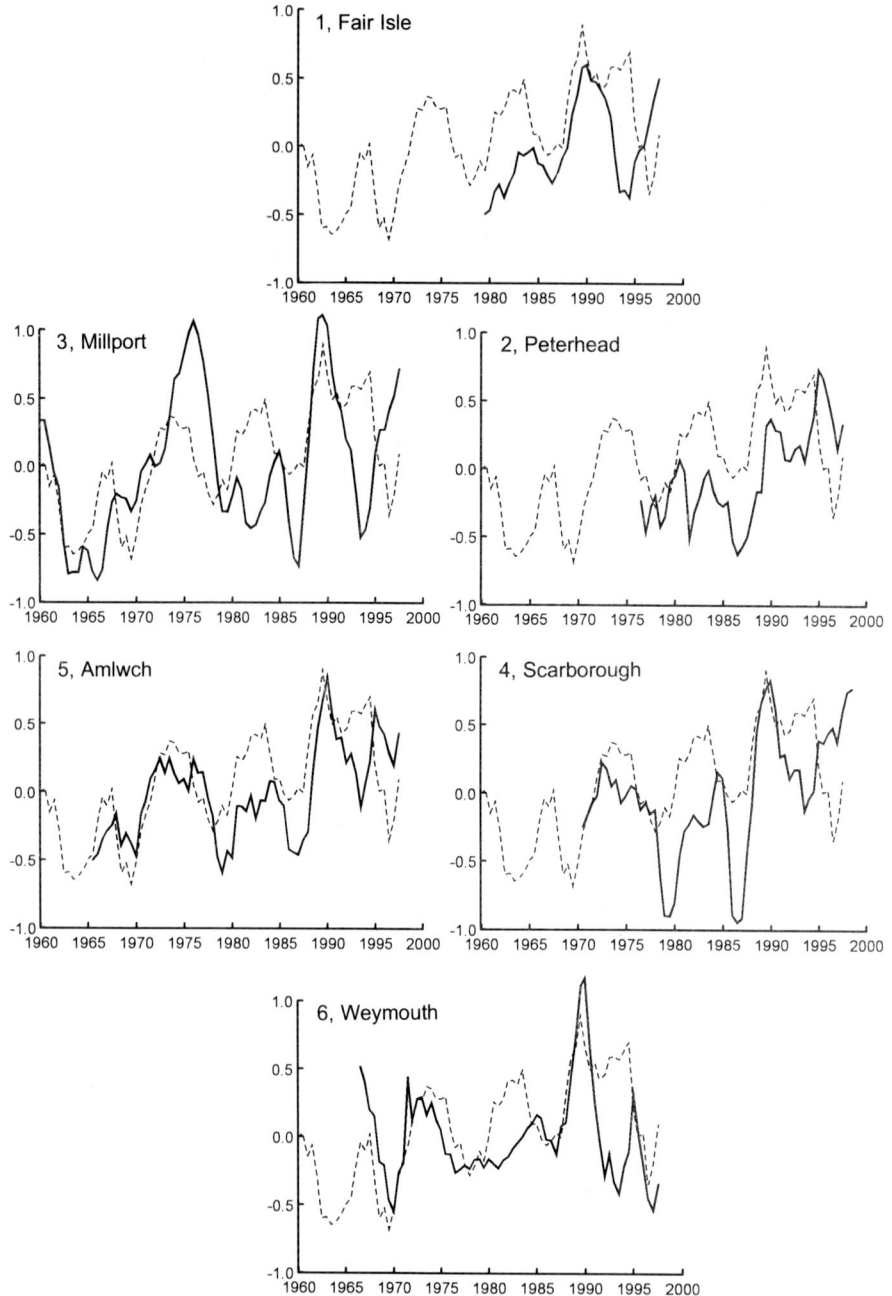

Fig. 9.4 Sea surface temperature from six representative sites around the UK (see Plate 9.1), characterising the north coast of Scotland (1, Fair Isle), the east and west coasts of Scotland (2, Peterhead, and 3, Millport), the east and west coast of England and Wales (4, Scarborough, and 5, Amlwch) and the English south coast (6, Weymouth). All data have been de-seasoned and smoothed using a 2-year running mean. The NAO index is also shown (broken line), and is an update of Hurrell's (1995) winter index, smoothed using a 2-year running mean.

trend in sea temperatures in all areas since that time, and the decadal scale variability is also seen to some extent in most time series. As noted by Friedland *et al.* (1998a), it may not be the mean annual temperature variation that is most relevant to salmon in home waters, but rather the temperature at first entry to sea. However, for UK salmon this normally means entry at some time between March and May, and the variability of coastal temperatures for these months is not very different from that of the smoothed annual mean temperatures (Fig. 9.5).

It may be not only in terms of temperature that the NAO has influenced the home waters of UK salmon, but also in terms of oceanic circulation, particularly in the North Sea. The typical circulation around the coasts of the UK is shown in Plate 9.1. The anticlockwise circulation around Scotland and in the North Sea is primarily driven by wind stress during the months March–May. Winds from the south-west in particular drive this circulation. The wind data recorded at Lerwick (Fig. 9.2) show that the strength of the wind over home waters has generally increased with the strengthening NAO. The decadal-scale variability of the NAO is also seen, and clearly influences the frequency of south-westerly winds over the northern North Sea. The strengthening winds will have increased the wind-driven transport of the area, and also affected factors such as vertical mixing and wave height. All of these may have had direct or indirect effects on the marine survival, growth and migration patterns of post-smolts during their first few weeks at sea in north European home waters.

Middle-distance waters

The changes observed within home waters may be placed into the context of wider changes within the north-east Atlantic as a whole. Here, we couch our description in terms of the conditions that prevailed during the positive NAO, particularly the protracted and extreme positive phase of the 1990s.

Spring migration through home waters

Based on data from the past 45 years at 50-mile grid-spacing, Plate 9.2 (B. Planque pers. comm.) illustrates the ocean-wide correlations between the NAO Index and two of the key variables which are likely to determine the environment of fish leaving the European shelf for middle and distant waters, SST and windiness.

Plate 9.2a conforms with Becker & Pauly (1996) in showing that abnormally warm SST in winter–spring typify the entire European Shelf during this positive phase of the NAO, yet fish would meet with anomalously cool conditions as they pass west across the Continental Slope to the deeper waters of the north-east Atlantic, and more generally across a broad mid-latitude band to west Greenland. This warm/cold contrast is also found in less amplified form to the west of Norway, and appears to extend as far north as the Barents Sea-Fram Strait.

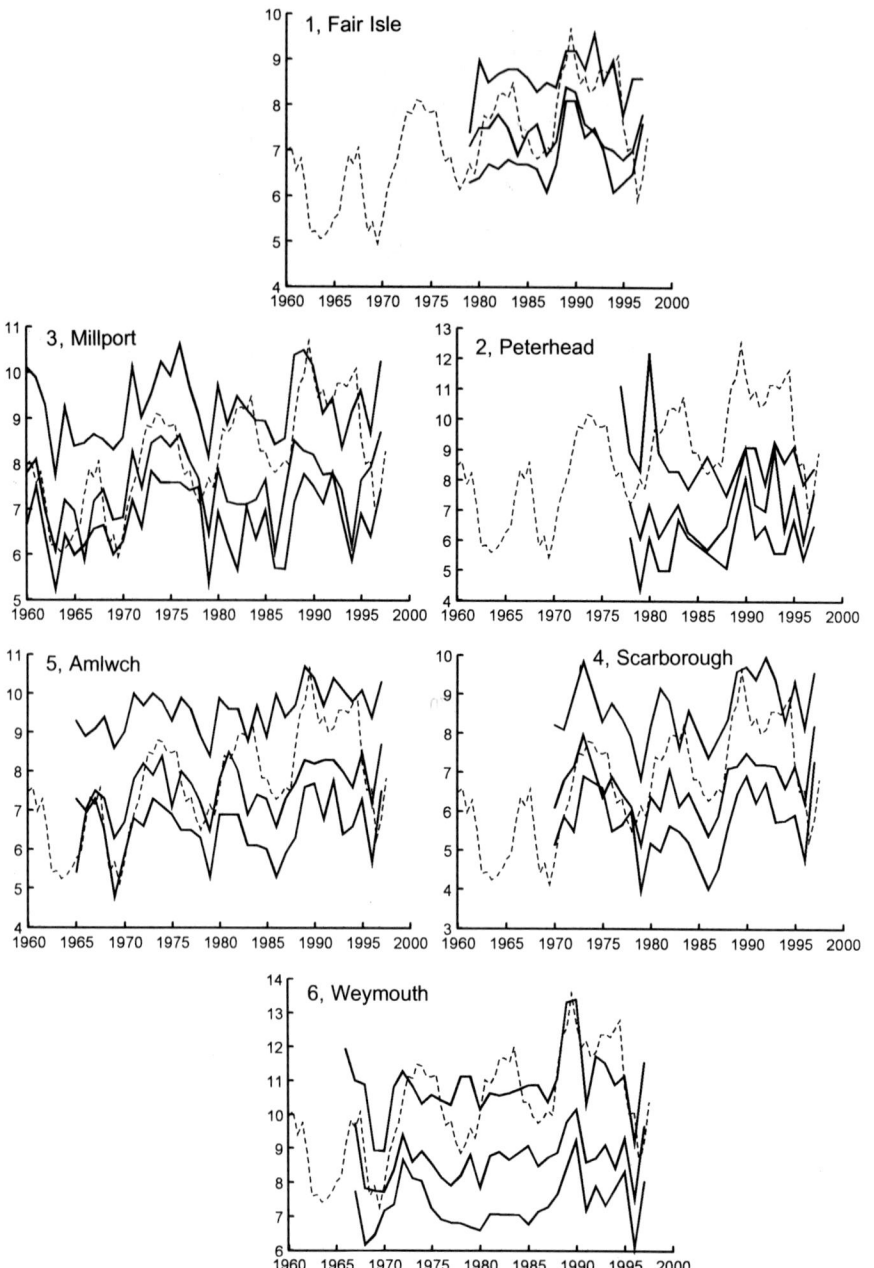

Fig. 9.5 Monthly mean values of SST at the five coastal monitoring sites. Months are progressively warmer – March (lower solid line), April (central solid line), May (upper solid line). No smoothing has been used. The NAO index is also shown (broken line), and is an update of Hurrell's (1995) winter index, smoothed using a 2-year running mean.

Migration north into the Nordic seas

It is now thought likely that part of the ocean's response to the amplifying NAO between the 1960s and 1990s was a progressive change in the width of the Norwegian Atlantic Current. Figure 9.6 shows the evidence (from Blindheim *et al.* in press) that the longitude of the 35 isohaline on a Russian hydrographic section through the southern Norwegian Sea at 65°45'N retracted eastward in three main steps over this period, in apparent synchrony with the NAO. Altogether, the change amounts to an eastward shift of > 300 km since the 1960s and, we would suppose, not irrelevant to the success and distribution of salmon passing into the Norwegian Sea and spreading through its current system. We do not yet know whether any resulting or accompanying change took place in the Norwegian Atlantic Current volume transport, although there is evidence that transport towards the Norwegian Atlantic Current from the north of Scotland increased from the 1960s to the 1990s (S.R. Dye, G.R. Bigg & W.R. Turrell pers. comm.)

A second factor of potential relevance to salmon stocks in middle-distance waters has been a change in the freshwater balance of the Nordic seas, once again as a result

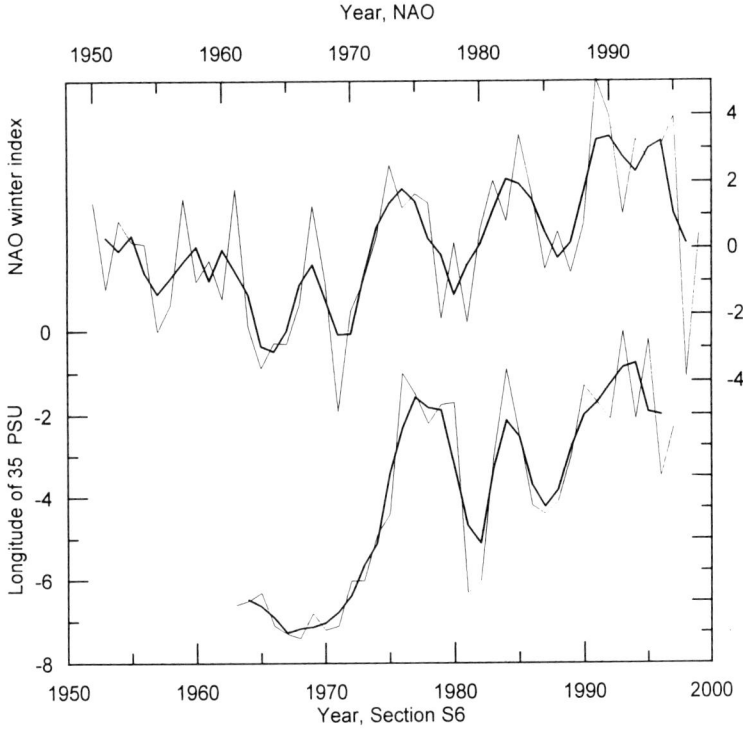

Fig. 9.6 Width of the Norwegian Atlantic Current, as measured by the longitude of the 35‰ isohaline at 65°45'N (Blindheim *et al.* in press) compared with the winter NAO index of Hurrell (1995a). Both bold curves are 3-year running means.

of the amplifying NAO. It has long been known (Hurrell 1995a) that the axis of maximum moisture transport over the Atlantic extends much further to the north and east at times of high NAO Index. Hurrell & van Loon (1997) illustrate just such a pattern in showing (their Fig. 14) the change in precipitation corresponding to a unit deviation of the NAO index in winter during 1900–94. Since Hurrell's report, Xie & Arkin (1996) have greatly improved the data coverage, using station data blended with satellite products and, where neither exists, data from an assimilating model to provide global, gridded precipitation at 2.5° resolution for every month from 1979 to 1995. From this new data set, Dickson et al. (in press) compiled composites of mid- to high-latitude precipitation for the contrasting positive and negative phases of the NAO. Though the difference in the distribution of precipitation between these two composites (Plate 9.3) will underestimate the precipitation change from low- to high-index conditions, it agrees qualitatively with Hurrell (1995b) in showing that the major precipitation increase (up to +15 cm per winter) during positive NAO conditions occurs in the 'conduit' of the Norwegian-Greenland Seas, especially over the Norwegian Atlantic Current, presumably reflecting the changes in the Atlantic storm track (see below).

Overwintering in the Nordic seas

Winter wind speeds (Fig. 9.7) were amplified during the 1990s period over almost the whole range occupied by 1SW and MSW fish. It is already well established that the NAO exerts a significant control on the track, prevalence and intensity of Atlantic storms (Rogers 1990, 1997, see also Hurrell 1995b). As the NAO index switches from negative to positive, so the centre of winter storm activity extends north-eastward into the north-east Atlantic and Norwegian-Greenland Sea. This is accompanied by a remarkable increase in the incidence of Atlantic storms deeper than 950 hPa (R. Francke, Deutsche Wetterdienst, Seewetteramt, pers. comm.) from near zero (mid 1960s) to around 15 per winter (1990s) (Dickson et al. in press). More recent studies have quantified the change in storm climate northwards into the Norwegian Sea. Serreze et al. (1997) apply an automated cyclone detection and tracking algorithm to twice-daily sea-level pressure fields over the period 1966–93 (i.e. covering much of the change from low- to high-index NAO activity) to describe a general increase in cold-season cyclone frequency over this period in the region north of 60°N.

Since it spans much of the geographical range of 1SW fish (the Aberdeen-Faroes-OWS M-Bergen rectangle) and covers the whole period since 1880, the storm index of Alexandersson (in press) gives perhaps the clearest indication of the extreme nature of this recent increase in storminess, as well as confirming its long-term general link with the NAO (Fig. 9.7). In general, this index suggests that winter storm activity over the north-east Atlantic-southern Norwegian Sea during the NAO peak of the 1990s was at its most intense for 100 years.

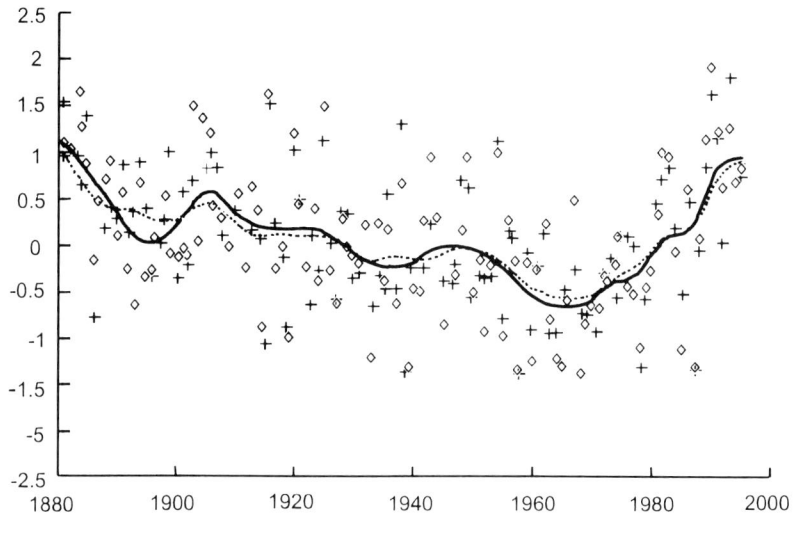

Fig. 9.7 Comparison of the winter NAO index of Hurrell (1995a) with a winter storm index constructed by Alexandersson et al. (in press) for the area Aberdeen-Torshavn-OWS M-Bergen, 1880–1995.

106   *The Ocean Life of Atlantic Salmon*

Distant waters

We identify three radical environmental changes in more distant waters that seem likely to have been experienced by MSW fish over recent decades.

West Greenland

First, the waters of the Greenland Sea and Iceland Sea affected by the cold, fresh influence of the East Greenland Current have experienced a parallel freshening to that of the Norwegian Atlantic Current further east, but in this location largely through an increase in the annual efflux of ice from Fram Strait. Ice volume-flux estimates of increasing reliability are available since 1976 (Vinje & Finnekasa 1986, Vinje *et al.* 1997, see also Kwok & Rothrock 1998). Since the annual mean ice thickness does not appear to alter by more than $\pm 10\%$, Vinje *et al.* conclude that changes in the efflux of ice are not primarily due to the origin of the ice but to changes in the local wind field over Fram Strait. In turn, Dickson *et al.* (in press) demonstrate a close correspondence between the ice flux and the NAO, hence an increasing ice flux into the 1990s, rising to a maximum of $4687\,km^3$ per year in 1994–95. Although, at its peak, the increase in the annual volume flux of ice was comparable to that associated with the 'Great Salinity Anomaly' of the 1960s ($\gg +2000\,km^3$; Dickson *et al.* 1988, Aagaard & Carmack 1989), a much greater proportion appears to have been retained within the Greenland-Iceland-Norwegian Seas in the case of the mid-1990s event. An analysis by K. Grotefendt (pers. comm.) suggests that the 1994–95 peak in the flux of ice through Fram Strait is the origin of a 100 m thick, low-salinity surface layer ($S < 34.65$) that is observed spreading eastwards into the southern Greenland Sea in April–May 1996.

Cooling north-west Atlantic

The second potential effect on distant water fish has already been introduced in Plate 9.2. The marked sea surface cooling that accompanied the amplifying NAO across a broad band of mid-latitudes from the British Isles to west Greenland. Both this broad-scale cooling and the accompanying warmth on the European Shelf are consistent with the patterns of sensible and latent heat flux described for NAO extrema by Cayan (1992c, his Figs 7a, 8a). Although restricted to years before 1986, and missing the most extreme positive development of the NAO, his analysis still shows an increase in total heat flux of up to 200 W/m^2 south and south-east of Greenland associated with positive versus negative extreme winter months of the NAO. The 1990s were very much more extreme than that. Figure 9.8 illustrates the time-dependence of the mean winter north-west wind stress for the Labrador Sea since 1947 (Dickson *et al.* 1996), showing that the strength of these chill north-westerlies, which drive the sea-to-air heat flux in this sector, has amplified steadily with the NAO from a minimum in the 1960s to record post-war intensity in the 1990s. We can expect

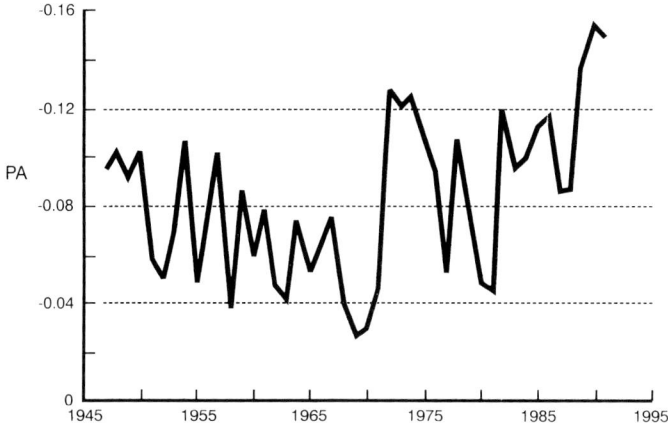

Fig. 9.8 The annual vector-averaged north-west wind-stress for the Labrador Sea (average of six sites). Negative stresses indicate winds from the north-west (from Drinkwater 1994).

effects of these chilling conditions both on the component of European young fish which will spread to west Greenland in the course of a MSW life cycle, as well as on the North American fish that occupy the Labrador Sea in numbers during their first winter.

Strengthening ocean circulation

One effect of the long increase in winter severity since the 1960s was a non-steady but progressive and at times explosive increase in the intensity of open-ocean deep convection in the Labrador Sea between the 1960s, when production was effectively capped, and the early 1990s, when convective overturn was reaching deeper than ever observed (to 2300 m), and the potential temperature, salinity and thickness of Labrador Sea Water (LSW) were at their coldest, freshest and thickest ever observed in a record extending back to the 1930s (Lazier 1995).

The production of these vertically-homogeneous 'mode waters' (such as LSW) at their three main Atlantic sites is observed to be co-ordinated by the NAO in such a way that as LSW becomes colder, fresher and denser, the production of '18-degree water' in the subtropical gyre dwindles and becomes less dense (Dickson *et al.* 1996). Curry & McCartney (in press) recognised the implication. That as the NAO amplified from the 1960s to the 1990s, the effect on the potential energy anomaly gradient between the subpolar and subtropical gyres would produce an increase in the strength of the Atlantic circulation along the gyre–gyre boundary (Fig. 9.9). This is the path of the main North Atlantic Current, and is therefore cited here as the final example of radical and recent environmental change of potential relevance to the Atlantic salmon. The link with the salmon is certainly speculative, as all of our examples have been, and depends on the extent to which salmon depend on the gyre circulation in making their outward or homeward migrations. However, the implied change is an

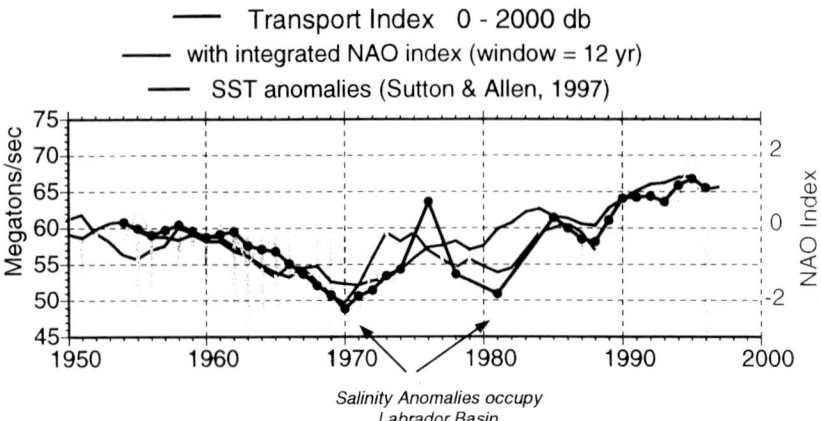

Fig. 9.9 Comparison between the weighted NAO index, the index of 0–2000 dbar transport along the subpolar-subtropical gyre boundary of the North Atlantic, and the SST anomaly in 'Area 2' of Sutton & Allen (1997). R. Curry & M. McCartney pers. comm. 1998.

important one, amounting to an increase of 30% in the strength of the North Atlantic Current since the late 1960s. If the linkage to the NAO proves valid, then the North Atlantic Current has been stronger in the 1990s than at any other time since the start of the instrumented record in 1865, as Curry & McCartney point out.

The NAO and previous observations of salmon population change

If the NAO and its effects are the dominant feature of the last decades in terms of ocean climate and its impacts on salmon stocks, then the extensive and detailed studies mentioned in the Introduction should be compatible with this overriding forcing.

General stock decline

Figure 9.10 demonstrates that the general decline in European salmon, as indicated using the total catch from Scottish rivers as an index, from the peak years of the 1960s and 1970s, does correspond with the strengthening NAO. The fall of the NAO index after 1995 was only temporary, and the 1997 winter index was again high.

Martin and Mitchell (1985) analysis

In the period from 1920 to the 1950s the NAO index was generally weakening towards its low values in the 1960s. Waters north of Iceland, influenced by areas of the subpolar gyre, which exhibit an inverse relationship between SST and the NAO (Plate 9.2a), were warming as the NAO weakened. Thus, following Martin & Mitchell's analysis, the northern boundary of the salmon's ocean habitat moved north, and a

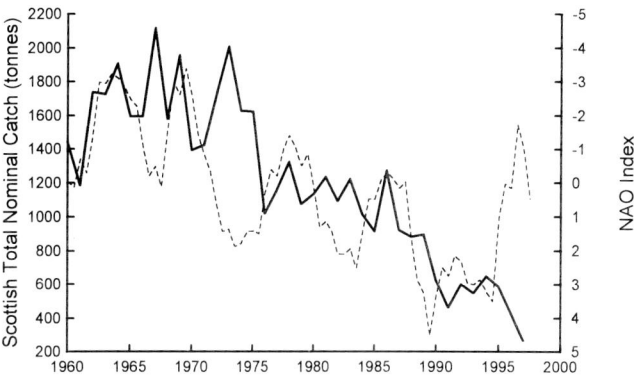

Fig. 9.10 The overall decline of European salmon catches, as illustrated using the total nominal Scottish catch (Anon 1998 – solid line) compared with the inverted NAO index (broken line). The inverted NAO index is an update of Hurrell's (1995) winter index, smoothed using a 2-year running mean.

greater proportion of the population returned as MSW fish. Since the 1970s, as the NAO rises and waters in the subpolar gyre cool, lower proportions of MSW can be expected in the catch, as has already been observed in many European rivers (Turrell & Shelton 1993).

North-east Atlantic thermal habitats and North Sea SST

The spring thermal habitat of European salmon described by Friedland *et al.* (1993), Friedland *et al.* (in press), and Friedland & Reddin (1993), lies between 0°W and 20°W in the north-east Atlantic. Its southern and northern boundaries are the 13°C and 7°C isotherms respectively, which in the spring lie approximately at latitudes of 40°N and 65°N (Turrell 1994). Hence the southern boundary is within the area of the subtropical gyre, which exhibits a positive response to a changing NAO index, while the northern boundary lies within areas exhibiting a negative response. As the NAO increased from its low index values in the 1960s to the high index values of the 1990s, the habitat area shrank, as the southern boundary moved north in water that became warmer, and the northern boundary moved south in waters which were cooling (Fig. 9.11).

The thermal habitat of European salmon was re-examined in greater detail by Friedland *et al.* (1998a,b). Their refined thermal habitat again has its southern and northern boundaries within waters exhibiting positive and negative responses, respectively. Thus the computed habitat area steadily decreased in size from 1971 to 1993 as the NAO strengthened (Fig. 9.11). They attributed changes in the habitat area primarily to changes within the North Sea, as years exhibiting particularly good index-river return rates (1971–74) had warm waters along the coast of Norway in May, while years with particularly poor return rates had cool Norwegian coastal waters. In Fig. 9.11 their analysis is confirmed using additional data obtained for a point on the Norwegian coast. The overall increasing trend in the NAO is reflected in

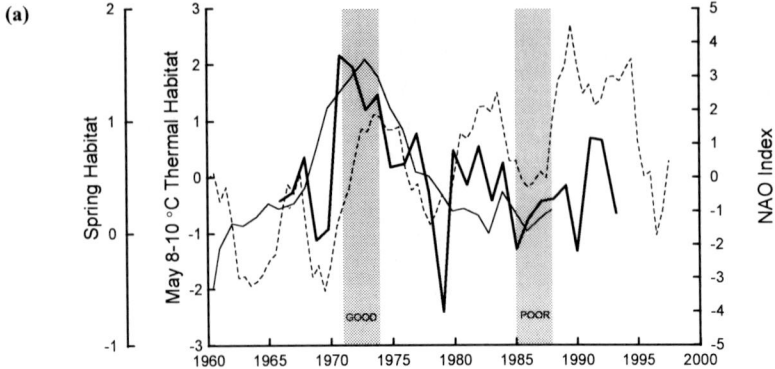

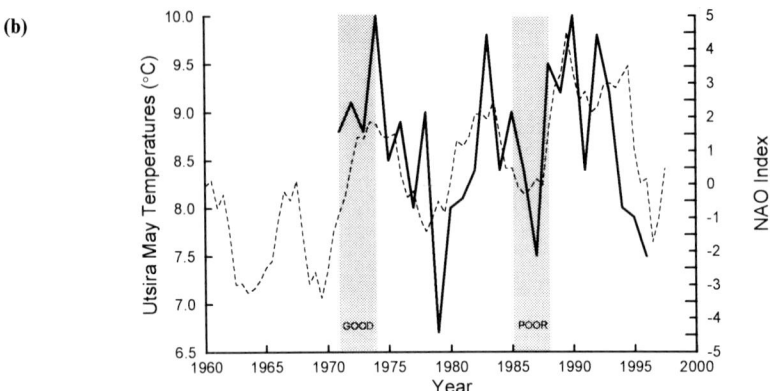

Fig. 9.11 (a) The spring habitat area of Friedland *et al.* (1993a,b) (thin solid line) and the normalised 8–10°C thermal habitat area of Friedland *et al.* (1998a) (thick solid line), compared with the NAO index. (b) Temperatures off the coast of Norway (Station 7, Plate 9.1 – solid line) compared with the NAO index. (SST data from http://www.bsh.de/daten/). The NAO index (broken line) is an update of Hurrell's (1995) winter index, smoothed using a 2-year running mean.

the overall decreasing size of the computed thermal habitat area. However, the difference between years of 'good' and 'poor' 1SW return rates is the result of the strong decadal scale variability seen in the NAO, and within coastal SST time series throughout the north-west European shelf (e.g. Figs 9.4, 9.5). The critical point is that while Norwegian coastal temperatures have again risen since the minima of 1985–88, 1SW return rates have not, and continue to decline.

Summary – the NAO and north-east Atlantic thermal habitat areas

The conclusion from this analysis is that calculated thermal habitat areas in the north-eastern Atlantic, which produce the most significant correlations with observed trends in catch or inferred marine survival, are obtained from large ocean areas spanning the subpolar/subtropical division, separating waters that experience opposite responses to the changing NAO. An increasing NAO shrinks these areas, and hence these correlate well with the reducing abundance and survival of stocks observed since the 1960s. The waters of the north-west European continental shelf have responded more to the decadal scale variability of the NAO during the last few decades, rather than just to its overall trend. This decadal scale change has not been observed in catch statistics. Hence it may be postulated that the key areas do not lie within home waters or the North Sea, but rather in the more distant Nordic Seas.

North-west Atlantic thermal habitat

Finally, the north-west Atlantic thermal habitat index currently used by the 1998 ICES Working Group on North Atlantic salmon to estimate pre-fishery abundance of salmon in the area and hence to provide advice on fishing quotas, is compared with the inverted NAO index in Fig. 9.12. Clearly there is a relationship, and as the NAO strengthens the Labrador Sea is cooled and the thermal habitat area shrinks. The habitat area responds not only to the general trend in the NAO, but also to the decadal scale variability. The correlation between the two indices increases if a lag of 1 year is introduced (r^2 values of 0.27, 0.63 and 0.58 for lags of 0, 1 and 2 years respectively). The delayed response of the Labrador Sea to a changing NAO implies that the NAO index provides some predictive capability, and hence may be of use to the ICES Working Group when they consider the provision of advice.

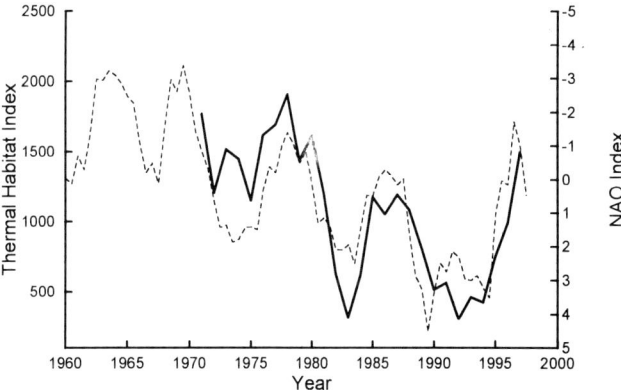

Fig. 9.12 The thermal habitat index (solid line) for February in the north-west Atlantic north of 41°N and west of 29°W (includes Davis Strait, Labrador Sea, Irminger Sea and the Grand Banks) which was used by the 1998 ICES Working Group on North Atlantic salmon to estimate pre-fishery abundance of salmon in the area, and hence to provide advice on fishing quotas. This is compared with the inverted NAO index (broken line).

Conclusions

- The North Atlantic Oscillation (NAO) is the dominant recurrent mode of atmospheric behaviour throughout the Atlantic Sector, and therefore acts as a dominant control on the environment of Atlantic salmon throughout their range. The NAO Index exhibits a considerable decadal variability, which appears to be amplifying with time. In a record extending back to 1865, the NAO evolved to its most persistent and extreme negative state in the 1960s, and thereafter to an equally extreme positive state in the late 1980s–early 1990s.
- This gradual evolution from low- to high-index conditions over the past 3–4 decades brought a north-eastward extension of the Atlantic storm track to the Greenland, Iceland and the Norwegian and Barents Seas, together with an increase in the numbers of deep Atlantic storms from near-zero during low-index conditions to around 15 per winter during the high index phase of the 1990s.
- The change from the 1960s to the 1990s in the winter NAO and storm track are associated with changes in many variables, which we suppose to be of relevance to the salmon at one stage or other in their migration/life cycle, and these are summarised in Table 9.1.

Table 9.1 Summary of climatic change which accompanied the high NAO index (low catch) years of the 1990s, compared with the low NAO index (high catch) years of the 1960s.

Salmon	Area	Temperature	Other effects
Early life	Fresh water	Warmer	Wetter, with higher flows in north-west UK. Drier in south-east UK.
First migration to sea	Coastal waters	Warmer	More winter storms. Windier. Increased wind-driven circulation in North Sea. Increased wave heights.
	European shelf edge	Warmer. Increased warm/cold contrast offshore	Stronger slope current.
Migration north	Norwegian Atlantic Current	Warmer	Narrower, shifted > 200 km towards east. Increased winter storms and precipitation (+15 cm). Lower salinities.
Overwintering	Nordic seas	Colder	More winter storms and of increased intensity. Increased Fram Strait ice-flux adds to widespread freshening of the Nordic Seas.
MSW fish	Labrador Sea, West Greenland	Colder	Fresher owing to increased precipitation and increased volume flux of ice from Fram Strait. Increase of winter deep convection. Increased NW wind stress enhancing cooling.
MSW return home	North Atlantic Current	Colder	Increased strength of the N Atlantic gyre circulation including the main North Atlantic Current along the sub-polar: sub-tropical gyre boundary

- When the overall decline in total catch and sea survival, as indicated by 1SW return rate to index rivers, are compared with the contrasting response of home waters (strong decadal scale response to NAO, currently warming) and distant waters (negative response to increasing NAO since 1960, currently cooling), then it is effects in middle and distant waters that may have the greatest impact on European salmon stocks.

References

Anon. (1998) Report of the Working Group on North Atlantic Salmon. International Council for the Exploration of the Sea CM 1998/ACFM:15.

Aagaard, K. & Carmack, E.C. (1989) The role of sea ice and other fresh water in the Arctic circulation. *Journal of Geophysical Research*, **94**, 14485–14489.

Alexandersson, H., Smith, T., Iden, K. & Tuomenvirta, H. (1998) Long term trend variation of the storm climate over NW Europe. *The Global Ocean Atmosphere System*, **6**, 97–120.

Bacon, S. & Carter, D.J.T. (1993) A connection between mean wave height and atmospheric pressure gradient in the North Atlantic. *International Journal of Climatology*, **13**, 423–36.

Becker, G.A. & Pauly, M. (1996) Sea surface temperature changes in the North Sea and their causes. *ICES Journal of Marine Science*, **53**, 887–98.

Blindheim, J., Borovkov, V., Hansen, B., Malmberg, W.R., Turrell, W.R. & Osterhus, S. (in press) Upper layer cooling and freshening in the Norwegian Sea in relation to atmospheric forcing. *Deep-Sea Research*.

Cayan, D.R. (1992a) Latent and sensible heat flux anomalies over the Northern Oceans: The connection to monthly atmospheric circulation. *Journal of Climate*, **5**, 354–69.

Cayan, D.R. (1992b) Variability of latent and sensible heat fluxes estimated using bulk formulae. *Atmosphere-Ocean*, **30**, 1–42.

Cayan, D.R. (1992c) Latent and sensible heat flux anomalies over the northern oceans: driving the sea surface temperature. *Journal of Physical Oceanography*, **22**, 859–81.

Cayan, D.R. & Reverdin, G. (1994) Monthly precipitation and evaporation variability estimated over the North Atlantic and North Pacific. *Atlantic Climate Change Program: Proceedings of PI's meeting*, Princeton, May 9–11, 1994, pp. 28–32.

Cook, E.R., D'Arrigo, R.D. & Briffa, K.R. (1998) A reconstruction of the North Atlantic Oscillation using tree-ring chronologies from North America and Europe. *The Holocene*, **8**, 9–17.

Cotton, P.D. & Carter, D.J.T. (1994) Interannual variability in global wave climate from satellite data. Proceedings of EOS/SPIE International Symposium on Satellites and Remote Sensing, Rome, 1994.

Cotton, P.D., Challenor, P.G. & Carter, D.J.T. (1997) Analysis of Inter-annual variability of altimeter measured global wave climate. 3rd ERS Symposium, Florence, Italy, March 1997.

Curry, R. & McCartney, M. (in press) Ocean gyre circulation changes associated with the North Atlantic Oscillation. *Journal of Physical Oceanography*. WOCE Conference, Halifax, NS, Abstracts volume, p.193.

Dempson, J.B., Myers, R.A. & Reddin, D.G. (1986) Age at first maturity of Atlantic Salmon (*Salmo salar*) – Influences of the Marine Environment. In: *Salmonid Age at Maturity* (ed. D.J. Meerburg) *Canadian Special Publication Fisheries and Aquatic Sciences*, **89**, 79–89.

Dickson, R.R., Meincke, J., Malmberg, S.-A. & Lee, A.J. (1988) The 'Great Salinity Anomaly' in the northern North Atlantic 1968–1982. *Progress in Oceanography*, **20**, 103–51.

Dickson, R., Lazier, J., Meinke, J., Rhines, P. & Swift, J. (1996) Long-term coordinated changes in the convective activity of the North Atlantic. *Progress in Oceanography*, **38**, 241–95.

Dickson, R.R. (1997) From the Labrador Sea to global change. *Nature*, **386**, 649–50.

Dickson, R.R., Osborn, T.J., Hurrell, J.W., Meincke, J., Blindheim, J., Adlandsvik, B., Vigne, T., Alekseev, G. & Maslowski, W. (in press) The Arctic Ocean response to the North Atlantic Oscillation. *Journal of Climate*.

Friedland, K.D. & Reddin, D.G. (1993) Marine survival of Atlantic salmon from indices of post-smolt growth and sea temperature. In: *Salmon in the Sea and new Enhancement Strategies* (ed. D. Mills) pp. 119–38. Fishing News Books, Oxford.

Friedland, K.D., Reddin, D.G. & Kocik, J.F. (1993a) Marine survival of North American and European Atlantic salmon: Effects of growth and environment. *ICES Journal of Marine Science*, **50**, 481–92.

Friedland, K.D., Reddin, D.G. & Kocik, J.F. (1993b) The production of North American and European Atlantic salmon: Effects of post-smolt growth and ocean environment. International Council for the Exploration of the Sea CM 1993/M:13.

Friedland, K.D., Hansen, L.P. & Dunkley, D.A. (1998) Marine temperatures experienced by postsmolts and the survival of Atlantic salmon, *Salmo salar* L., in the North Sea. *Fisheries Oceanography*, **7**, 22–34.

Friedland, K.D., Hansen, L.P., Dunkley, D.A. & MacLean, J.C. (in press) Linkage between ocean climate, post-smolt growth, and survival of Atlantic salmon (*Salmo salar* L.) in the North Sea area. *ICES Journal of Marine Science*,

Fromentin, J.-M. & Planque, B. (1996) *Calanus* and environment in the eastern North Atlantic. II. Influence of the North Atlantic Oscillation on *C. finmarchicus* and *C. helgolandicus*. *Marine Ecology Progress Series*, **134**, 111–18.

Holm, V.M.E., Holst, J.C. & Hansen, L.P. (1996) Salmon in the Norwegian Sea. Fishing experiments and recorded captures of salmon in the Norwegian Sea and adjacent areas, July 1991–August 1995. (In Norwegian with English summary and figure legends). *Fisken Havet*, **1**, 13 pp.

Hurrell, J.W. (1995a) Decadal trends in the North Atlantic Oscillation: regional temperatures and precipitation. *Science*, **269**, 676–9.

Hurrell, J.W. (1995b) An evaluation of the transient eddy forced vorticity balance during northern winter. *Journal of Atmospheric Science*, **52**, 2286–301.

Hurrell, J.W. & van Loon, H. (1997) Decadal variations in Climate associated with the North Atlantic Oscillation. *Climate Change*, **36**, 301–26.

Kushnir, Y., Cardone, V.J., Greenwood, J.G. & Cane, M. (1997) On the recent increase in North Atlantic wave heights. *Journal of Climate*, **10**, 2107–13.

Kwok, R., & Rothrock, D.A. (1998) Variability of Fram Strait Ice Flux and the North Atlantic Oscillation. *Journal of Geophysical Research (Oceans)*, **104**, 5177–89.

Lazier, J.R.N. (1995) The salinity decrease in the Labrador Sea over the past thirty years. In: *Natural Climate Variability on Decade-to-Century Time Scales* (eds D.G. Martinson, K. Bryan, M. Ghil, M.M. Hall, T.M. Karl, E.S. Sarachik, S. Sorooshian & L.D. Talley) pp. 295–304. National Academy Press, Washington, DC.

Martin, J.H.A. & Mitchell, K.A. (1985) influence of sea temperature upon the numbers of grilse and multi-sea-winter Atlantic salmon (*Salmo salar*) caught in the vicinity of the River Dee (Aberdeenshire). *Canadian Journal of Fisheries and Aquatic Sciences*, **42**, 1513–21.

McCartney, M.S., Curry, R.G. & Bezdek, H.F. (1996) North Atlantic's Transformation pipeline chills and redistributes subtropical water. *Oceanus*, **39** (2), 19–23.

McCartney, M.S., Curry, R.G. & Bezdek, H.F. (1997) The interdecadal warming and cooling of Labrador sea water. *ACCP Notes*, **4**(1), 1–11.

Myers, R.A., Helbig, J. & Holland, D. (1989) Seasonal and interannual variability of the Labrador Current and West Greenland Current. International Council for the Exploration of the Sea CM 1989/C:16.

Planque, B. & Fox, C.J. (1998) Interannual variability in temperature and the recruitment of Irish Sea cod. *Marine Ecology Progress Series*, **172**, 101–05.

Reddin, D.G. & Friedland, K.D. (1993a) Marine environmental factors influencing the movement and survival of Atlantic salmon. International Council for the Exploration of the Sea CM1993/M:42.

Reddin, D.G. & Friedland, K.D. (1993b) Marine environmental factors influencing the movement and survival of Atlantic salmon. In: *Salmon in the Sea and New Enhancement Strategies* (ed. D.H. Mills), pp. 79–103. Fishing News Books, London.

Reddin, D.G. & Shearer, W. (1987) Sea-surface temperature and distribution of Atlantic salmon in the Northwest Atlantic Ocean. American Fisheries Society Symposium on Common Strategies in Anadromous/Catadromous Fishes, **1**, 262–75.

Reddin, D.G., Friedland, K.D., Rago, P.J., Dunkley, D.A., Karlsson, L. & Meerburg, D.M. (1993) Forecasting the abundance of North American two-sea winter salmon stocks and the provision of catch advice for the West Greenland salmon fishery. International Council for the Exploration of the Sea CM1993/M:43.

Rhines, P. (1994) Climate change in the Labrador Sea, its convection and circulation. In: *Atlantic Climate Change Program*. Proceedings of the PI's Meeting, Princeton, May 9–11 1994, pp. 85–96.

Rogers, J.C. (1990) Patterns of low-frequency monthly sea-level pressure variability (1899–1986) and associated wave cyclone frequencies. *Journal of Climate*, **3**, 1364–79.

Rogers, J.C. (1997) North Atlantic storm track variability and its association to the North Atlantic Oscillation and climate variability of Northern Europe. *Journal of Climate*, **10**, 1635–47.

Saunders, R.L. (1986) The thermal biology of Atlantic salmon: influence of temperature on salmon culture with particular reference to constraints imposed by low temperature. *Report of the Institute of Freshwater Research Drottningholm*, **63**, 68–81.

Serreze, M.C., Carse, F., Barry, R.G. & Rogers, J.C. (1997) Icelandic Low cyclone activity: climatological features, linkages with the NAO, and relationships with recent changes in the Northern Hemisphere circulation. *Journal of Climate*, **10**, 453–64.

Shelton, R.G.J., Turrell, W.R., Macdonald, A., McLaren, I.S. & Nicoll, N. (1997) Records of post-smolt Atlantic salmon, *Salmo salar* L., in the Faroe-Shetland Channel in June 1996. *Fisheries Research*, **31**, 159–62.

Sutton, R.T. & Allen, M.R. (1997) Decadal predictability of North Atlantic sea surface temperature and climate. *Nature*, **388**, 563–7.

Turrell, W.R. (1994) Oceanographic influences on Scottish west coast salmon and sea trout. In: *Problems with Sea Trout and Salmon in the Western Highlands*, pp. 244–41. Atlantic Salmon Trust, Pitlochry, Perthshire.

Turrell, W.R. & Shelton, R.G.J. (1993) Climatic Change in the North-east Atlantic and its impacts on Salmon Stocks. In: *Salmon in the Sea and New Management Strategies* (ed. D.H. Mills), pp. 40–78. Fishing News Books, Oxford.

Vinje, T. & Finnekasa, O. (1986) The ice transport through the Fram Strait. *Norsk Polarinstitut Skrifter*, **186**, 37–39.

Vinje, T., Nordlund, N. & Vambekk, A.K. (1997) Monitoring ice thickness in Fram Strait. *Journal of Geophysical Research*, **103**, 10437–49.

WASA Group (1997) '*Changing Waves and Storms in the Northeast Atlantic*?' GKSS Forschungszentrum Geesthacht GmbH, Report 97/E/46, Geesthacht, Germany.

Welch, D.W., Chigirinsky, A.I. & Ishida, Y. (1995) Upper thermal limits on the oceanic distribution of Pacific salmon (*Oncorhynchus* spp.) in the spring. *Canadian Journal of Fisheries and Aquatic Sciences*, **52**, 489–503.

Wilby, R.L., O'Hare, G.O. & Barnsley, N. (1997) The North Atlantic Oscillation and British Isles climate variability, 1865–1996. *Weather*, **52**, (9), 266–76.

Xie, P. & Arkin, P.A. (1996) Analyses of global monthly precipitation using gauge observations, satellite estimates and numerical model prediction. *Journal of Climate*, **9**, 840–58.

Chapter 10
Changes in Ocean Climate and its General Effect on Fisheries: Examples from the North-west Atlantic

K.F. DRINKWATER

Introduction

Fish and shellfish respond directly to climate fluctuations as well as to changes in their biological environment (predators, prey, species interactions, disease) and to fishing pressures. While this multi-forcing sometimes makes it difficult to establish unequivocal linkages between changes in the physical environment and the response of fish or shellfish stocks, some effects are clear (see reviews by Cushing & Dickson 1976, Bakun et al. 1982, Cushing 1982, Sissenwine 1984, Shepherd et al. 1984, Sharp 1987). These include effects on the growth and reproduction of individual fish, as well as on the distribution and abundance of some fish populations. The influence on abundance occurs principally through recruitment (how many young survive long enough to potentially enter the fishery), but in some cases may also be due to direct mortality of adult fish.

In this paper, the general response of fish to climate variability is discussed. The following section provides examples of environmental influences on fish stocks, principally derived from groundfish stocks in the north-west Atlantic Ocean (Fig. 10.1). The third section focuses specifically upon some environmental influences on salmon. The fourth section provides a brief description of recent climate changes in the Labrador Sea and their biological effects. Some concluding remarks appear in the final section.

Environmental/fisheries linkages: some examples

Growth

Environmental conditions have a marked effect on the growth of many fish species. For example, mean bottom temperatures account for 90% of the observed (tenfold) difference in growth rates between different Atlantic cod (*Gadus morhua*) stocks in the North Atlantic (Brander 1994, 1995). Warmer temperatures lead to faster growth rates (Fig. 10.2). Regional studies have shown similar results (Fleming 1960, Shackell et al. 1995). In the north-west Atlantic, the largest cod are found on Georges Bank with a 4-year-old fish being, on average, five times bigger than one off Labrador and Newfoundland. Temperature accounts not only for differences in growth rates between cod stocks but also for year-to-year changes in growth rates within a stock. For example, approximately 50% of the observed changes in size-at-age of Atlantic

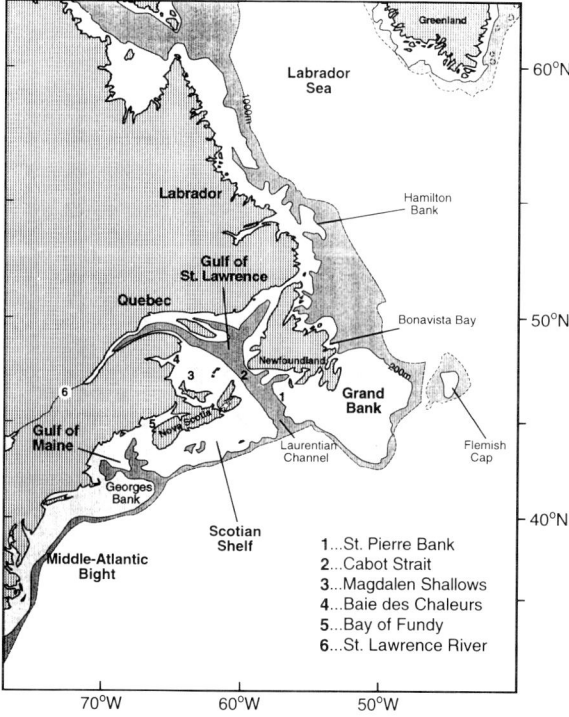

Fig. 10.1 The north-west Atlantic.

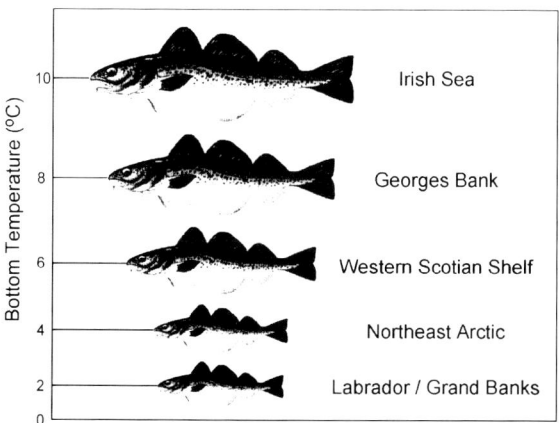

Fig. 10.2 The relative size of an age 4 cod from different areas around the North Atlantic showing increasing size with increasing bottom temperature. The data were taken from Brander (1995).

cod on the north-eastern Scotian Shelf (Campana 1995) and off Newfoundland (de Cárdenas 1996, Shelton *et al.* 1996) have been attributed to the variability in sea temperature. This is particularly important given that 50–75% of the decline in the spawning stock biomass of the Newfoundland, Gulf of St Lawrence and north-eastern Scotian Shelf cod stocks from the 1980s to the mid-1990s were caused by reduced weights-at-age (Sinclair 1996). Krohn & Kerr (1996) found that 50% of the variability in weights-at-age of the 4–8-year-old northern cod could be accounted for by the weight at age 3, indicating that size at an early age affects future fish production. They further noted, however, that from 1979 to the mid-1990s, the weight of northern cod older than 4 years was lower than expected based on predictions from weight at age 3, suggesting that the growth environment for these older fish had worsened.

Temperature-dependent growth rates are not restricted to cod. Cold bottom temperatures resulted in decreased growth rates of adult American plaice (*Hippoglossoides platessoides*) on the Grand Bank during the 1980s (Brodie 1987). Reduced length-at-age and weight-at-age for ages 3 and 4 capelin (*Mallotus villosus*) off Newfoundland during the 1990s have been shown to be a direct response to cold ocean temperatures (Nakashima 1996). Fifty per cent of the interannual variations of growth of herring (*Clupea harengus harengus*) aged 3–7 in two large bays on the south-east coast of Newfoundland were accounted for by the March to December water temperatures (Winters *et al.* 1986). Growth rates of larvae have also been found to be temperature dependent for many species in the north-west Atlantic including Atlantic cod, haddock (*Melanogrammus aeglefinus*) and winter flounder (*Pseudopleuronectes americanus*) (Morse 1989).

Reduced growth rates at lower temperatures are, in part, due to changes in feeding rates. Laboratory experiments by McKenzie (1934, 1938) found that Atlantic cod from the Bay of Fundy and Scotian Shelf ate well at temperatures within their normal tolerance range but ceased feeding at very high (>17°C) or very low temperatures (<0°C). He also noted that at low temperatures cod had difficulty swallowing and the size of the food particles consumed decreased as the cod were unable to open their mouths as wide as in warmer water (McKenzie 1938). Kohler (1964) found feeding rates for Atlantic cod increased with temperature over the range 4° to 13°C, a result consistent with that of McKenzie (1934, 1938). Reduced feeding with subsequent weight loss at low temperatures for adult American plaice from the Grand Banks has also been measured in the laboratory by Morgan (1992). Besides lower feeding rates, reduced growth may also arise through delayed spawning (see below), initially causing a short growing season, and subsequently smaller size later in life.

Spawning and reproduction

In addition to growth, the environment affects the reproductive cycle of fish and shellfish. For example, the age of sexual maturity of certain fish species, such as Atlantic cod, is determined by ambient temperatures (Fig. 10.3) (Myers *et al.* 1996a).

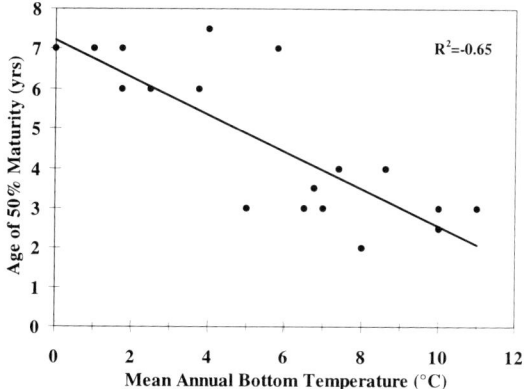

Fig. 10.3 The relationship between age of 50% maturity and the mean annual bottom temperature for North Atlantic cod stocks (based upon data presented in Hutchings & Myers 1994a).

Off Labrador and the northern Grand Bank, cod mature at age 7 and in the northern Gulf of St Lawrence and the eastern Scotian Shelf at age 6, while in the warmer waters off south-west Nova Scotia and on Georges Bank they mature at 3.5 and 2 years, respectively.

Spawning times, too, are influenced by temperature. Cold temperatures typically result in delayed spawning through slow gonad development, as has been observed in Atlantic cod on the northern Grand Bank (Hutchings & Myers 1994a). While warm temperatures promote gonad development, resulting in earlier spawning, the relationship between temperature at the spawning site and time of spawning depends on local hydrography and fish distribution. In contrast to the positive relationship between local temperatures and time of spawning on the northern Grand Banks, cold temperatures lead to earlier spawning of cod off southern Newfoundland (Hutchings & Myers 1994a). However, these fish reside in warm offshore waters and move onto St Pierre Bank prior to spawning. In very cold years on the Bank, they appear to delay migration onto the Bank, thereby remaining in the warm offshore waters longer, resulting in faster gonad development and an earlier readiness to spawn.

Temperature-dependent spawning again is not limited to cod. In the early 1990s, extremely low temperatures during spring off Newfoundland delayed capelin spawning by over a month, which led to slow growth rates and poor condition (Nakashima 1996). Marak & Livingston (1970) found a 1.5° to 2°C temperature change produced a difference of a month in the spawning time of haddock on Georges Bank, with earlier spawning in warm years. In these warm years the duration of spawning also was longer than in colder years. From studies in the Baie des Chaleurs within the Gulf of St Lawrence, spawning of the giant scallop (*Placopecten magellanicus*) has been shown to be associated with rapid temperature changes caused by wind-induced upwelling (Bonardelli *et al.* 1996). It therefore follows that interannual variations in timing of spawning probably depend upon the frequency and extent of wind forcing.

Miller *et al.* (1995) found that 52% and 70% of the seasonal variance of egg and larval size at hatch, respectively, of Atlantic cod on the Scotian Shelf are temperature dependent over the range 2°–14°C with size decreasing as temperature increases. Similar dependence of egg size on temperature was found by Ware (1977) for Atlantic mackerel (*Scomber scombrus*) in the Gulf of St Lawrence. This is believed to be in part an ecological advantage in order to match available prey size at the time of hatching, as the latter is temperature dependent (Ware 1977).

Incubation times of cod eggs are also temperature dependent (Pepin *et al.* 1997). Page & Frank (1989) found that they varied from 8 to 42 days at 14° to 1°C, respectively, for Atlantic cod on the Scotian Shelf. Thus eggs in colder water are more vulnerable to predation owing to longer exposure time and may therefore experience lower survival.

Distribution and migration

Temperature is one of the primary factors, together with food availability and suitable spawning grounds, in determining the large-scale distribution pattern of fish and shellfish. Because most fish species or stocks tend to prefer a specific temperature range (Coutant 1977, Scott 1982), long-term changes in temperature can lead to expansion or contraction of the distribution range of certain species. These are generally most evident near the northern or southern boundaries of the species range; warming results in a distributional shift northward and cooling draws species southward.

Many studies have documented distributional changes off West Greenland. Jensen & Hansen (1931) and Hansen (1949) reviewed the response to the warming in the 1920s. It included the introduction of new species that came from Iceland via the Irminger Current; the extension northward of warm water species such as cod, herring and halibut; fish such as herring and redfish being found in areas outside their traditional grounds; and overwintering of some species in northerly regions. In addition, the southward migration of white whales in autumn was delayed and the abundance of cold water species such as capelin increased in the north while decreasing in the south.

Capelin, the major food source of Atlantic cod off Newfoundland and Labrador, spread southward as far as the Bay of Fundy when temperatures declined south of Newfoundland in the mid-1960s and retracted northward as temperatures rose in the 1970s (Tibbo & Humphreys 1966, Colton 1972, Frank *et al.* 1996). During cooling in the second half of the 1980s and into the 1990s, capelin again extended their range, eastward to Flemish Cap and southward onto the north-eastern Scotian Shelf off Nova Scotia (Frank *et al.* 1996, Nakashima 1996). For example, small quantities of capelin began to appear in the groundfish trawl surveys on the Scotian Shelf in the mid- to late 1980s, and since then numbers have increased dramatically (Fig. 10.4) (Frank *et al.* 1996). Initially only adult capelin were caught, but in recent years juveniles have appeared, suggesting that capelin are successfully spawning. At the

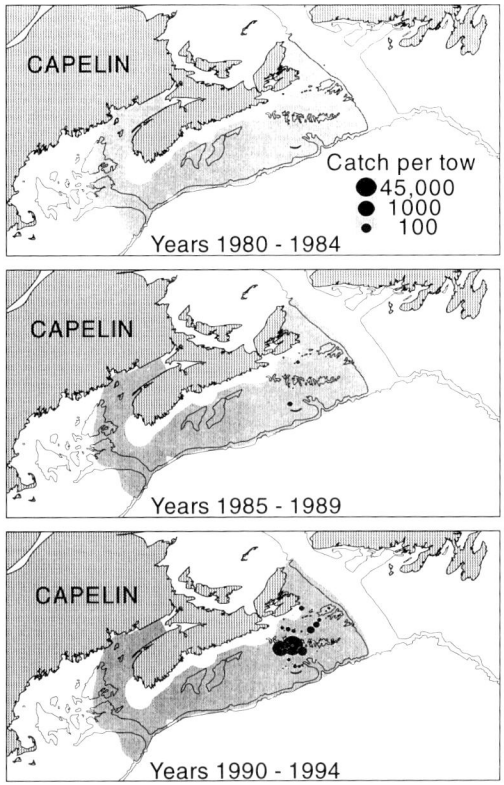

Fig. 10.4 The catch per tow of capelin during the summer groundfish surveys of the Scotian Shelf grouped by 5-year periods. The surveyed area is shaded.

same time, another cold water species, the snow crab (*Chionoecetes opilio*), extended its distribution and abundance on the north-eastern Scotian Shelf. This was interpreted as expansion due to an increase in the areal extent of their preferred thermal habitat (Tremblay 1997).

While capelin and snow crab were spreading onto the Scotian Shelf, similar range extensions were observed further north off Newfoundland. The distribution of Arctic cod (*Boreogadus saida*), a small cold-water pelagic fish whose primary grounds have traditionally been the Labrador Shelf, stretched southward to northern Newfoundland in the late 1980s and early 1990s. By the mid-1990s these fish had pushed southward to the Grand Banks and into the Gulf of St Lawrence in large numbers. This southward movement was suggested by Gomes *et al.* (1995) and substantiated from annual autumn ground surveys off Newfoundland through the 1990s (Lilly *et al.* 1994, K. Zwanenburg & G. Howell, Bedford Institute of Oceanography, pers. comm.). Southward shifts in the distribution of groundfish species off Newfoundland and Labrador at this time have also been documented; in Atlantic cod by deYoung and Rose (1993), Rose *et al.* (1994), Taggart *et al.* (1994), and Atkinson *et al.* (1997), and in fish assemblages consisting of both commercial (e.g. Greenland halibut (*Reinhardtius hippoglossoides*), and American plaice) and non-commercial species by Gomes *et al.* (1995).

Changes in distribution were also observed during the warming trend in the 1940s in the Gulf of Maine which produced a northward shift in abundance and distribution of Atlantic mackerel, American lobster (*Homarus americanus*), yellowtail flounder (*Limanda ferruginea*), Atlantic menhaden (*Brevoortia tyrannus*), and whiting (*Merluccius bilinearis*), as well as the range extension of more southern species such as the green crab (*Carcinus maenas*) (Taylor *et al.* 1957). In contrast, during the cooling trend in the Gulf of Maine from 1953 to 1967, American plaice extended their range southward white butterfish (*Peprilus triacanthus*) retracted southward to their more traditional distributional range prior to the warm 1950s (Colton 1972). Mountain & Murawski (1992) have documented north-south shifts in distribution as a function of temperature within the Gulf of Maine. The latitudinal position of the weighted mean catch for 14 out of 30 stocks investigated from groundfish surveys conducted during 1968–1989 was found to increase northward with increasing temperature. This relationship was found to be strongest for Atlantic mackerel, Atlantic herring and silver hake (*Merluccius bilinearis*).

Many species appear to use environmental conditions as cues during their migrations. For example, Atlantic mackerel migrate from their overwintering grounds off the Middle Atlantic Bight across the Gulf of Maine, along the Atlantic coast of Nova Scotia and into the Gulf of St Lawrence. Their arrival at any location along their route requires temperatures warmer than 7–8°C (Sette 1950). Similarly, the north-south migrations of American shad along the Atlantic coast of North America are regulated by the seasonal movement of waters in the 13–18°C range (Leggett & Whitney 1972). April sea surface temperatures and ice conditions in the southern Gulf of St Lawrence determine the average arrival time of Atlantic herring on their spawning grounds (Lauzier & Tibbo 1965, Messieh 1986). Ice conditions also appear to control the arrival time in spring of Atlantic cod onto the Magdalen Shallows in the Gulf of St Lawrence (Sinclair & Currie 1994). This is in contrast to their return migration in the autumn to the deep waters in the Laurentian Channel south of Cabot Strait, which appears to be unrelated to environmental conditions.

Abundance and recruitment

Understanding recruitment variability has arguably been the number one issue in fisheries science this century. Evidence of past changes in fish abundance in the absence of fishing suggests the likelihood of environmental causes. Since the advent of intensive fishing, it has become increasingly difficult to sort out the relative importance of fishing versus environment as the cause of recruitment variability. Still, recruitment levels have frequently been associated with variations in temperature during the first years of life of the fish (Drinkwater & Myers 1987). The recruitment levels of Atlantic cod off West Greenland, Labrador and Newfoundland have generally been high when ocean temperatures are warm and decrease when temperatures are cold (Taggart *et al.* 1994), but the warm periods were also those in which the spawning stock biomass was high, and thus temperature as the main cause of recruitment decline can not be confirmed. During the last 10 years of extremely cold temperatures in the northern regions, cod recruitment from Labrador to the Grand Bank has been poor. At the same time, as previously mentioned, cod populations have moved further southward. A recent hypothesis suggests these two features are related; in cold years, spawning tends to occur at southerly locations where larval retention and hence survival is poor (deYoung & Rose 1993). Other studies have suggested that the collapse of the cod stocks was not caused by a low larval survival index (recruitment/spawning stock biomass) and attributes the poor state of the fish stocks to overfishing (Myers *et al.* 1996b). In the Gulf of St Lawrence, the survival index of Atlantic cod is weakly related to the freshwater runoff from the St Lawrence River system (Chouinard & Frechet 1994). High survival occurs only when runoff is above normal, although years of high runoff do not always lead to high survival, implying that other factors are also important. The discharge is not considered to have a direct effect on the cod, but may be a proxy for food resources through influences on nutrient levels, phytoplankton production or zooplankton abundance.

American lobster landings in Canada and the United States increased steadily during the 1980s and into the 1990s to all time historic highs in most regions (Fig. 10.5). This is due primarily to higher recruitment rather than increased fishing effort (Drinkwater *et al.* 1996). Relationships between temperature and lobster landings had been established in several areas (e.g. from the Gulf of Maine to the Gulf of St Lawrence) prior to the large increase in landings, showing higher landings during warm temperatures. This suggested that perhaps the recent high landings may have been due to a large-scale warming trend. However, examination of the data showed

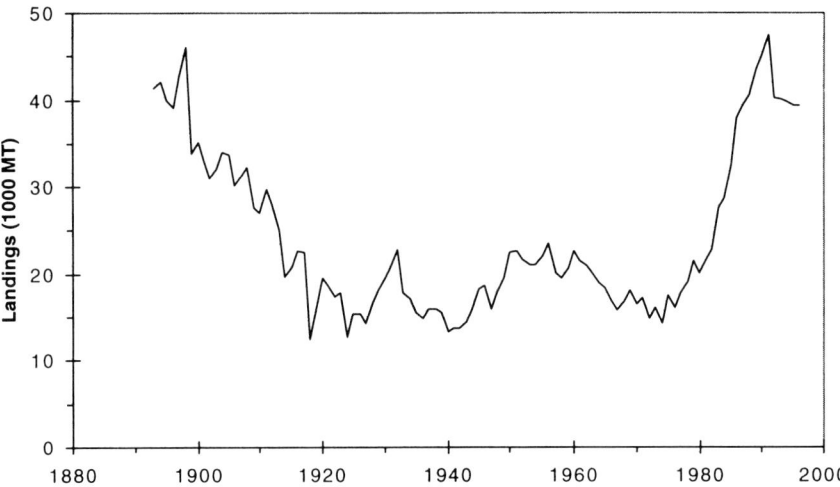

Fig. 10.5 Lobster landings in Atlantic Canada.

no such warming, and using recent temperature data the temperature/landing relationships were unable to predict any significant rise in landings during the 1980s and 1990s (Drinkwater *et al.* 1996). This is an example of a 'failed' relationship, one in which a linear regression between an environmental variable and fish or shellfish abundance was established, only to find that it was unable to explain the observed abundance in later years. These can arise either because the original correlations were spurious or the abundance is controlled by more than one variable. In the latter case, one variable may be dominant for a period of time, only to be replaced later by another variable or variables.

In a recent re-examination of previously published environment-recruitment correlations, Myers (1998) found that most relationships failed when tested with new data. Those relationships that did remain statistically significant upon re-examination tended to be for stocks that were near the geographical limits of a species range, i.e. at the northern or southern boundaries of their distribution.

While most of the examples mentioned above have involved changes in temperature, currents can also affect recruitment through transport of eggs and larvae. The 1987 haddock year class on eastern Georges Bank, which appeared to have been spawned normally in early spring, was located almost entirely in the Middle Atlantic Bight by June. This unusually large south-westward displacement was the result of an enhanced transport of water from the bank (Polochek *et al.* 1992). Recent improvements in numerical models of the currents over the continental shelves have allowed scientists to study the potential drift patterns of eggs and larvae (Werner *et al.* 1993, Lough *et al.* 1994). Advection into unfavourable sites leads to reduced recruitment if the fish die or if they can not make it back to the parent stock to reproduce (Sinclair 1988). One example of transport-related effects on recruitment involves Gulf Stream rings off north-eastern United States and eastern Canada. Large meanders in the Gulf Stream will sometimes pinch off and separate from the stream to form Gulf Stream rings or eddies. Eddies on the north side of the stream rotate clockwise and tend to trap warm Sargasso Sea water in their centre, giving rise to the terminology 'warm-core' rings. Those rings that approach the shelf often entrain large amounts of shelf water, transporting them off the shelf into the adjacent deeper slope water region. Greater numbers of Gulf Stream rings close to the continental shelf during the spawning or larval periods have been shown to lead to reduced recruitment in 14 of 17 groundfish stocks, including Atlantic cod, redfish, haddock, pollock and yellowtail flounder (Fig. 10.6) (Myers & Drinkwater 1989). The leading hypothesis is that rings entrain shelf waters laden with eggs and larvae, transporting them off the shelf, where they may die because they cannot find appropriate habitat. Death can also occur if they encounter temperatures that are too high, as observed in the waters off Georges Bank by Colton (1959).

The history of the West Greenland cod stock has been discussed by Hóvgard & Buch (1990) and Dickson & Brander (1993). Cod stocks at the turn of the century were very low, but abundance began to increase with the warming in the 1920s. Landings peaked in the 1950s and 1960s, but fell rapidly when significant cooling

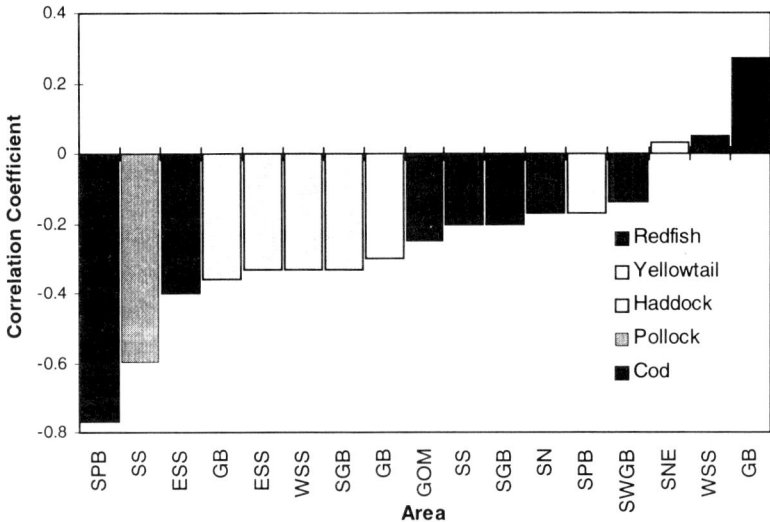

Fig. 10.6 Correlations between Gulf Stream rings and recruitment for various groundfish stocks in the Gulf of Maine, Scotian Shelf and off Newfoundland. The stock areas include St Pierre Bank (SPB), Scotian Shelf (SS), eastern and western Scotian Shelf (ESS, WSS), Georges Bank (GB), southern Grand Bank (SGB), Gulf of Maine (GOM), southern Newfoundland shelf (SN), south-west Grand Bank (SWGB), and southern New England shelf (SNE). The data were obtained from Myers & Drinkwater (1989).

started around 1970. Landings declined almost to zero during a cold period in the early 1980s. While intensive fishing is believed to have reduced the stock size, the actual collapse is considered to have been caused by recruitment failure (Hovgård & Buch 1990). It has long been known that part of the recruitment for West Greenland comes from Iceland (Tånning 1937). This has led to the hypothesis that variations in recruitment may be caused by variable inflow of larvae from Iceland (Hovgård & Buch 1990). Dickson & Brander (1993), based on work by Kushnir (1994), noted the contrast in winds off southern Greenland and around Iceland during the warm and cold periods. They found that during the warm years, when recruitment was generally high, winds were favourable for enhanced drift from Iceland to West Greenland, although they stated that the actual effects on the currents were unknown.

Increased transport of water around the Labrador Sea was found by Myers *et al.* (1989) during warm years (e.g. in the 1960s). Similar results were found by Petrie & Drinkwater (1993) when they examined transports on the eastern edge of the Grand Banks. These observations support the advective hypothesis and suggest increased transport of larvae from Iceland to West Greenland during the 1960s, when the recruitment off Greenland was good. Similarly, the low transport of water during cold years may explain in part the lower recruitment since 1970. As noted above, during the 1930s and 1940s the cod distribution along West Greenland also changed. It moved northward relative to its location in the colder years. Hovgård & Buch (1990) found that in the 1980s, as temperatures decreased, the cod fishery was dis-

126   *The Ocean Life of Atlantic Salmon*

placed southward. At the same time, the size-at-age decreased. This may be due to reduced food production or less feeding under cold conditions.

Catchability and availability

The ocean climate can also affect the fishery through influences upon availability and catchability. Availability is how many fish there are for the fishermen to catch and catchability is how easy it is for the fishermen to catch them. Availability and catchability depend not only upon the total abundance of fish but upon when and how they are distributed. If migrating fish such as herring are abundant, but do not arrive on the fishing grounds during the time the fishery is permitted to fish, then the availability is low. Also, if fish are abundant but widely distributed, so that concentrations are low, then catchability is likely to be low. Similarly, if the fish are not very abundant but highly concentrated then catch rates in those areas containing fish are good.

There are numerous examples of the environment affecting catchability. For example, when temperatures are low, American lobsters are known to move slowly, reducing the potential for encountering lobster traps, and hence cause reduced catchability (McLeese & Wilder 1958). The landings in Atlantic cod traps off Newfoundland have also been shown to depend upon temperature variability (Templeman 1966). If the traps are located in waters that are too cold, catches are low. Only when the temperatures are warm enough do catches increase. Similar results were observed in cod traps from Quebec off the north shore of the Gulf of St Lawrence (Rose & Leggett 1988).

Catch rates in the groundfish surveys conducted by the Canadian Government have also been shown to be influenced by environment conditions. For example, the number of age 4 Atlantic cod on the eastern Scotian Shelf caught in the annual spring surveys is greater during years when a larger proportion of ocean bottom is covered by temperatures less than $5°C$ and salinities of 32–33.5‰, known locally as cold intermediate layer (CIL) waters (Smith *et al.* 1991). This may result from cod seeking preferred conditions or an inability to avoid the trawls owing to reduced swimming speeds at lower temperatures (Smith & Page 1996).

Salmon

Most of the Atlantic salmon stocks of eastern North America are thought to overwinter in the Labrador Sea and northern Grand Banks (Reddin & Shearer 1987). There is a large variability in the numbers of salmon returning to the rivers of eastern Canada each year. The similarity in the interannual variability from different rivers over widely separated regions suggests that the numbers of returning salmon are most likely determined in the marine environment. A winter index of the areal extent of sea surface temperatures in the Labrador Sea conducive for salmon has been developed, which shows a high positive correlation with the number of salmon returning to

North America during the following spring and summer (Friedland *et al.* 1993, Reddin & Friedland 1993). This winter index is now used to predict prefishery abundance of salmon entering the rivers during the following late spring or summer. It is one of few examples where the state of the environment is used to predict fish abundance for fisheries assessment purposes.

Other studies have suggested that the number of returning salmon is linked to environmental change. Narayanan *et al.* (1995) suggested that the abundance of the salmon off Newfoundland and Labrador was linked to the amount of water of $<0°C$ on the shelf in summer. In years when the amount of cold water was large and the marine climate tended to be cold, there were fewer salmon returning to coastal waters. Recently, a statistically significant relationship has been found between the areal extent of ice off Labrador and northern Newfoundland and the number of returning salmon in one of the major rivers on the Atlantic coast of Nova Scotia (Harvie & Amiro 1998).

Salmon migration on the east coast of North America is also influenced by environmental conditions. The timing and geographical distribution of Atlantic salmon (*Salmo salar*) along the Newfoundland and Labrador coasts have been shown to be dependent upon the arrival of the 4°C water (Narayanan *et al.* 1995). Salmon arrive earlier during warmer years.

Labrador Sea: biological responses to recent climate changes

Ocean climate

The Labrador Sea is extremely important as an overwintering area for Atlantic salmon. Thus it is useful to describe some of the changes in ocean climate in the region and the biological responses. In spite of the reported increase in global and northern hemisphere air temperatures during the past 100 years, the Labrador Sea has undergone significant cooling from the 1960s to at least the mid-1990s.

The large-scale atmospheric pressure patterns over the North Atlantic Ocean are dominated by the Icelandic Low, centred between southern Greenland and Iceland, and the Azores High, centred roughly above the Azores. This pattern occurs all the year round but is most intense in winter. The strengths of the Low and High vary from year to year with the tendency for both pressure systems to intensify (or weaken) in the same year. This tendency is known as the NAO (North Atlantic Oscillation) and is discussed in more detail by Dickson & Turrell in Chapter 9. Rogers (1984) defined an NAO index as the winter (December, January, February) sea surface pressure at the Azores minus that at Iceland. A high index corresponds to a deep Icelandic Low and an intense Azores High. The index is a latitudinal pressure gradient and therefore its increase results in corresponding increases in the strength of the westerly winds across the northern North Atlantic. It also means stronger north-west winds over the Labrador Sea. In such years, these north-west winds carry cold Arctic air masses further south, causing winter air temperatures to decrease. This in turn results in earlier and more ice formation, and the stronger north-west winds push the ice further south, leading to more extensive ice coverage. The cold, windy conditions in winter also result in high air-sea heat exchanges leading to extensive cooling of the waters over the shelf and in the Labrador Sea (Cayan 1992). In years of low NAO, the opposite tends to occur, i.e. weakened Icelandic Low, reduced north-west winds, warmer-than-normal winter air temperatures, later and less extensive ice, and reduced heat exchange leading to warmer ocean temperatures.

The above relationships are clearly seen in time series plots of the NAO, the winter time air temperatures at Cartwright on the Labrador coast, the NW wind stress over the Labrador Sea, the area of ice off Labrador and Newfoundland south of 55°N, the temperature at a hydrographic monitoring site just off St John's, Newfoundland, and the area of cold ($<0°C$) water (termed the cold intermediate layer or CIL) in summer along a transect off Bonavista Bay in northern Newfoundland (Fig. 10.7). When the NAO is high (intensified Icelandic Low), the air temperatures are cold, the wind stress is high, there is lots of ice, the ocean temperatures are cold and there is more than normal CIL waters. Of particular interest is the period since the 1960s. From the low NAO years (warm air temperatures over the Labrador Sea) in the 1960s to the mid-1990s, the NAO signal consisted of a steady rise superimposed upon near decadal oscillations with peaks in the early 1970s, the early 1980s and the early 1990s. Correspondingly, there were negative air and ocean temperature trends and increasing winds, ice area and CIL area, with evidence of the near-decadal variations in all variables.

Biological responses

A large number of changes were recorded in the fish stocks around the Labrador Sea during the 1980s and early 1990s, a period that coincides with the extremely low temperatures. A review of trends in the northern cod fisheries located off southern Labrador and Newfoundland has been written by Taggart *et al.* (1994). Landings were typically 100–400 × 10^3 t from 1800 to 1950 (Fig. 10.8). In the late 1950s and early 1960s landings rose to almost 1200 × 10^3 t as technology improved and large distant-water foreign fleets arrived. By the late 1960s landings were dropping. They reached a minimum in 1977 around 200 × 10^3 t, rose through to the mid-1980s, but fell dramatically to levels of less than 100 × 10^3 t in the 1990s. A fishing moratorium was imposed in 1992 and remains in place.

Spawning stock biomass declined from the mid-1960s to the mid-1970s, rose slightly into the early 1980s but fell again in the 1990s to very low levels. The weights-at-age also fell steadily through the 1980s into the 1990s and contributed over 50% of the decrease in biomass (Sinclair 1996). The decrease in weight has been linked to the presence of colder waters, as discussed previously. Recruitment declined along with the spawning stock biomass through the 1960s, but since then there have been three distinct oscillations. These were found to co-vary with salinity oscillations by Sutcliffe *et al.* (1983) and confirmed by Myers *et al.* (1993) although a reanalysis by Hutchings

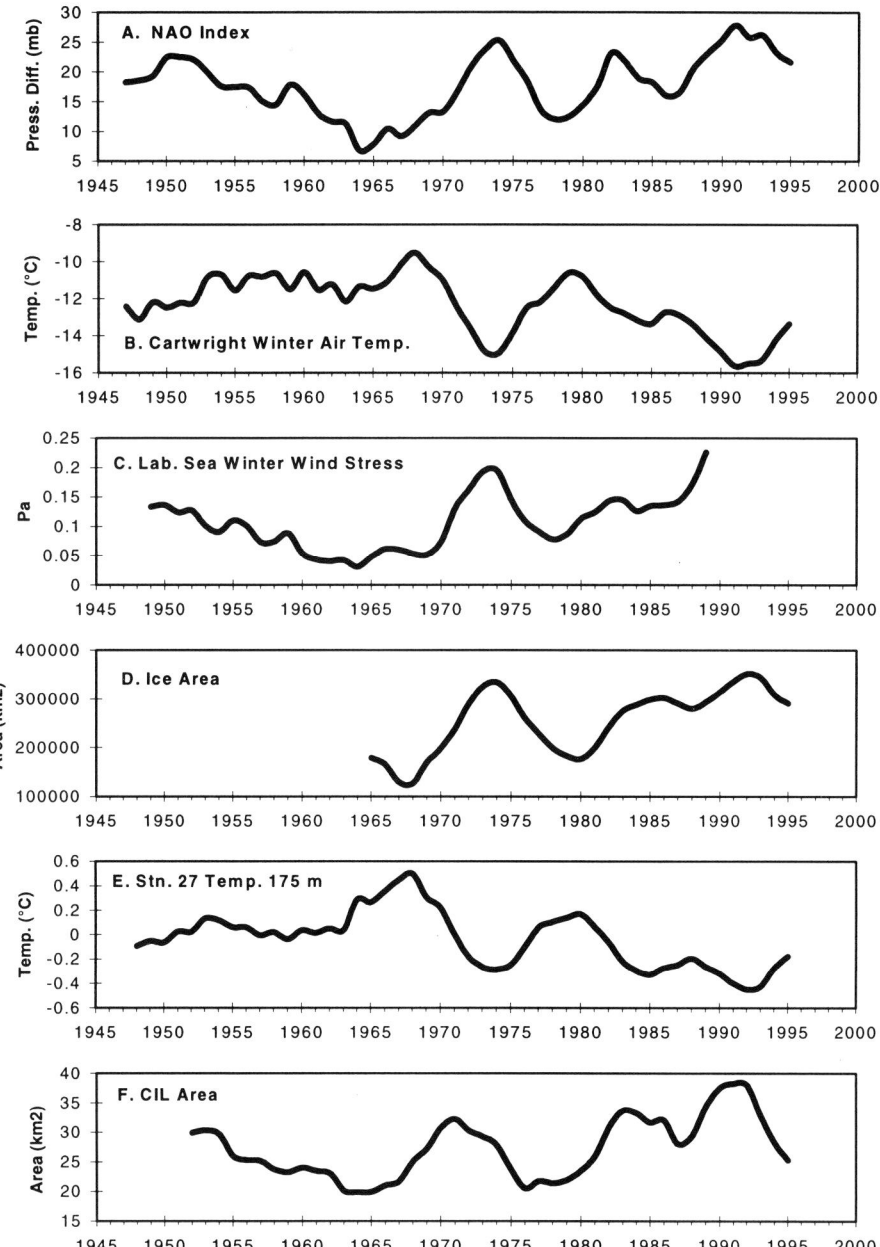

Fig. 10.7 Five-year running means of (A) the North Atlantic Oscillation (NAO) Index, (B) winter air temperatures at Cartwright on the Labrador coast, (C) north-west wind stresses over the Labrador Sea, (D) the ice extent south of 55°N during February, (E) the near-bottom (175 m deep) temperature anomalies at station 27 off St John's Newfoundland, and (F) the areal extent of the cold intermediate layer (CIL) off Bonavista, Newfoundland during the summer.

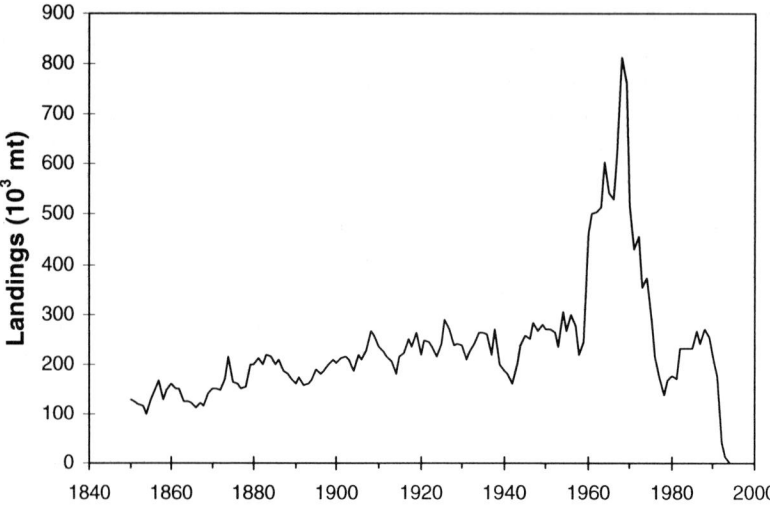

Fig. 10.8 The landings of cod off southern Labrador and northern Newfoundland (NAFO subareas 2J3KL).

& Myers (1994b) questions any relationship. DeYoung & Rose (1993) showed a southward shift in the cod distributions coincident with the decline in abundance, and proposed that egg and larval retention and survival are spatially dependent, so that in cold years spawning tends to occur at more southerly locations where larval retention, and hence recruitment, is poor.

Atkinson (1993) has discussed trends in the biomass and numbers of demersal fish caught off southern Labrador and northern Newfoundland in the Canadian fall groundfish surveys between 1981 and 1991. He found a general decrease in biomass for all species, many of the changes exceeding that experienced by cod. Decreases were also observed in the abundance of most species. These general declines corresponded in time to the decline of sea temperatures. In addition to, or perhaps as a consequence of, these declines, the groundfish concentrated along the slope edge.

Morozova (1993) examined the distribution of yellowtail flounder on the Grand Bank based upon Russian surveys between 1971 and 1991. She found a southward shift during this period. In recent years young fish have been distributed only on the southern Grand Bank. This is in contrast to the 1970s when they were found on the northern Grand Bank. Although the Grand Bank is the northern limit of the yellowtail, they can tolerate wide temperature variations and low temperatures (Walsh 1992). Morozova (1993) suggested that the southerly distribution of young reflects the general decline in abundance. When stock size increases (or decreases) the areal distribution is pushed northward (or southward). Therefore, while the decrease in temperature may not have contributed directly to the southward movement, and this can not be ruled out entirely, it is thought to have played a role in the general decline in the abundance of yellowtail and hence its southward displacement.

American plaice is another of the groundfish species on the Grand Bank whose biomass and abundance has declined steadily through the 1980s (Brodie *et al.* 1993). An earlier study by Brodie (1987) had shown a significant relationship between bottom temperatures and weight of fish caught per tow in the period 1971–86. This suggested that the overall decline may be in large part temperature-induced. Morgan (1992) conducted laboratory experiments and found that plaice were tolerant to the observed temperatures, but that at the low bottom temperatures of recent years they became more active, did not eat and lost weight. She suggested that weakened condition or starvation may have led to reproduction failure.

Nakashima (1996) has discussed changes in capelin during the 1990s off Newfoundland in response to extreme cold conditions. He found reduced length-at-age and weight-at-age for ages 3 and 4 fish. In addition during the 1990s there was a late arrival of mature capelin inshore owing to extreme cold and late ice retreat. This caused delayed spawning, extended spawning times and later emergence of larval capelin. This is believed to have led to smaller fish and lower survival.

Ratz (1993) documented significant changes in the demersal fish species off West Greenland (Cape Farewell to 67°N) during 1982–92 from German groundfish surveys. The combined abundance and biomass of the demersal species caught in the surveys were low in the early 1980s, rose in the mid-1980s to a peak in 1987, after which the abundance fell by 85% and the biomass by 98%. Commercially important species such as cod, plaice, redfish, wolffish and skate contributed most to the decline. Length distributions reveal that small individuals now dominate the biomass. The low abundance in the early 1980s and the increase in the mid-1980s correspond with temperature changes. Ratz (1993) states that the decline in the late 1980s into the early 1990s did not correspond to temperature trends observed in November at a slope station on the Fyllas Bank section. However, data for this station in June show strong cooling during this period and are more consistent with air temperature records than the November data.

It is clear that significant ecosystem changes occurred in the Labrador Sea and particularily on the continental shelves off West Greenland and Labrador/Newfoundland during the 1980s and 1990s. These included distributional shifts of cold water species southward, a movement of many species into deep waters, a decline in growth rates, reduced recruitment, low abundance of most groundfish species, and delayed spawning times. This was also a time of generally reduced salmon returns to North American rivers. While overfishing may be the cause of certain of the observations, environmental changes have also played a major role.

Summary

This chapter has provided several examples of the effects of environment and environmental variability on fish stocks. As stated by Frank *et al.* (1990), taking the next step to then predict *a priori* the response of local marine organisms to possible climate variability scenarios becomes a highly speculative exercise. However, observations of changes to the fish stocks due to climate variability in the past allow us to predict some general responses. Climate change, and here we will not deal with either the amplitude or the sign of the change, can be expected to result in distributional shifts in species, with the most obvious changes occurring near the northern or southern boundaries of their range. Migration patterns will shift, causing changes in arrival times along the migration route. Growth rates are expected to vary, with the amplitude and direction being species-dependent. Recruitment success could be affected owing to changes in time of spawning, fecundity rates, survival rate of larvae, and food availability. Qualitative predictions will require improved knowledge of the life histories of the fish and further understanding of the relative roles that environment, species interactions and fishing play in determining the variability of growth, reproduction, distribution and abundance of fish stocks. This multi-forcing and numerous past examples of 'failed' environment/fish relationships indicate the difficulty fisheries scientists face in providing reliable predictions of the response of fish to climate change.

References

Atkinson, D.B. (1993) Some observations on the biomass and abundance of fish captured during stratified random bottom trawl surveys in NAFO Divisions 2J3KL, fall 1981–1991. *NAFO Scientific Council Research Document*, 93/29.

Atkinson, D.B., Rose, G.A., Murphy, E.F. & Bishop, C.A. (1997) Distribution changes and abundance of northern cod (*Gadus morhua*), 1981–1993. *Canadian Journal of Fisheries and Aquatic Sciences*, **54** (Supplement 1), 132–8.

Bakun, A., Beyer, J., Pauly, D., Pope, J.G. & Sharp, G.D. (1982) Ocean sciences in relation to living marine resources: a report. *Canadian Journal of Fisheries and Aquatic Sciences*, **39**, 1059–70.

Bonardelli, J.C., Himmelman, J.H. & Drinkwater, K. (1996) Relation to spawning of the giant scallop, *Placopecten magellanicus*, to temperature fluctuations during downdwelling events. *Marine Biology*, **124**, 637–49.

Brander, K.M. (1994) Patterns of distribution, spawning, and growth in North Atlantic cod: the utility of inter-regional comparisons. *ICES Marine Science Symposium*, **198**, 406–13.

Brander, K.M. (1995) The effects of temperature on growth of Atlantic cod (*Gadus morhua* L.) *ICES Journal of Marine Science*, **52**, 1–10.

Brodie, W.B. (1987) American plaice in divisions 3LNO – an assessment update. *NAFO Scientific Council Research Document*, 87/40.

Brodie, W.B., Power, D. & Morgan, M.J. (1993) An assessment of the American plaice stock in NAFO Divisions 3LNO. *NAFO Scientific Council Research Document*, 93/91.

Campana, S.E., Mohn, R.K., Smith, S.J. & Chouinard, G. (1995) Spatial visualization of a temperature-based growth model for Atlantic cod (*Gadus morhua*) off the eastern coast of Canada. *Canadian Journal of Fisheries and Aquatic Sciences*, **52**, 2445–6.

Cayan, D.R. (1992) Latent and sensible heat flux anomalies over the northern oceans: the connection to monthly atmospheric circulation. *Journal of Climate*, **5**, 354–369.

Chouinard, G.A. & Fréchet, A. (1994) Fluctuations in the cod stocks of the Gulf of St Lawrence. *ICES Marine Science Symposium*, **198**, 121–39.

Colton, J.B. Jr (1959) A field observation of mortality of marine fish larvae due to warming. *Limnology and Oceanography*, **4**, 219–22.

Colton, J.B. Jr (1972) Temperature trends and the distribution of groundfish in continental shelf waters, Nova Scotia to Long Island. *Fisheries Bulletin*, **70**, 637–57.

Coutant, C.C. (1977) Compilation of temperature preference data. *Journal of the Fisheries Research Board of Canada*, **34**, 739–45.

Cushing, D.H. (1982) *Climate and Fisheries*. Academic Press, London.
Cushing, D.H. and Dickson, R.R. (1976) The biological response in the sea to climatic changes. *Advances in Marine Biology*, **14**, 1–122.
de Cárdenas, E. (1996) Some considerations about annual growth rate variations in cod stocks. *NAFO Science Council Studies*, **24**, 97–107.
deYoung, B. & Rose, G.A. (1993) On recruitment and distribution of Atlantic cod (*Gadus morhua*) off Newfoundland. *Canadian Journal of Fisheries and Aquatic Sciences*, **501**, 2729–41.
Dickson, R.R. & Brander, K.M. (1993) Effects of a changing windfield on cod stocks of the North Atlantic. *Fisheries Oceanography*, **21**, 124–53.
Drinkwater, K.F. & Myers, R.A. (1987) Testing predictions of marine fish and shellfish landings from environmental variables. *Canadian Journal of Fisheries and Aquatic Sciences*, **44**, 1568–73.
Drinkwater, K.F., Harding, G.C., Mann, K.H. & Tanner, N. (1996) Temperature as a possible factor in the increased abundance of American lobster, *Homarus americanus*, during the 1980s and early 1990s. *Fisheries Oceanography*, **5**, 176–93.
Fleming, A.M. (1960) Age, growth and sexual maturity of cod (*Gadus morhua* L.) in the Newfoundland area, 1947–1950. *Journal of the Fisheries Research Board of Canada*, **17**, 775–809.
Friedland, K.D., Reddin, D.G. & Kocik, J.K. (1993) Marine survival of North American and European Atlantic salmon: effects of growth and environment. *ICES Journal of Marine Science*, **50**, 481–92.
Frank, K.T., Carscadden, J.E. & Simon, J.E. (1996) Recent excursions of capelin (*Mallotus villosus*) to the Scotian Shelf and Flemish Cap during anomalous hydrographic conditions. *Canadian Journal of Fisheries and Aquatic Sciences*, **53**, 1473–86.
Frank, K.T., Perry, R.I. & Drinkwater, K.F. (1990) Predicted response of northwest Atlantic invertebrate and fish stocks to CO_2-induced climate change. *Transactions of the American Fisheries Society*, **119**, 353–65.
Gomes, M.C., Haedrich, R.L. & Villagarcia, M.G. (1995) Spatial and temporal changes in the groundfish assemblages on the north-east Newfoundland/Labrador Shelf, north-west Atlantic, 1978–91. *Fisheries Oceanography*, **4**, 85–101.
Hansen, P.M. (1949) Studies on the biology of the cod in Greenland waters. *Rapports et Procès-Verbaux des Réunions, Conseil International pour l'Exploration de la Mer*, **123**, 1–83.
Harvie, C.J. & Amiro, P.G. (1998) Area of ice over the Newfoundland Shelf as a variable to reduce the variance of inseason forecasts of Atlantic salmon at Morgan Falls, LaHave River. *DFO Canadian Stock Assessment Secretariat Research Document*, 98/57.
Hovgård, H. & Buch, E. (1990) Fluctuation in the cod biomass of the West Greenland Sea ecosystem in relation to climate. p. 36–43. In: *Large Marine Ecosystems, Patterns, Processes and Yields* (eds K. Sherman, L.M. Alexander & B.D. Gold), pp. 36–43. American Association for the Advancement of Science.
Hutchings, J.A. & Myers, R.A. (1994a) Timing of cod reproduction: interannual variability and the influence of temperature. *Marine Ecology Progress Series*, **108**, 21–31.
Hutchings, J.A. & Myers, R.A. (1994b) What can be learned from the collapse of a renewable resource: Atlantic cod, *Gadus morhua*, of Newfoundland and Labrador. *Canadian Journal of Fisheries and Aquatic Sciences*, **51**, 2126–46.
Jensen, Ad.S. & Hansen, P.M. (1931) Investigations on the Greenland cod (*Gadus callarias* L.). *Rapports et Procès-Verbaux des Réunions, Conseil International pour l'Exploration de la Mer*, **72**, 1–41.
Kohler, A.C. (1964) Variations in the growth of Atlantic cod (*Gadus morhua* L.). *Journal of the Fisheries Research Board of Canada*, **21**, 57–100.
Krohn, M. & Kerr, S. (1996) Declining weight-at-age in northern cod and the potential importance of the early-years and size-selective fishing mortality. *NAFO Scientific Council Research Document*, 96/56.
Kushnir, Y. (1994) Interdecadal variations in North Atlantic sea surface temperature and associated atmospheric conditions. *Journal of Climate*, **7**, 141–57.
Lauzier, L.M. & Tibbo, S.N. (1965) Water temperature and the herring fishery of Magdalen Islands, Quebec. *ICNAF Special Publication*, **6**, 591–6.
Leggett, W.C. & Whitney, R.R. (1972) Water temperature and the migrations of American shad. *Fishery Bulletin*, **70**, 659–70.
Lilly, G.R., Hop, H., Stansbury, D.E. & Bishop, C.A. (1994) Distribution and abundance of polar cod

(*Boreogadus saida*) off southern Labrador and eastern Newfoundland. International Council for the Exploration of the Sea CM 1994/O:6.

Lough, R.G., Smith, W.G., Werner, F.E., Loder, J.W., Page, F.H., Hannah, C.G., Naimie, C.E., Perry, R.I., Sinclair, M. & Lynch, D.R. (1994) Influence of wind-driven advection on interannual variability in cod egg and larval distributions on Georges Bank: 1982 vs 1985. *ICES Marine Science Symposium*, **198**, 356–78.

McKenzie, R.A. (1934) Cod and water temperature. *Biological Board of Canada, Atlantic Progress Report*, **12**, 3–6.

McKenzie, R.A. (1938) Cod take smaller bites in ice-cold water. *Fisheries Research Board of Canada, Atlantic Progress Report*, **22**, 12–14.

McLeese, D.W. & Wilder, D.G. (1958) The activity and catchability of the lobster (*Homarus americanus*) in relation to temperature. *Journal of the Fisheries Research Board of Canada*, **15**, 1345–54.

Marak, R.R. & Livingstone, R. Jr (1970) Spawning date of Georges Bank haddock. *ICNAF Research Bulletin*, **7**, 56–8.

Messieh, S.N. (1986) The enigma of Gulf herring recruitment. *NAFO Scientific Council Research Document* 86/103, Serial No. N1230.

Miller, T.J., Herra, T. & Leggett, W.C. (1995) An individual-based analysis of the variability of eggs and their newly hatched larvae of Atlantic cod (*Gadus morhua*) on the Scotian Shelf. *Canadian Journal of Fisheries and Aquatic Sciences*, **52**, 1088–93.

Morgan, M.J. (1992) Low-temperature tolerance of American plaice in relation to declines in abundance. *Transactions of the American Fisheries Society*, **121**, 399–402.

Morozova, G.N. (1993) Distribution of yellowtail flounder (*Limanda ferruginea*) on the Grand Bank of Newfoundland by the data from Russian surveys 1971–91. *NAFO Scientific Council Research Document*, 93/10.

Morse, W.W. (1989) Catchability, growth, and mortality of larval fishes. *Fishery Bulletin*, **87**, 417–46.

Mountain, D.B. & Murawski, S.A. (1992) Variation in the distribution of fish stocks on the northeast continental shelf in relation to their environment, 1980–1989. *ICES Marine Science Symposium*, **195**, 424–32.

Myers, R.A. (1998) When do environment-recruitment correlations work? *Reviews in Fish Biology and Fisheries*, **8**, 285–305.

Myers, R.A. & Drinkwater, K.F. (1989) The influence of Gulf Stream warm core rings on recruitment of fish in the northwest Atlantic. *Journal of Marine Research*, **47**, 635–56.

Myers, R.A., Drinkwater, K.F., Barrowman, N.J. & Baird, J.W. (1993) Salinity and recruitment of Atlantic cod (*Gadus morhua*) in the Newfoundland region. *Canadian Journal of Fisheries and Aquatic Sciences*, **50**, 1599–609.

Myers, R.A., Helbig, J. & Holland, D. (1989) Seasonal and international variability of the Labrador Current and West Greenland Current. International Council for the Exploration of the Sea, CM 1989/C:16.

Myers, R.A., Hutchings, J.A. & Barrowman, N.J. (1996a) Hypotheses for the decline of cod in the North Atlantic. *Marine Ecology Progress Series*, **138**, 293–308.

Myers, R.A., Mertz, G. & Fowlow, P.S. (1996b) The population growth rate of Atlantic cod (*Gadus morhua*) at low abundance. *NAFO Scientific Council Research Document*, 96/40.

Nakashima, B.S. (1996) The relationship between oceanographic conditions in the 1990s and changes in spawning behaviour, growth and early life history of capelin (*Mallotus villosus*). *NAFO Scientific Council Studies*, **24**, 55–68.

Narayanan, S., Carscadden, J., Dempson, J.B., O'Connell, M.F., Prinsenberg, S., Reddin, D.G. & Shackell, N. (1995) Marine climate off Newfoundland and its influence on Atlantic salmon (*Salmo salar*) and capelin (*Mallotus villosus*). In: *Climate Change and Northern Fish Populations* (ed. J. Beamish), pp. 461–474. Canadian Special Publication of Fisheries and Aquatic Sciences, **121**.

Page, F.H. & Frank, K.T. (1989) Spawning time and egg stage duration in northwest Atlantic haddock (*Melanogrammus aeglefinus*) stocks with emphasis on Georges and Brown Bank. *Canadian Journal of Fisheries and Aquatic Sciences*, **46** (Supplement 1), 68–81.

Pepin, P., Orr, D.C. & Anderson, J.T. (1997) Time to hatch and larval size in relation to temperature and egg size in Atlantic cod (*Gadus morhua*). *Canadian Journal of Fisheries and Aquatic Sciences*, **54** (Supplement 1), 2–10.

Petrie, B. & Drinkwater, K. (1993) Temperature and salinity variability on the Scotian Shelf and in the Gulf of Maine 1945–1990. *Journal of Geophysical Research*, **98**, 20079–89.

Polocheck, T., Mountain, D., McMillan, D., Smith, W. & Berrien, P. (1992) Recruitment of the 1987 year class of Georges Bank Haddock (*Melanogrammus aeglefinus*): the influence of unusual larval transport. *Canadian Journal of Fisheries and Aquatic Sciences*, **49**, 484–96.

Ratz, H.-J. (1993) Abundance and present length structure of demersal fish stocks off West Greenland (Divisions 1B–F, 0–400 m). *NAFO Scientific Council Research Document*, 93/26.

Reddin, D.G. & Friedland, K.D. (1993) Marine environmental factors influencing the movement and survival of Atlantic salmon. In: *Salmon in the Sea and New Enhancement Strategies* (ed. D. Mills), pp. 79–103. Fishing News Books, Oxford.

Reddin, D.G. & Shearer, W.M. (1987) Sea-surface temperature and distribution of Atlantic salmon in the Northwest Altantic Ocean. *American Fisheries Society Symposium on Common Strategies in Anadromous/Catadromous Fishes*, **1**, 262–75.

Rogers, J.C. (1984) The association between the North Atlantic Oscillation and the Southern Oscillation in the Northern Hemisphere. *Monthly Weather Review*, **112**, 1999–2015.

Rose, G.A. & Leggett, W.C. (1988) Atmosphere-ocean coupling and Atlantic cod migrations: the effects of wind forced variations in sea temperatures and currents on nearshore distributions and catch rates of *Gadus morhua*. *Canadian Journal of Fisheries and Aquatic Sciences*, **45**, 1234–43.

Rose, G.A., Atkinson, B.A., Baird, J., Bishop, C.A. & Kulka, D.W. (1994) Changes in distribution of Atlantic cod and thermal variations in Newfoundland waters, 1980–1992. *ICES Marine Science Symposium*, **198**, 542–52.

Scott, J.S. (1982) Depth, temperature and salinity preferences of common fishes of the Scotian Shelf. *Journal of Northwest Atlantic Fishery Science*, **3**, 29–39.

Sette, O.E. (1950) Biology of the Atlantic Mackerel (*Scomber scombrus*) of North America, Part II – migrations and habits. *Fishery Bulletin*, **49**, 251–358.

Shackell, N.L., Frank, K.T., Stobo, W.T. & Brickman, D. (1995) Cod (*Gadus morhua*) growth between 1956 and 1966 compared with growth between 1978 to 1985, on the Scotian Shelf and adjacent areas. International Council for the Exploration of the Sea CM 1995/P:1.

Sharp, G.D. (1987) Climate and fisheries: cause and effect or managing the long and short of it all. In: *South African Journal of Marine Science* (eds A.I.L. Payne, J.A. Gulland & K.H. Brink) **5**, 811–38.

Shelton, P.A., Lilly, G.R. & Colbourne, E. (1996) Patterns in the annual weight increment for 2J3KL cod and possible prediction for stock projection. *NAFO Scientific Council Research Document*, 96/47.

Shepherd, J.G., Pope, J.G. & Cousens, R.D. (1984) Variations in fish stocks and hypotheses concerning their links with climate. *Rapports et Procès-Verbaux des Réunions, Conseil International pour l'Exploration de la Mer*, **185**, 255–67.

Sinclair, A. (1996) Recent declines in cod species stocks in the northwest Atlantic. *NAFO Scientific Council Studies*, **24**, 41–52.

Sinclair, A. & Currie, L. (1994) Timing of cod migration into and out of the Gulf of St Lawrence based on commerical fisheries, 1986–1993. *DFO Atlantic Fisheries Research Document*, 94/47.

Sinclair, M. (1988) *Marine Populations: an Essay on Population Regulation and Speciation*. University of Washington Press, Seattle.

Sissenwine, M.P. (1984) Why do fish populations vary? In: *Exploitation of Marine Communities* (ed. R.M. May). Dahlem Konferenzen 1984, Spring-Verlag, Berlin, Germany.

Smith, S.J. & Page, F. (1996) Associations between Atlantic cod (*Gadus morhua*) and hydrographic variables: implications for the management of the 4VsW cod stock. *ICES Journal of Marine Science*, **53**, 597–614.

Smith, S.J., Perry, R.I. & Fanning, L.P. (1991) Relationships between water mass characteristics and estimates of fish population abundance from trawl surveys. *Environmental Monitoring and Assessment*, **17**, 227–45.

Sutcliffe, W.H., Loucks, R.H., Drinkwater, K.F. & Coote, A.R. (1983) Nutrient flux onto the Labrador shelf from Hudson Strait and its biological consequences. *Canadian Journal of Fisheries and Aquatic Sciences*, **40**, 1692–701.

Taggart, C.T., Anderson, J., Bishop, C., Colbourne, E., Hutchings, J., Lilly, G., Morgan, J., Murphy, E., Myers, R., Rose, G. & Shelton, P. (1994) Overview of cod stocks, biology, and environment in the

Northwest Atlantic region of Newfoundland, with emphasis on northern cod. *ICES Marine Science Symposium*, **198**, 140–57.

Tånning, A.V. (1937) Some features in the migration of cod. *Journal du Conseil Permanent International pour l'Exploration de la Mer*, **12**, 1–35.

Taylor, C.C. Bigelow, H.B. & Graham, H.W. (1957) Climate trends and the distribution of marine animals in New England. *Fisheries Bulletin*, **57**, 293–345.

Templeman, W. (1966) Marine resources of Newfoundland. *Fisheries Research Board of Canada Bulletin*, **154**, 1–170.

Tibbo, S.N. & Humphreys, R.D. (1966) An occurrence of capelin (*Mallotus villosus*) in the Bay of Fundy. *Journal of the Fisheries Research Board of Canada*, **23**, 463–7.

Tremblay, M.J. (1997) Snow crab (*Chionoecetes opilio*) distribution limits and abundance trends on the Scotian Shelf. *Journal of Northwest Atlantic Fishery Science*, **21**, 7–22.

Walsh, S.J. (1992) Factors influencing distribution of juvenile yellowtail flounder (*Limanda ferruginea*) on the Grand Bank of Newfoundland. *Netherlands Journal of Sea Research*, **29**, 193–203.

Ware, D.M. (1977) Spawning time and egg size of Atlantic mackerel, *Scomber scombrus*, in relation to the plankton. *Journal of the Fisheries Research Board of Canada*, **34**, 2308–15.

Werner, F.E., Page, F.H., Lynch, D.R., Loder, J.W., Lough, R.G., Perry, R.I., Greenberg, D.A. & Sinclair, M.M. (1993) Influences of mean advection and simple behaviour on the distribution of cod and haddock early life stages on Georges Bank. *Fisheries Oceanography*, **2**, 43–64.

Winters, G.H., Wheeler, J.P. & Dalley, E.L. (1986) Survival of a herring stock subjected to a catastrophic event and fluctuation environmental conditions. *Journal du Conseil International pour l'Exploration de la Mer*, **43**, 26–43.

Xie, L. & Hsieh, W.W. (1989) Predicting the return migration routes of the Fraser River sockeye salmon (*Oncorhynchus nerka*). *Canadian Journal of Fisheries and Aquatic Sciences*, **46**, 1287–92.

Chapter 11
Historical and Potential Long-term Climatic Change in the North Atlantic

GRANT R. BIGG

Introduction

Salmon inhabit, throughout their lifetime, fluid environments that change both seasonally and from one year to the next. They exploit these environments, for instance by using currents in the oceanic gyres to take them to the rich feeding grounds within the sub-polar gyres. They are also critically dependent on predictable seasonal changes, such as river water temperature thresholds, to trigger smolt migration (Mills 1989) and salmon return to their home river (Power 1981). Atlantic salmon breed over a wide geographical range of rivers, from sub-polar Labrador, Iceland and Siberia to warm temperate Connecticut and northern Spain. This apparent environmental tolerance, however, conceals the existence of a number of different races (Mills 1989), with further genetic specialisation to the level of river origin (Ståhl, 1987).

Genetic modification by environmental influences means that salmon populations have the potential to be severely disrupted by rapid climatic change. In this chapter we will examine the past occurrence and future likelihood of such change. We begin by discussing the oceanic and freshwater characteristics of the North Atlantic region that have the potential for rapid change and play a role in the salmon life cycle. We then look at the historical record for evidence of such change in the past, before turning to current predictions of the course of anthropogenically driven climatic change over the next century. This will bring us to consider the uncertainties in such predictions and the instabilities and feedback mechanisms in the climate system that can lead to such rapid change.

Sensitive characteristics of the North Atlantic climate

Deep water formation

Perhaps the major determinant in why the North Atlantic climate is so much milder than one would expect of this latitude band (Bigg 1996) is the existence of open ocean convection and consequent deep water formation in the sub-polar Atlantic. This convection transports 15–18 Sv ($10^6 \, m^3/s$) (Schmitz & McCartney 1993) of relatively warm, salty water into the deep Atlantic to form North Atlantic Deep Water (NADW) that then spreads out at depth over much of the global ocean. To maintain this substantial volume transport branches of the North Atlantic Drift are drawn into the Norwegian and Labrador Seas. The strength of the wind-driven Gulf Stream and sub-tropical gyre are thus enhanced by a significant thermohaline component. The resulting additional meridional transport of heat leads to greater latent and sensible heat input to the overlying atmosphere than there would otherwise be.

Deep water that contributes to NADW currently forms in either the Norwegian-Greenland Sea or the Labrador Sea. There are two theories for how this occurs in the former. The classical theory (Reid 1979) assumes that warm, salty water from the North Atlantic Drift and the Slope Current passes across the Iceland-Scotland Ridge to form the Norwegian Coastal Current. Basin-scale gyres (Hebbeln *et al.* 1998) then transport some of this water into the interior of the Norwegian and Greenland Seas (Fig. 11.1). Winter cooling, combined with some brine rejection from sea-ice formation, increases the surface density sufficiently for convection to occur.

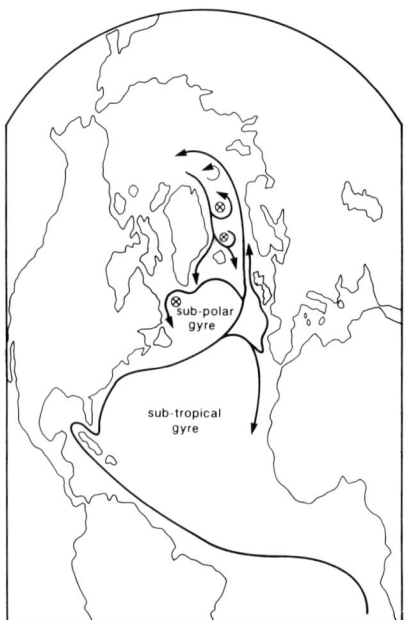

Fig. 11.1 Schematic of the upper ocean circulation of the North Atlantic Ocean and Nordic Seas. Convection sites are shown by ⊗.

This convection occurs spasmodically over rather small areas called chimneys rather than whole basins (Killworth 1983). Prior to the mid-1970s the convection penetrated to great depths within the Norwegian-Greenland Sea but more recently the water has been cooled only sufficiently for convection to reach about 1000 m (Dickson *et al.* 1996). This dense water then spills over the various sills in the Greenland-Iceland-Scotland Ridge into the deep North Atlantic. More recently, Mauritzen (1996) has proposed that only limited convection occurs in the Norwegian-Greenland Sea and that most of the overflow water derives from the Arctic. As the

Norwegian Coastal Current flows past North Cape some of the water passes through the shallow Barents Sea and, because it is relatively warm, then sinks upon leaving the extensive continental shelf for the High Arctic. This mid-level (1000–1500 m, Schlosser *et al.* 1996) Atlantic water then recirculates through the Fram Strait, being cooled en route by direct or convective mixing. Much of this water then passes back into the Atlantic through the Denmark Strait, while the remainder is entrained with the waters of the Norwegian-Greenland Sea to supplement the overflow across the Iceland-Scotland Ridge.

Deep convection also occurs in the Labrador Sea. A branch of the North Atlantic Drift, the Irminger Current, moves north-east to form the northern arm of the sub-polar gyre. Within the Labrador Sea similar wintertime processes to those occurring in the Norwegian-Greenland Sea again lead to the formation of dense water that convects. Prior to the mid-1970s the cooling was less here and the water thus convected to shallower depths, but this region has recently seemed to be the major source of NADW (Dickson *et al.* 1996). Note, however, that there is some indication that the situation has reverted since 1995 (McCartney *et al.* 1998). This exchange in the location of densest winter water may be linked to the North Atlantic Oscillation.

Storms, oceanic gyres and surface temperatures

The sub-tropical and sub-polar gyres of the North Atlantic are driven by the wind (Brown *et al.* 1990). Ekman convergence or divergence respectively forces variation in the sea level across these gyres, which then leads to pressure differences driving geostrophic currents around the gyres. The latitudinal positioning of the gyres is fundamentally linked to meridional changes in mean zonal wind direction. Thus a long-term change in the position of the westerlies, for example, leads to a consequent shift in gyre location.

The ocean also affects the atmosphere (Bigg 1996). Part of the heat implicit in the Gulf Stream and North Atlantic Drift is transferred to the atmosphere through sensible and latent heating, the latter occurring at heights and locations well away from the initial source of the evaporated water vapour. This heating both affects the atmospheric pressure distribution directly, and, more importantly, provides energy for the generation of synoptic cyclones. The position of the North Atlantic Drift therefore influences the positioning of the Atlantic storm tracks.

There are feedbacks between these ocean and atmosphere interactions. An increase in local storminess leads to a strengthening of the mean wind, which shifts the position of the westerlies and also increases evaporation, hence enhancing storm generation. However, prolonged evaporation cools the sea surface, which in turn cools the atmosphere and reduces evaporation. Storm generation thus decreases and may switch to another region which remained warm, but out of the main storm belt, during this process. There is thus a tendency for a north-south oscillation in the Atlantic storm tracks and gyre circulation, the North Atlantic Oscillation (Dickson *et al.* 1996). We shall see, in the section 'Salmon and instabilities in the climate system', that this oscillation may occur over a few weeks, or as long as a few decades. It may also be linked into larger, pan-Atlantic, climatic oscillations (Xie & Tanimoto 1998).

Terrestrial component of the hydrological cycle

The volume flux and health of the rivers discharging into the North Atlantic are intimately linked to climatic variation. Changes in storm tracks, as discussed above, will lead to variation in the net precipitation over catchments, and through associated changes in temperature and cloudiness evaporation rates will also alter. The net water balance, and hence discharge, of rivers will not, therefore, be a simple function of changing precipitation patterns. There are further, anthropogenic, effects which can modify any climatic variability in river states. Evolving land use throughout the course of a river will affect run-off characteristics and a river's water quality. For example, run-off can be dramatically increased by deforestation, particularly in the head waters, and through urbanisation.

There is abundant evidence of long term variation in rainfall around the North Atlantic. The rainfall of Britain shows a stable annual average, but a shifting seasonal distribution, with winter rainfall increasing at the expense of that in summer (Smith 1995). Many coastal regions of north-western Europe and eastern Canada are showing trends towards increasing rainfall of up to 5% per decade (Nicholls et al. 1996). There also seems to be a tendency towards reduced evaporation. It also needs to be remembered that short term variability can be greater than these multi-decadal trends. The drought in England during the early 1990s drew groundwater and river levels to historic lows despite relatively modest precipitation shortfalls (Marsh & Monkhouse 1993).

Historical change to the North Atlantic circulation

Over the last hundred thousand years salmon have had to cope with major rearrangements of the climate of the North Atlantic Ocean, some of which have occurred over only a decade or two. The former view of relatively stable climates, both in the atmosphere and more particularly in the ocean, during interglacial and glacial periods, with a smooth transition between these two extremes, is now changing. We will briefly examine some of the evidence for rapid change in each of these geological regimes before proceeding, in the section on 'Climate change at the end of the second millennium', to examine how the next 100 years may change.

Glacial variability

The oxygen isotope record in planktonic remains within deep-sea sediments is a proxy for global ice volume, and hence temperature. As water molecules with the ^{18}O isotope of oxygen are heavier than the standard $H_2^{16}O$, the latter, lighter, isotope water molecules evaporate preferentially, thus making the water vapour in the atmosphere depleted, relative to sea water, in the $H_2^{18}O$ water molecule. There is a further fractionation when the vapour condenses, with the end result that snow that falls on icecaps tends to be very depleted in ^{18}O. The water that remains in the sea is thus enriched in ^{18}O, and this signal, modified by the temperature of the water, is seen within the skeletons of hard-bodied plankton. Figure 11.2 shows significant variation in ^{18}O, and hence ice volume and global temperature, even over the peak glacial period from 50 000 to 15 000 years BP (before the present). Sarnthein et al. (1995) have examined this, and allied records in detail to reconstruct probable major changes in North Atlantic circulation over the last 30 000 years. Over much of the last 10 000 years, and prior to 30 000 years BP, the circulation seems to have been similar to today, while at other times the boundary between the sub-tropical and sub-polar gyres, the Polar Front, was much further south, at about 40°N (e.g. from 25 000 to 15 000 years BP).

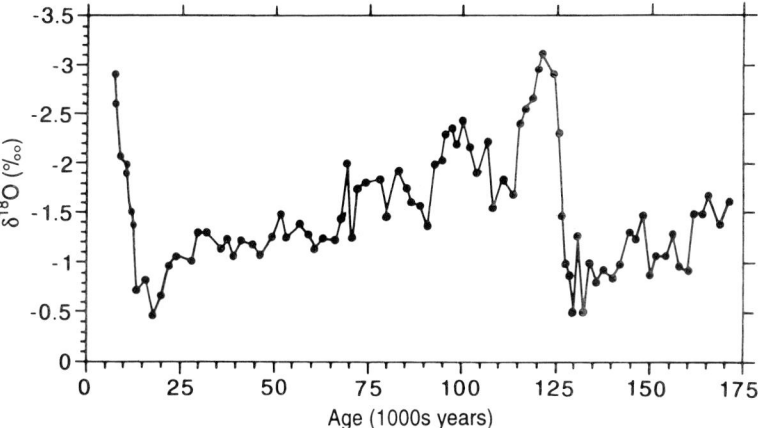

Fig. 11.2 Oxygen isotope measure from the remains of *Globigerinoides ruber* from a sediment core in the equatorial Indian Ocean. High values correspond to low ice volume or interglacial periods. (After Bigg 1996.)

Interspersed with these longer-term changes to the circulation there is evidence for rapid circulation change associated with occasional massive discharges of icebergs from both the Laurentide (Bond et al. 1992) and the Fennoscandinavian ice sheets (Hebbeln et al. 1998), known as Heinrich events. The accompanying injections of fresh water from iceberg melting capped the North Atlantic with fresh water that prevented convection. This resulted in a southward diversion, and weakening, of the Gulf Stream, as mass conservation no longer required significant northward transport of heat and salt to close the thermohaline circulation. This major change may have occurred within the last 1000 years (Dokken & Hald 1996), and has the potential to have become a new stable state for the ocean circulation (Bigg et al. 1998), although interaction between the ocean and atmosphere is likely to have slowly removed the surface freshening.

142 The Ocean Life of Atlantic Salmon

Possible surface circulation patterns for these two alternative modes of circulation during the last glacial period are shown in Fig. 11.3. Also shown in this figure is an indication of the geographical limits for rivers along the Atlantic seaboard of North America and Europe that were in the same climatic zones at the last glacial maximum as salmon are found in today. Comparing the circulation patterns salmon would have experienced in either state shows potentially large differences in migration routes. During periods with enhanced deep-water production (Fig. 11.3a) post-smolt salmon from North America and Europe north of Spain would have been easily advected northwards into rich sub-polar waters, but would have had strong currents to counter when returning to breed. In contrast, in the reduced convection state (Fig. 11.3b) salmon would need to have been active swimmers, and so not have used advection as

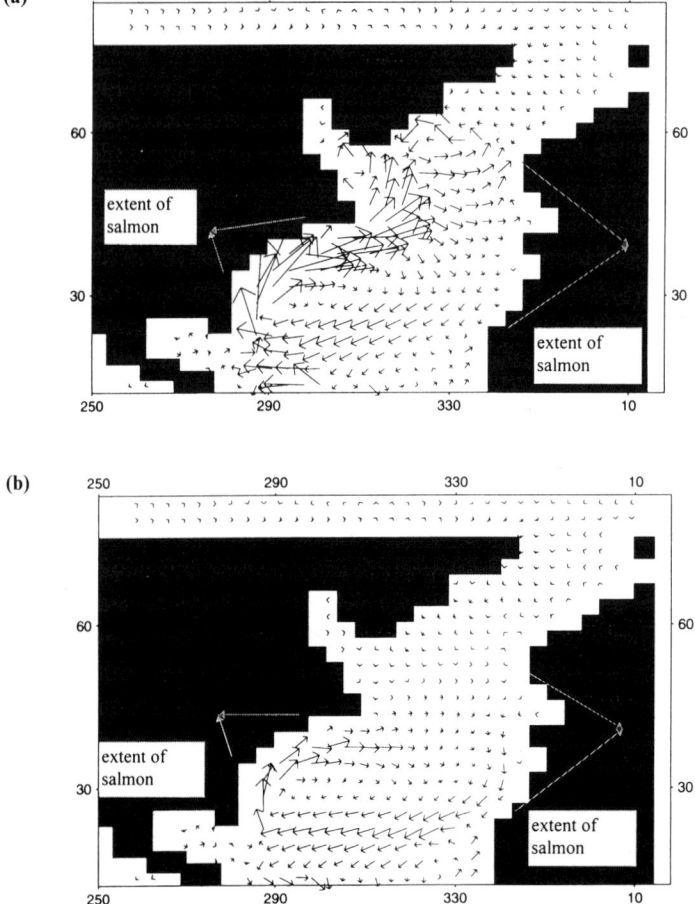

Fig. 11.3 Modelled upper ocean circulation at the Last Glacial Maximum (LGM) with (a) enhanced production of North Atlantic Deep Water, (b) reduced production of North Atlantic Deep Water. The possible salmon river ranges are based on the LGM air temperatures over coastal regions. Note that the eastern Atlantic range extends from southern Ireland to northern Morocco.

a transport mechanism, to travel north of 50°N, as the currents are very weak in this area. This region was, however, much colder than today, and is therefore likely to have been more productive, obviating the need for migration even further northward as today.

Glacial termination

The ending of the last glacial period was accompanied by large additions of fresh water to the North Atlantic, principally concentrated at two times, around 15 000 BP (mostly from Europe) and 11 000 BP (largely from North America). The first of these was accompanied by rapid atmospheric surface-ocean and deep water climate changes (Adkins *et al.* 1998). However, it was the second injection of fresh water, known as the Younger Dryas event, that had the most dramatic impact on the North Atlantic, as it occurred as the global climate had begun to warm. The addition of fresh water shut down the North Atlantic convection, resulting in a circulation similar to the second state in Fig. 11.3 being re-established for several hundred years. The global climate also cooled by several degrees over a few decades (Kroon *et al.* 1997, Smith *et al.* 1997) as a result of this change, with the maximum change around the North Atlantic.

Natural variability during interglacial periods

Both the last interglacial (Fronval & Jansen 1996) and the present one have experienced significant variability in the North Atlantic circulation. Lamb (1995) reconstructed the historical positions of the storm belt and the sub-tropical high over the eastern Atlantic, suggesting more northerly latitude for both between 6000 and 2000 BP, with a stronger meridional pressure gradient. This would have pushed the oceanic polar front north, probably with a similar impact on salmon migration routes. Millennial scale variation in the strength of the deep water leaving the North Atlantic has been reported by Marchitto *et al.* (1998), and similar variability is seen in the surface circulation of the northern Atlantic (Bond *et al.* 1997). This is well known in recent millennia over western Europe as seen in the warmer climate experienced around 0, AD 1000 and today, but cooler climates experienced during the Dark Ages and AD 1400–1900.

This last cold period, the Little Ice Age, began with an increase in storminess in western Europe (AD 1250–1350), colder water leaving the Arctic in the East Greenland Current but enhanced inflow of Atlantic waters into the occurring Norwegian Sea (Lamb 1995). Eventually the southward spread of sea-ice led to the failure of fisheries in the Nordic Seas as far south as the Faroes, with ice at the coast of Iceland for half the year. Such changes are probably associated with changes in sea surface temperature of less than 2°C (Bond *et al.* 1997). While these conditions do not suggest such a radical change in circulation as at the Younger Dryas they at least show a decrease in the strength of the North Atlantic Drift and a southward shift in its mean position. The latter is associated with an increased frequency and strength of storms further south, because of the physical effects described in the section on 'Storms, oceanic gyres and surface temperatures' (Lamb 1995). However, it should be noted that there was decadal-scale modulation in the severity of these conditions. Presumably the North Atlantic Oscillation was operating within this climate to cause such variation, while some more global ocean-atmosphere interaction was the cause of the millennial-scale variability. This ocean-atmosphere link is suggested by the close correlation between the signal of the millennial-scale events in the Greenland ice cores and the North Atlantic sediments, particularly in the change in ice-rafted debris pattern (Bond et al. 1997).

Climatic change at the end of the second millennium

Elsewhere in this volume, Drinkwater (Chapter 10) and Dickson and Turrell (Chapter 9) examine recent climatic change and its effect on fisheries. Predictions for the evolution of the climate over the North Atlantic for the next century are examined here, paying attention to the development of recent change, before discussion of mechanisms involved in this potential change in the section on 'Salmon and instabilities in the climate system'.

The general behaviour of the response of global temperature to projected increases in greenhouse gases is similar in most coupled models of the ocean-atmosphere climate system. There is obvious uncertainty in the actual rate at which these gases will increase in the atmosphere, but most experiments using a typical value of about 1% increase per annum lead to a global temperature increase of 1–2°C by the middle of the next century (Kattenberg et al. 1996, Mitchell & Johns 1997). These models are not good at reproducing observed global trends in natural variability, such as the cool period around AD 1910 and the warm period in the 1940s. However, all contain the tendency of the last 30 years for a steady increase in global temperature of about 0.4°C (Fig. 11.4). What is not so well known is that most of the models suggest that the temperature impact over the North Atlantic will be less than elsewhere, with the possibility of no change at all immediately south of Greenland. A simulation example is shown in Fig. 11.5. Some models show more warming over the North Atlantic (Mitchell & Johns 1997), but the general pattern in Fig. 11.5 is common to all simulations.

The principal reason for this apparent inertia is contained in the sensitivity of the North Atlantic thermohaline circulation, and so the strength of the North Atlantic Drift, to perturbation in the surface density. Most models show an increase in precipitation of about 5% for an increase in global temperature of 2.5°C (Mitchell et al. 1990). In the North Atlantic the increases are predicted to occur mostly in the subpolar seas, thus decreasing the surface density in convection zones. This makes convection harder to initiate in either the Nordic or Labrador Seas and thus tends to slow the thermohaline circulation down by 10–30% by AD 2050 (Kattenberg et al. 1996). In addition, the preferential warming of the polar regions (Fig. 11.5) compared

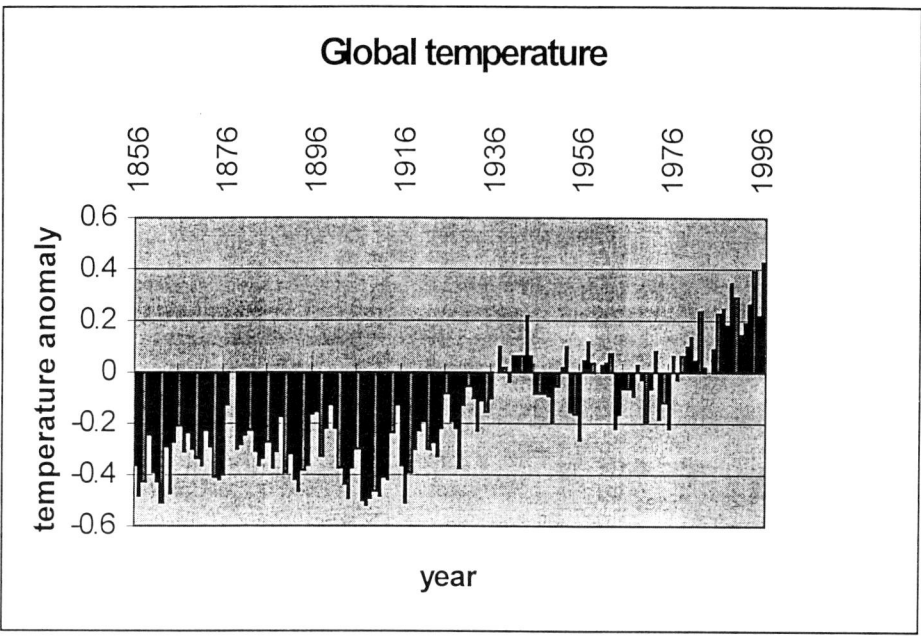

Fig. 11.4 Observed near-surface global temperature variation since 1860. Values are in °C relative to the 1961–90 mean. Bars show individual years and the solid line is a running mean. (After Parker *et al.* 1998.)

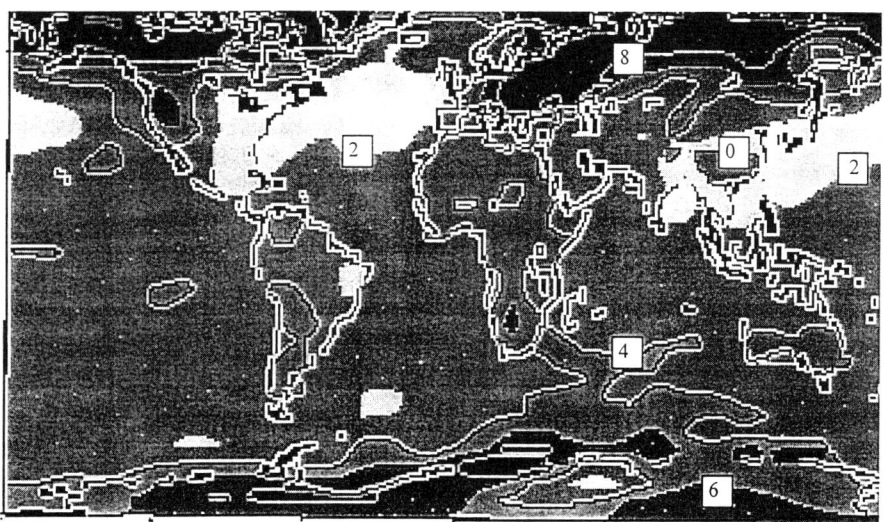

Fig. 11.5 One projection of the global temperature pattern, relative to 1961–90, in winter of the decade 2040–2049 (after Kattenberg *et al.* 1996). Each grey scale is an anomaly 1°C warmer, with the boundary of the lightest grey being an anomaly of 1°C. Small regions of the North Atlantic and China are cooler than today.

with the tropics reduces the atmospheric meridional temperature gradient and so reduces the need for pole-ward heat transport, further reducing the oceanic transport. One model saw a complete collapse of the thermohaline circulation after a further 100 years of atmospheric CO_2 increase (Manabe & Stouffer 1994).

Another forecast change which occurs in most models is for land surface temperatures around the North Atlantic rim to warm more in winter than in summer, with enhanced precipitation during the winter but a decrease in the summer (Johanessen *et al.* 1995, Kattenberg *et al.* 1996, Mitchell & Johns 1997). These will affect the hydrological conditions experienced by salmon in their breeding streams. Migration of the smolts may occur earlier in the warmer river waters, but potentially low and warm summertime river flows may discourage breeding salmon from return and allow disease and anoxia to become more widespread. Both migration and return run timing may thus change in the future. An additional hazard is possible if the freshwater and coastal sea surface temperatures become uncoupled. Warmer terrestrial temperatures, but lesser change in oceanic conditions, would mean that smolt enter less hospitable seas upon beginning their migration.

The potential change in suitable river environments towards the end of the twenty-first century is shown in Fig. 11.6. This has been constructed by considering the change in coastal temperature distribution predicted by the Hadley Centre climate model, with some consideration also being paid to summer rainfall predictions. The two major potential changes are the possibly considerable extension of salmon-bearing rivers into the Arctic, with a retreat from those rivers currently at the southern limit of their range in Spain, France and the eastern United States. With sea-ice likely to retreat because of the polar warming, but the North Atlantic Drift

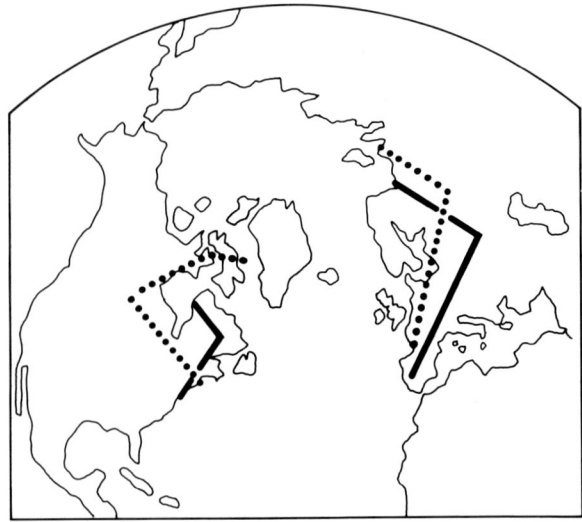

Fig. 11.6 Potential changes in the salmon range in rivers around the North Atlantic in 2050. The thick black lines delineate the present boundaries, while the dotted line shows the projected boundary in 2050.

possibly weakening, there may be an increase in productive sub-polar waters for salmon to feed from in their ocean phase.

Uncertainties

Before leaving this discussion it is worth considering the possible uncertainties in the predictions discussed above. We have already seen that there is a range of 1–2°C in model predictions for the global temperature increase over the next century, and that regional variation in the predictions occur (Kattenberg *et al.* 1996) even though large-scale patterns are common to most models. What is not clear from the discussion, however, is how well the models perform in reproducing the current climate, and whether all the processes that cause climatic change are included within them. Gates *et al.* (1996) have considered the former factor, and show that climate models are good at simulating the large scale components of the current climate, but that major problems still exist in model representations of clouds and their radiative effects, the hydrological cycle over land and the exchange of heat between the ocean and atmosphere. The impact of this was seen in Fig. 11.4, where the shorter time-scale variability of the twentieth-century global temperature was not reproduced although the current warming trend was.

A more problematic issue is the extent to which factors causing climatic change are in existing models. Until a few years ago the cooling effect of sulphate aerosols was not represented, and including this has reduced estimates of global warming by about a degree over the next 50 years (Mitchell & Johns 1997). However, these aerosols have both a direct radiative effect on climate and indirect effects. Only the former has been fully implemented in models to date. There are also only very preliminary attempts to include the biological feedback between CO_2 uptake by phytoplankton, atmospheric levels of CO_2 and dimethylsulphide, and the interaction of phytoplankton with light transmission within the ocean. It is therefore possible that predictions will be modified considerably in the future, although a very major effect would have to be undiscovered for a change in the predicted trend to occur.

Salmon and instabilities in the climate system

We have seen several ways in which significant, sometimes dramatic, change has occurred in the North Atlantic circulation in the past, or may occur in the future. These mechanisms will now be examined in more detail and related to potential impact on salmon migration patterns.

Freshwater catastrophes

The North Atlantic is very sensitive to the supply of fresh water to its surface. Rahmstorf (1995) has demonstrated that an increase of only about 0.1 Sv (roughly one third of the present annual average rainfall over the North Atlantic) would cause almost complete collapse of the basin's thermohaline circulation. The transition, when modelled, occurs over a small change in freshwater flux and not as a linear decline. The circulation would appear to have a hysteresis character in that a decrease in freshwater flux two to three times that causing the collapse would be needed before a circulation with today's strength would be reinstated. It is worth noting that such a change in the freshwater flux can take place in one, or a combination, of several ways.

Precipitation can increase, the most likely scenario in the next hundred years. Evaporation can decrease if temperatures lower or wind speeds decline. The transport of fresher water from the Pacific, through the Arctic into the Atlantic (approximately 1 Sv, Reason & Power 1994) could increase because of freshening in the North Pacific. Increased runoff, or glacial melting, from the surrounding land masses would also increase the freshwater flux into the Atlantic. Decreases in the flux or salinity of the intermediate depth outflow from the Mediterranean could also freshen the surface waters, through winter mixing (Reid 1979, Rahmstorf 1998).

The consequences for the sea surface temperature are shown by the contrast between two states in Fig. 11.7, adapted from Rahmstorf (1995). As continuity of mass no longer requires extensive transport of warm water into northern convection regions the North Atlantic Drift would be pushed further south, leading to an enhanced sub-polar gyre. Less warm water, if any, would enter the Norwegian Sea through the Faroe-Shetland Channel, and the Slope Current may reverse sign. Such changes have implications for the atmospheric circulation, pulling storm tracks south and cooling north-western Europe, both of which would tend to accentuate the oceanically-induced change. Some of these feedbacks do, however, tend to stabilise the thermohaline circulation (Rahmstorf & Willebrand 1995). They would also affect salmon migration routes, assuming that these have a component of advective

Fig. 11.7 Sea surface temperature anomaly of the present North Atlantic compared with the state with fresh water capping and no convection. Such a state is in excess of 6°C colder off south-eastern Greenland. (After Rahmstorf 1995.)

transport. A larger region of the northern Atlantic would become productive feeding ground, although displaced southwards somewhat. It may, however, be more difficult for fish from either side of the Atlantic to reach these feeding grounds because of the changes to coastal currents into which they first enter.

We have seen that such a transition is likely to have occurred a number of times in the past 30 000 years as a result of iceberg or melt-water discharge. Some tendency for a thermohaline circulation decline is also found in model simulations of the enhanced greenhouse climate of the next century, owing to precipitation increases. Recent evidence (Kroon *et al.* 1997) shows that the transition can occur in a few decades, leaving little time for the salmon to adjust to the new environment.

The North Atlantic Oscillation

One of the major signals in the sub-century variability of the North Atlantic is the North Atlantic Oscillation (NAO). This is an oscillation in the meridional atmospheric pressure pattern of the Icelandic Low and the Azores High between an accentuation of the average, the positive phase, and a lessening of the meridional pressure gradient, the negative phase (Barnston & Livezey 1987). The positive phase leads to more precipitation and storminess over northern Europe and less further south. While the strength of this oscillation varies from week to week there is a strong tendency for the mean circulation to reside in one or the other of these two states for more than a decade. For 15 years before 1995 the positive phase was dominant, but a very sudden change occurred, over only two months, in the autumn of 1994 to a prolonged negative phase (Harold *et al.* 1998), which is still in place.

Dickson *et al.* (1996) have suggested that the change in the principal location of North Atlantic deep water formation from the Nordic Seas to the Labrador Sea in the 1970s is associated with the NAO. They also show links with earlier oscillations this century. Such changes, which climate models predict to continue in an enhanced greenhouse world, lead to changes in the distribution of upwelled nutrients and ocean currents supplying the convection sites. The state of the NAO will thus affect salmon migration patterns, and as the phase is relatively stable could provide a means of pathway prediction.

Multi-decadal trends

The climate over the North Atlantic also evolves over slower scales, but still sub-century. Some of this change will increasingly be due to anthropogenic effects, but there is natural climate variation on this scale too due to solar variability, ocean-atmosphere interaction and changes in the frequencies of volcanic eruptions. One of the most notable changes over the last few decades has been a general increase in the strength of the surface wind and wave fields over the North Atlantic (Carretero *et al.* 1998). This has also been seen in the tropics (Bigg 1993). This change is partly related to atmospheric pressure tendencies seen in the positive phase of the NAO, but is possibly part of a longer, more geographically extensive variation (Xie & Tanimoto 1998). For instance, Carretero *et al.* (1998) show that the current wave and wind field is similar to that at the turn of the century.

Volcanic eruptions affect the climate over 1–3 years through injecting particulates into the stratosphere that reflect solar radiation. Major eruptions with a particularly high stratospheric input tend to lead to stronger westerlies and more storms across the northern Atlantic, as shown by Dawson *et al.* (1997) from a 200-year record of gales in Edinburgh. The frequency of such eruptions varies significantly from century to century. The nineteenth century experienced rather more such events, and hence experienced climatic perturbation more often than the twentieth century (Lamb 1995).

Conclusions

The North Atlantic is one of the most climatically sensitive regions of the world because of the interaction between oceanic and atmospheric processes that can occur. Atlantic salmon have had to adjust to large, and temporally dramatic, changes in ocean circulation and climate since the last glacial period. Such changes will continue in the near future, and may accelerate. Left to themselves salmon would most probably readily adjust to the coming climate, but the challenge will be for our management of this resource to be as flexible as the salmon themselves.

References

Adkins, J.F., Cvheng, H., Boyle, E.A., Druffel, E.R.M. & Edwards, R.L. (1998) Deep-sea coral evidence for rapid change in ventilation of the deep North Atlantic 15 400 years ago. *Science*, **280**, 725–8.

Barnston, A.G. & Livezey, R.E. (1987) Classification, seasonality and persistence of low-frequency atmospheric circulation patterns. *Monthly Weather Review*, **115**, 1083–126.

Bigg, G.R. (1993) Comparison of coastal wind and pressure trends over the tropical Atlantic: 1946–87. *International Journal of Climatology*, **13**, 411–21.

Bigg, G.R. (1996) *The Oceans and Climate*. Cambridge University Press, Cambridge.

Bigg, G.R., Wadley, M.R., Stevens, D.P. & Johnson, J.A. (1998) Simulations of two Last Glacial Maximum ocean states. *Paleoceanography*, **13**, 340–51.

Bond, G., Heinrich, H., Broecker, W., Labeyrie, L., McManus, J., Andrews, J., Huon, S., Jantschik, R., Clasen, S., Simet, C., Tedesco, K., Klas, M., Bonani, G. & Ivy, S. (1992) Evidence for massive discharge of icebergs into the North Atlantic ocean during the last glacial period. *Nature*, **360**, 245–9.

Bond, G., Showers, W., Cheseby, M., Lotti, R., Almasi, P., deMenocal, P., Priore, P., Cullen, H., Hajdas, I. & Bonani, G. (1997) A pervasive millennial-scale cycle in North Atlantic Holocene and glacial climates. *Science*, **278**, 1257–66.

Brown, J., Collins, A., Park, D., Phillips, J., Rothey, D. & Wright, J. (1990) *Ocean Circulation*, Pergamon, Oxford.

Carretero, J.C., Gomez, M., Lozano, I. *et al.* (1998) Changing waves and storms in the northeast Atlantic. *Bulletin of the American Meteorological Society*, **79**, 741–60.

Dawson, A.G., Hickey, K., McKenna, J. & Foster, I.D.L. (1997) A 200-year record of gale frequency, Edinburgh, Scotland: possible link with high-magnitude volcanic eruptions. *Holocene*, **7**, 337–41.

Dickson, R., Lazier, J., Meincke, J., Rhines, P. & Swift, J. (1996) Long-term coordinated changes in the convective activity of the North Atlantic. *Progress in Oceanography*, **38**, 241–95.

Dokken, T.M. & Hald, M. (1996) Rapid climatic shifts during isotope stages-2-4 in the polar North-Atlantic. *Geology*, **24**, 599–602.

Fronval, T. & Jansen, E. (1996) Rapid changes in ocean circulation and heat-flux in the Nordic Seas during the last interglacial period. *Nature*, **383**, 806–10.

Gates, W.L., Henderson-Sellers, A., Boer, G.J., Folland, C.K., Kirkby, A. & McAvaney, B.J. (1996) Climate models – evaluation. In: *Climate Change 1995* (eds J.T. Houghton, L.G. Meira Filho, B.A. Callander, N. Harris, A. Kattenberg & K. Maskell), pp. 229–84. Cambridge University Press, Cambridge.

Harold, J.M., Bigg, G.R. & Turner, J. (in press) Mesocyclone activity over the North-east Atlantic: Part 2, links with the large scale circulation. *International Journal of Climatology*.

Habbeln, D., Henrich, R. & Baumann, K.-H. (1998) Paleoceanography of the last interglacial/glacial cycle in the polar North Atlantic. *Quaternary Science Reviews*, **17**, 125–53.

Johanessen, T., Jonsson, T., Kallen, E. & Kaas, E. (1995) Climate-change scenarios for the Nordic countries. *Climate Research*, **5**, 181–95.

Kattenberg, A., Giorgi, F., Grassl, H. *et al.* (1996) Climate models – projections of future climate. In: *Climate Change 1995* (eds J.T. Houghton, L.G. Meira Filho, B.A. Callander, N. Harris, A. Kattenberg & K. Maskell), pp. 285–357. Cambridge University Press, Cambridge.

Killworth, P.D. (1983) Deep convection in the world ocean. *Reviews in Geophysics*, **21**, 1–26.

Kroon, D., Austin, W.E.N., Chapman, M.R. & Ganssen, G.M. (1997) Deglacial surface circulation changes in the northeastern Atlantic: temperature and salinity records off NW Scotland on a century scale. *Paleoceanography*, **12**, 755–63.

Lamb, H.H. (1995) *Climate, History and the Modern World*, 2nd edn. Routledge, London.

McCartney, M., Donohue, K., Curry, R., Mauritzen, C. & Bacon, S. (1998) Did the overflow from the Nordic Seas intensify in 1996–1997? *WOCE Newsletter*, **31**, 3–7.

Manabe, S. & Stouffer, R.J. (1994) Multiple century response of a coupled ocean-atmosphere model to an increase of atmospheric carbon dioxide, *Journal of Climate*, **7**, 5–23.

Marchitto, T.M., Curry, W.B. & Oppo, D.W. (1998) Millennial-scale changes in North Atlantic circulation since the last glaciation. *Nature*, **393**, 557–61.

Marsh, T.J. & Monkhouse, R.A. (1993) Drought in the United Kingdom, 1988–92. *Weather*, **48**, 15–22.

Mauritzen, C. (1996) Production of dense overflow waters feeding the North Atlantic across the Greenland-Scotland Ridge, Part 1: Evidence for a revised circulation scheme. *Deep-Sea Research*, **43**, 769–806.

Mills, D. (1989) *Ecology and Management of Atlantic Salmon*. Chapman and Hall, London.

Mitchell, J.F.B. & Johns, T.C. (1997) On modification of global warming by sulfate aerosols. *Journal of Climate*, **10**, 245–67.

Mitchell, J.F.B., Manabe, S., Tokioka, T. & Meleshko, V. (1990) Equilibrium climate change – and its implications for the future. In: *Climate Change: the IPCC Scientific Assessment* (eds J.T. Houghton, G.J. Jenkinss & J.J. Ephraums), pp. 131–172. Cambridge University Press, Cambridge.

Nicholls, N., Gruza, G.V., Jouzel, J., Karl, T.R., Ogallo, L.A. & Parker, D.E. (1996) Observed climate variability and change. In: *Climate Change 1995* (eds J.T. Houghton, L.G. Meira Filho, B.A. Callander, N. Harris, A. Kattenberg & K. Maskell), pp. 133–92. Cambridge University Press, Cambridge.

Parker, D.E., Horton, E.B. & Gordon, M. (1998) Global and regional climate in 1997. *Weather*, **53**, 166–75.

Power, G. (1981) Stock characteristics and catches of Atlantic salmon (*Salmo salar*) in Quebec and Newfoundland and Labrador in relation to environmental variables. *Canadian Journal of Fisheries and Aquatic Sciences*, **38**, 1601–11.

Rahmstorf, S. (1995) Bifurcations of the Atlantic thermohaline circulation in response to changes in the hydrological cycle. *Nature*, **378**, 145–9.

Rahmstorf, S. (1998) The effect of the Mediterranean Outflow on the North Atlantic. *EOS*, **79**, 281–2.

Rahmstorf, S. & Willebrand, J. (1995) The role of temperature feedback in stabilizing the thermohaline circulation. *Journal of Physical Oceanography*, **25**, 787–805.

Reason, C.J.C. & Power, S.B. (1994) The influence of the Bering Strait on the circulation in a coarse resolution global model. *Climate Dynamics*, **9**, 363–9.

Reid, J.L. (1979) On the contribution of the Mediterranean Sea outflow to the Norwegian-Greenland Sea. *Deep-Sea Research*, **26**, 1199–223.

Sarnthein, M., Jansen, E., Weinelt, M. *et al.* (1995) Variations in Atlantic surface ocean paleoceanography, 50°–80°N: a time-slice record of the last 30 000 years. *Paleoceanography*, **10**, 1063–94.

Schlosser, P., Swift, J.H., Lewis, D. & Pfirman, S.L. (1996) The role of the large-scale Arctic Ocean in the transport of contaminants. *Deep-Sea Research*, **43**, 1341–67.

Schmitz, W.J. & McCartney, M.S. (1993) On the North Atlantic circulation. *Reviews of Geophysics*, **31**, 29–49.

Smith, K. (1995) Precipitation over Scotland 1757–1992: some aspects of temporal variability. *International Journal of Climatology*, **15**, 543–56.

Smith, J.E., Risk, M.J., Schwarcz, H.P. & McConnaughey, T.A. (1997) Rapid climate change in the North Atlantic during the Younger Dryas recorded by deep-sea corals. *Nature*, **386**, 818–20.

Ståhl, G. (1987) Genetic population structure of Atlantic salmon. In: *Population Genetics and Fisheries Management* (eds N. Ryman & F. Utter), pp. 121–40. University of Washington Press, Seattle.

Xie, S.P. & Tanimoto, Y. (1998) A pan-Atlantic decadal climate oscillation. *Geophysical Research Letters*, **25**, 2185–8.

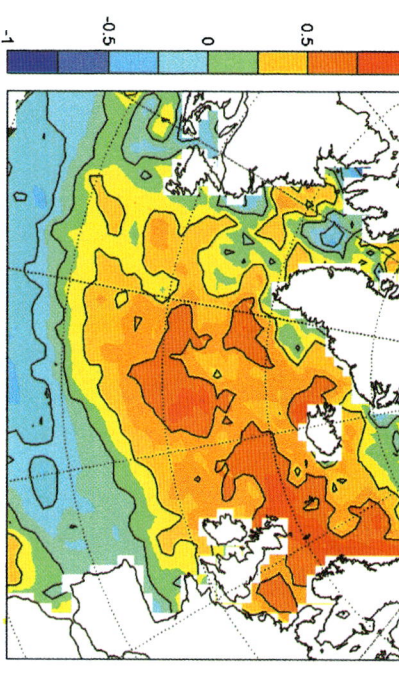

Plate 12.1 Correlations between the NAO index and year-to-year changes in scalar wind for the winter months (January–March). The value of the correlation is indicated by the colour scale.

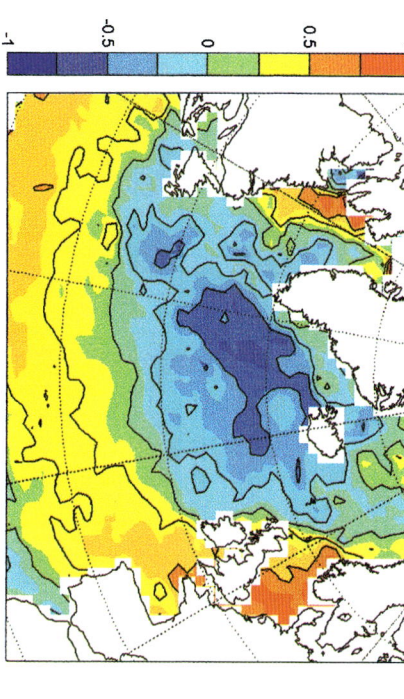

Plate 12.2 Correlations between the NAO index and year-to-year changes in SST for the winter months (January–March). The value of the correlation is indicated by the colour scale.

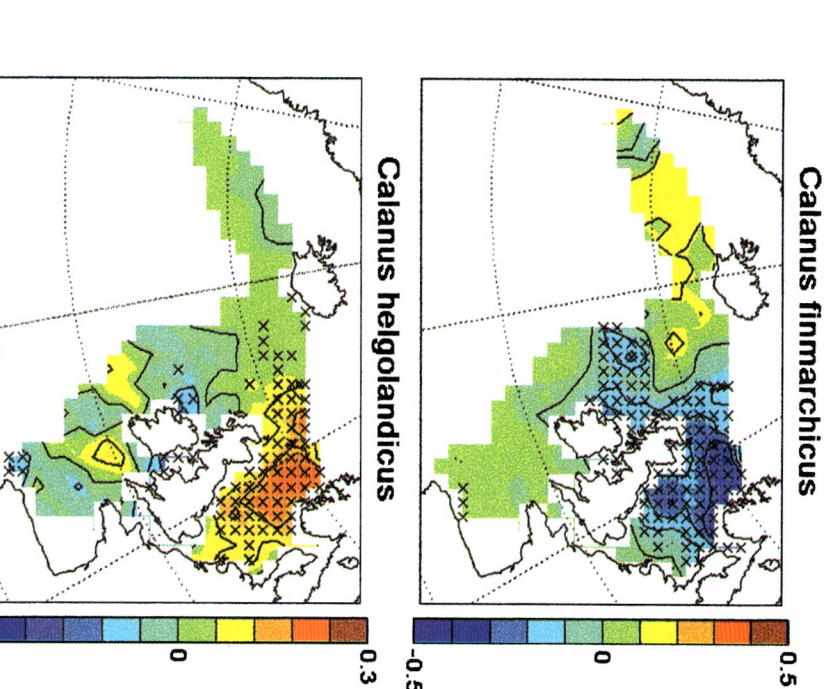

Plate 12.3 Differences between log-abundance of *Calanus finmarchicus* and *C. helgolandicus* during periods of high and low NAO. Positive values correspond to higher abundance during years of high NAO. Crosses indicate a statistically significant difference (t-test, $p < 0.05$). High and low NAO years are defined as in Fig. 12.3 (text p. 157).

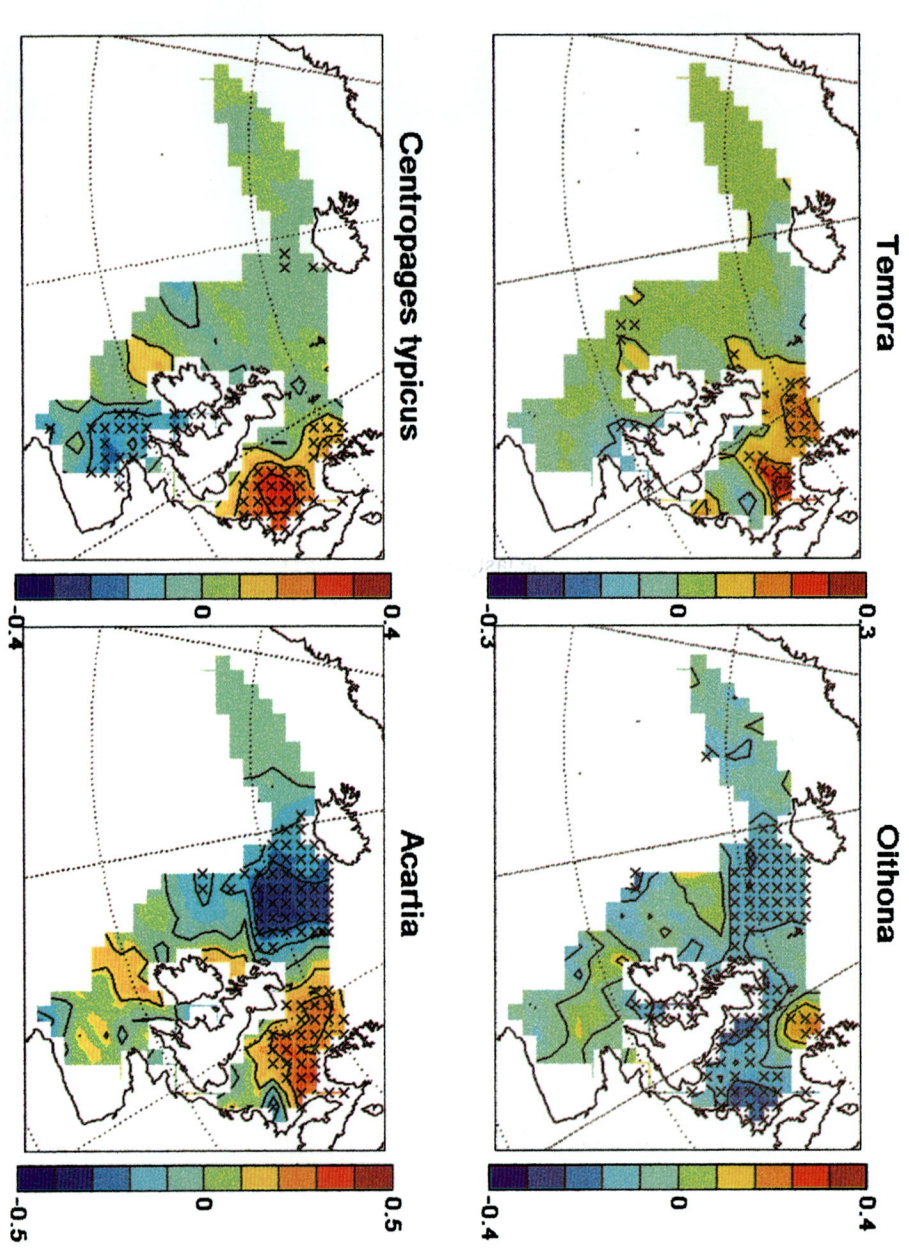

Plate 12.4 Differences between log-abundance of *Temora* spp., *Oithona* spp., *Centropages typicus* and *Acartia* spp. During periods of high and low NAO. Positive values correspond to higher abundance during years of high NAO. Crosses indicate a statistically significant difference (*t*-test, $p < 0.05$). High and low NAO years are defined as in Fig. 12.3 (text p. 157).

Chapter 12
Long-term Planktonic Variations and the Climate of the North Atlantic

PHILIP C. REID and BENJAMIN PLANQUE

Introduction

It is now well documented that runs of Pacific salmon in rivers along the western margin of North America stretching from Alaska to California vary on a decadal scale. When catches of a species are high in one region (e.g. Oregon) they may be low in another (e.g. Alaska). These changes are in part caused by marked interdecadal changes in the size and distribution of salmon stocks in the north-east Pacific (Brodeur & Ware 1995), which are in turn associated with important ecosystem shifts forced by hydroclimatic changes linked to El Niño and possibly climate change (Beamish & Bouillon 1995, Francis & Hare 1994).

In the North Atlantic major changes have been measured in the planktonic ecosystem and marine climate over the last few decades that are equal in magnitude to those observed in the Pacific. The planktonic changes have been measured from samples taken by the Continuous Plankton Recorder (CPR) survey which operates over an extensive area that is coincident with the migratory and forage area of the Atlantic salmon. In recent years salmon stocks and catches over much of the North Atlantic, and especially in UK waters, have shown a marked decline (Figs 12.1, 12.2). Rod catches in many, though not all, UK rivers (e.g. some rivers in Northern Ireland) have also dropped. Part of the reduction in total landings since 1988 may be attributed to closure of some net fisheries.

By analogy with the North Pacific it is likely that the changing fortunes of the salmon fishery in the north-east Atlantic in recent decades is linked to alterations in plankton productivity and/or structural changes in trophic (food chain) transfer, each forced by hydrometeorological variability and possibly climate change. Reductions in salmon runs in some UK rivers may be linked to pollution or deterioration in headwater spawning sites. The CPR survey operates offshore in waters > 20 m deep that are relatively pristine and only marginally affected by pollution or eutrophication. Therefore the major causes of observed plankton variation are likely to be environmentally induced. There is a possibility that density-driven, top-down control from fish, as seen in some freshwater lakes, might also contribute to plankton variability in the sea, but this has so far not been demonstrated.

This chapter outlines the changes that have been observed in the plankton of the North Atlantic and discusses the results in relation to a mode of climatic variability, the North Atlantic Oscillation (NAO) and the related Gulf Stream Index (GSI). It

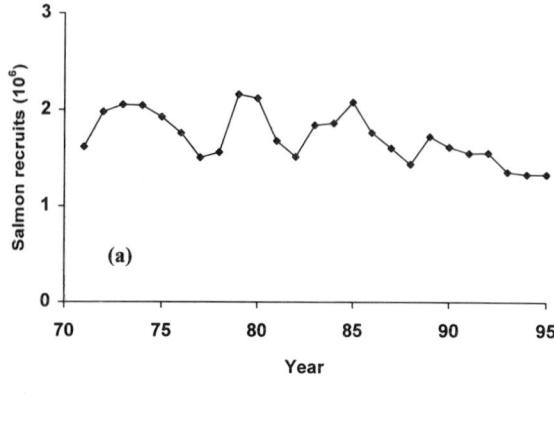
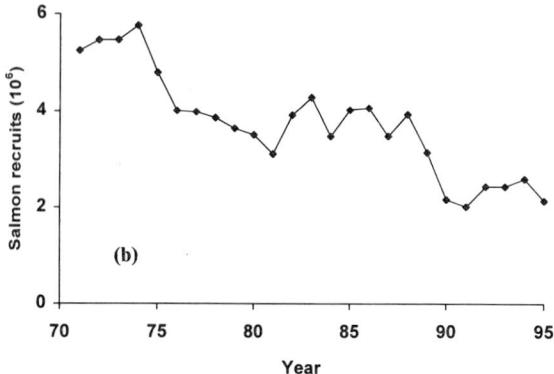

Fig. 12.1 Estimated pre-fishing abundance (1971–95) summed for maturing and non-maturing 1SW recruits for:
(a) the North European area
(b) the South European area
Data taken from Table 3.7.2.1 (ICES 1997).

concludes with a commentary on the relevance of the observed ecosystem and hydroclimatic changes to the sustainability of the salmon fishery.

Data sources

The Continuous Plankton Recorder (CPR) survey (Warner & Hays 1994) is an upper layer plankton monitoring programme that was initiated in the North Sea in 1931 and expanded into the eastern Atlantic in 1939 and since 1946 after World War II. The survey uses merchant ships on their normal routes of passage to tow CPRs that sample plankton on a moving band of silk. This silk moves forwards in the machine at a rate that is proportional to the speed of the ship so that for example a typical 400 nautical mile tow would provide a silk length of approximately 4 m long. On return to the laboratory the silk is cut into samples representing 10 miles of tow that are

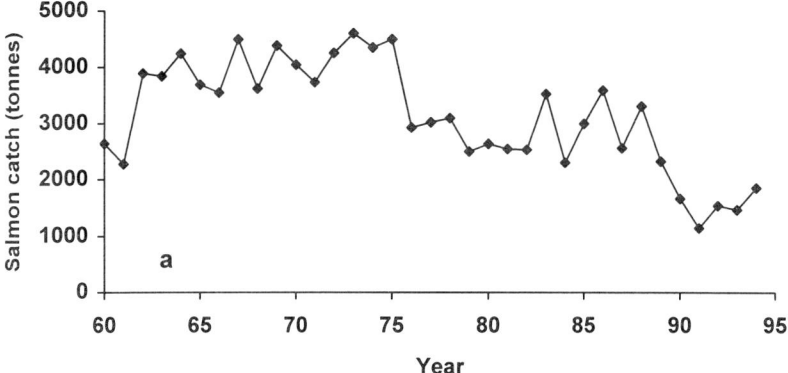
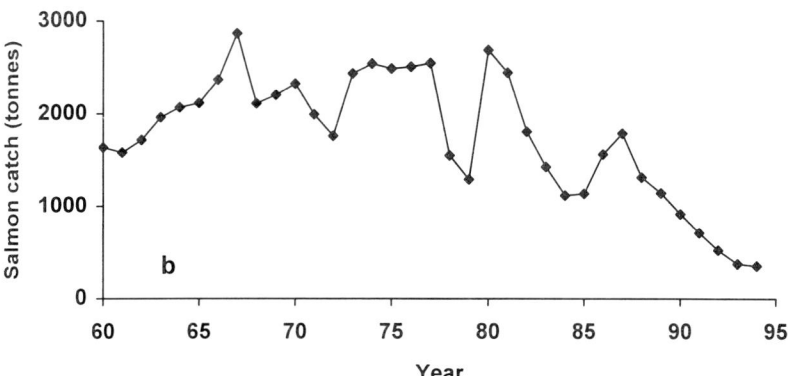
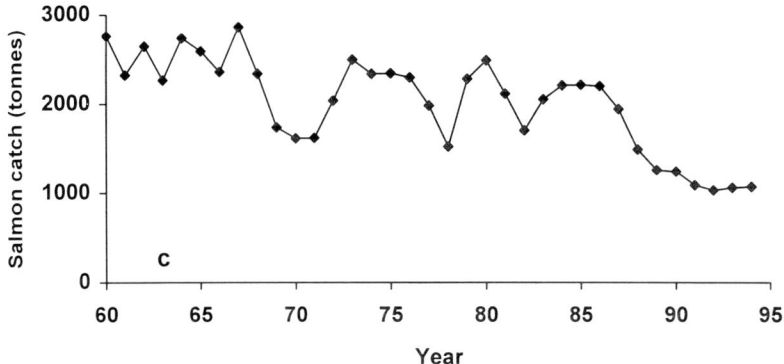

Fig. 12.2 Nominal catch of salmon (1960–94):
(a) totalled for United Kingdom, Ireland, France and Spain
(b) totalled for Canada, USA
(c) Norway and Russia
Data taken from Table 2.1.1 (ICES 1995).

examined under a microscope to identify and count the filtered plankton. Up to 400 different species or taxa are identified, and since the survey started approximately 4 million nautical miles have been towed and more than 200 000 samples analysed.

Statistics on salmon stocks and catches have been obtained from tables in the Report of the Working Group on North Atlantic salmon (ICES 1995).

Data on scalar wind speed and sea surface temperature were provided by the Comprehensive Ocean Data Set (COADS) (Woodruff *et al.* 1993). COADS provides wind and temperature monthly statistics in standard boxes of $2° \times 2°$ latitude and longitude. These data were interpolated on a new grid that matches the one used for the CPR plankton data.

The North Atlantic Oscillation (NAO) is primarily a winter atmospheric oscillation which can have oceanic consequences for a longer period of time. An index of its variability is defined as the difference between the normalised winter (December–March) sea level pressures measured at Iceland to represent the Iceland low and at Lisbon, Portugal, to approximate the Azores High pressure cell.

The Gulf Stream Index (GSI) is a measure of the latitudinal position of the north wall of the Gulf Stream where it breaks away from the east coast of North America (for details see Taylor 1995).

Regional climatic framework

The North Atlantic oscillation

A detailed description of the regional effects of the NAO will not be given here, but some key features in the hydrographic response to this oscillation need to be described before discussing the response of plankton to atmospheric variability. The principal aspects of the NAO discussed are related to physical processes known to drive plankton production in the ocean, especially effects of wind and sea surface temperature (SST) on seasonal timing. Additional effects on circulation at regional and ocean-wide scales will also be briefly presented as they may help to understand some of the recent and earlier climate-plankton links observed. The index has strengthened over the last 30 years to a period with high positive indices from 1988 to the present with a marked negative reversal in the winter of 1996 (Fig. 12.3).

The relationship between the NAO and the winter geostrophic wind field across the North Atlantic is evident from Plate 12.1 (Colour Section between pp. 152 & 153). Most of the central and eastern North Atlantic is undergoing stronger wind conditions during periods of higher NAO index, whilst in the western Atlantic the situation is to a lesser extent reversed. It is important to note that the increase in scalar wind during periods with a high NAO is perceptible only during winter months. The link between the NAO index and scalar wind for other seasons appears weak and patchy.

The SST response to the NAO is rather different (Plate 12.2). Both the eastern and western North Atlantic display increases in temperature during years of positive NAO, while temperature anomalies are negative in the central North Atlantic and in

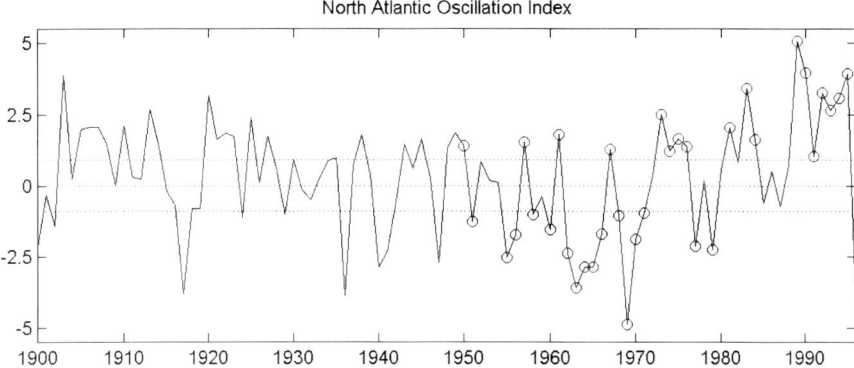

Fig. 12.3 The North Atlantic Oscillation Index calculated as the difference in standardised sea level pressure between Portugal and Iceland for the winter period (December–March). Years of high or low NAO are circled. Dotted lines indicates mean NAO (middle line) and mean NAO ± one Standard Deviation. Data were kindly provided by James W. Hurrell at the National Center for Atmospheric Research (Boulder, CO).

the Labrador Basin. An inverse pattern is also seen between the North and Celtic Seas and the North Atlantic. The sign and geographical distribution of the temperature anomalies is persistent for several months and can still be seen in summer. We will discuss later how this persistence in temperature anomalies may affect *Calanus* species when located at the limit of their temperature range.

Because of its large impact on wind and temperature anomalies, the NAO plays a critical role in the changes in circulation and convection in the North Atlantic. The formation of North Atlantic Deep Water in the Greenland Sea, and of Labrador Sea Intermediate Water, as well as the intensity of the subpolar gyre and the subtropical gyres, appear to be directly linked to changing phases of the NAO. Dickson & Turrell (see Chapter 9) have related the anomaly gradient between the sub-polar and sub-tropical gyres to the flow of the North Atlantic Current. Reid *et al.* (1998) have outlined a different scenario for the response of the North Atlantic Current and its extension into the Norwegian Sea, including a strengthening in the eastern boundary/ shelf edge current to the west of the UK (Fig. 12.4) during prolonged positive phases of the NAO.

The Gulf Stream Index

The Gulf Stream and its extension, the North Atlantic Current, play a key role in the distribution of heat in the northern hemisphere. It must however, be remembered that the primary source of warming for Europe is not the Gulf Stream, but extraction of heat by convection from west of the British Isles where the density gradient is weak and winds are strong. Ellett (1993) has likened this process to a 'fan-assisted storage heater'. Some potential climatic implications that might arise from changes in the course and volume flow of currents in the North Atlantic are discussed by Reid *et al.*

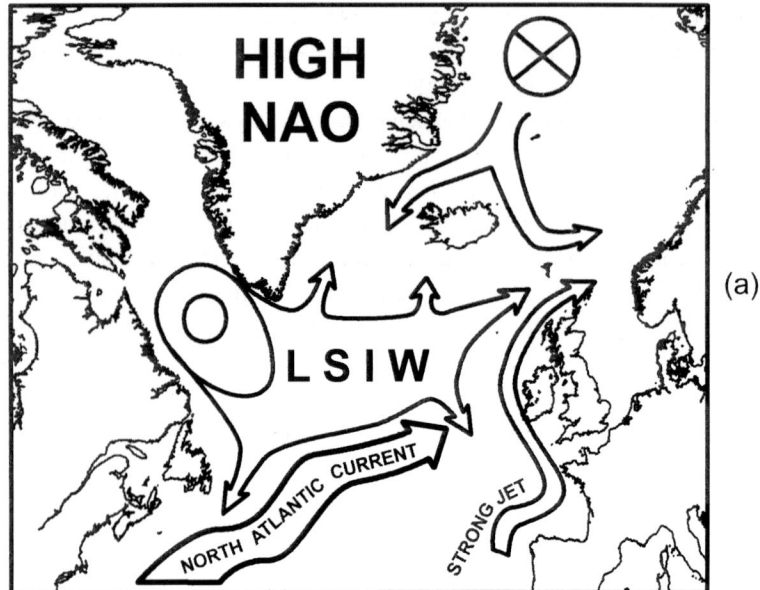
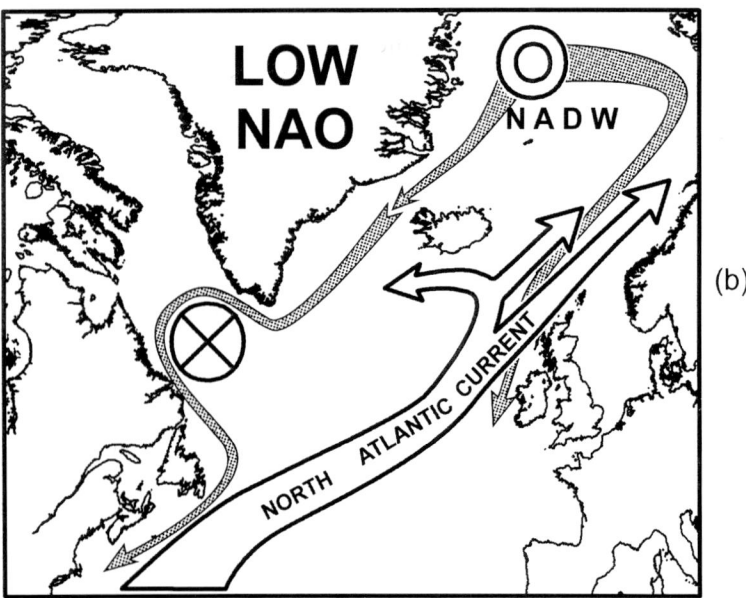

Fig. 12.4 Hypothesised changes in the circulation of the North Atlantic in the alternative scenarios of a prolonged period of (a) high or (b) low NAO index. The bifurcating surface flow north of Iceland in (a) indicates outflow from the Fram Strait, LSIW (Labrador Sea intermediate water); NADW (North Atlantic deep water). The shaded lines indicate approximate directions of deep water flow. From Reid et al. 1998b.

(1998a). Associated changes, linked to the NAO, in the strength of wind and the density, extent and depth of winter mixing to the west of the British Isles are likely to reinforce alterations in Europe's climate.

Taylor and Stephens (1998) have demonstrated a two-year lagged relationship between the NAO and the Gulf Stream Index (GSI), such that 60% of the variance in the position of the Gulf Stream can be predicted from the NAO. High NAO values correspond with stronger westerlies and trade winds which, with a two-year adjustment lag for ocean circulation, appear to force a more northerly position of the Gulf Stream. Through a teleconnection the trade winds are in turn strongly influenced by the El Niño Southern Oscillation (ENSO) events in the Pacific. Taylor et al. (1998) have shown that ENSO may also contribute to the GSI since it explains, with the NAO, 77% of the GSI variance.

Work by Taylor and colleagues (e.g. Taylor et al. 1992, Frid & Huliselan 1996) has shown that the position of the north wall of the Gulf Stream, where it separates from the shelf off the USA, is correlated with the trend in 'Total (small) Copepods' and other zooplankton measured by the CPR in the North Sea, and more recently with North Sea stratification and the timing of the spring bloom. A similar empirical relationship has been found between the GSI, zooplankton and the timing of spring stratification in Lake Windermere (George & Taylor 1995). Because it is a freshwater lake the link between the GSI and the plankton must be via the atmosphere. By analogy important climatic processes thus appear to be controlling water column stability and the biological production of the North Sea where the plankton appears to be integrating a climatic signal that has so far not been observed directly in meteorological observations.

Phytoplankton

At the base of the food chain, changes in phytoplankton as levels of standing stock, production or in community structure are likely to affect higher trophic levels and the fishery. There are no long-term and spatially widespread measurements of primary production, although Joint & Pomroy (1993) calculated monthly primary production for different areas of the southern North Sea in 1988 from measurements taken during the NERC North Sea Project. Sathyendranath et al. (1995) have calculated mean primary production for the North Atlantic using satellite imagery and a model with inputs for example on mean radiation and cloud cover. For the present the only available measurement of phytoplankton growth that can be used to reflect major patterns of variation in time and space in the north-east Atlantic and North Sea is sourced from the CPR survey.

Phytoplankton colour as measured on CPR silks (Fig. 12.5) has shown a pronounced and progressively increasing trend and longer growing season over the last five decades in the North Sea and in oceanic waters to the west of the British Isles south of 58°N (Reid et al. 1998b). The phytoplankton increase is not ubiquitous, as an inverse pattern to the trend is evident in eastern Atlantic waters between latitudes

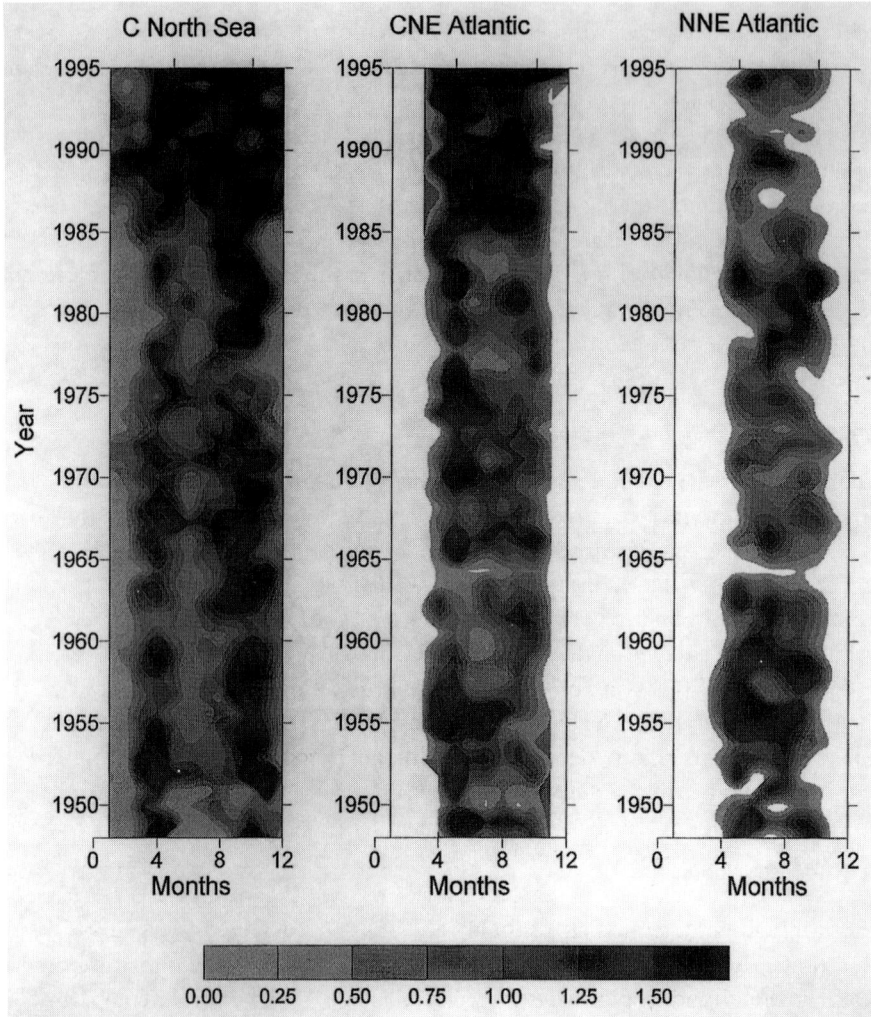

Fig. 12.5 Contour plots of mean monthly Phytoplankton Colour during 1948–95 for the central (C) North Sea, central north-east (CNE) Atlantic and northern north-east (NNE) Atlantic. From Reid et al. 1998a.

59°N and 63°N. It is not known whether the visual colour index provides an accurate estimate of chlorophyll a or the patterns observed also reflect changes in production. Nevertheless, the patterns of change shown by the index are consistent over wide areas of sea in both oceanic and shelf water, and thus must be providing a signal of real change. The mechanisms behind the changes are not understood; the spatial patterns show some relationship with the NAO signals described earlier for temperature and scalar wind, but there is only a weak correlation between the colour index and NAO time series. It is likely that the colour changes are integrating a wide range of meteorological signals, including cloud cover and wind strength and direction (Dickson et al. 1988).

Similar changes to those observed in CPR colour have been described over a shorter period (1981–91) from satellite measurements of the growth of terrestrial vegetation over a large area of the northern hemisphere north of 50°N (Myneni *et al.* 1997). This satellite derived index of plant growth also provided evidence for an increase in the active growing season of plants in the spring and autumn. The increases described by Myneni *et al.* (1997) were attributed to higher temperatures as a consequence of global warming. It is possible that the CPR colour index may be reflecting a similar long-term biological response in the North Atlantic to increasing global temperature (Reid *et al.* 1998b), which appears to be causing melting of ice and permafrost in the Arctic.

Zooplankton (*Calanus*)

The copepod genus *Calanus* is widely distributed in the oceans of the world. Four species are found in the North Atlantic, *C. helgolandicus*, *C. finmarchicus*, *C. glacialis* and *C. hyperboreus*. *C. finmarchicus* is found in temperate waters, between the warm waters of Gulf Stream influence and the cold waters of Arctic influence, where it often dominates the zooplankton community in terms of biomass and numbers. It constitutes a large part of the diet of small pelagic fish and fish larvae and is therefore a good indicator of the availability of zooplankton as a resource for higher trophic levels in this region. In the eastern North Atlantic, a large population of *C. finmarchicus* overwinters at depth, mainly in the Faroe-Shetland Channel, and is seeded into the North Sea each year (Backhaus *et al.* 1994). Fromentin & Planque (1996) have demonstrated that the abundance of *C. finmarchicus* is inversely and highly significantly correlated with the NAO signal (Plate 12.3 & Fig. 12.6).

The previously described increase in North Sea winter-spring temperature associated with high NAO periods is probably one of the main factors responsible for the decline of *C. finmarchicus* populations during the last three decades. The parallel increase in the abundance of *C. helgolandicus* (Plate 12.3), a direct competitor for food resources in the North Sea (Fromentin & Planque 1996), is a likely additional contributory factor to the decline in *C. finmarchicus*. Additionally, changes in the Atlantic water inflow into the North Sea associated with the NAO (Planque & Taylor 1998, Reid *et al.* in press, Stephens *et al.* 1998) may have adversely changed the migratory route from overwintering sites into the North Sea. However, because of the location of the sections along which the flows were estimated in the 2D model used by Stephens *et al.* (1998) it is not clear whether the total inflow of Atlantic water follows a similar pattern to that predicted by Reid *et al.* (in press) using a 3D model.

Fromentin & Planque (1996) suggested that the increase in late winter winds and their expected negative effect on the development of the spring phytoplankton bloom could have contributed to the observed decline in *C. finmarchicus* owing to reduced food availability. In contrast, based on the phytoplankton colour evidence from the North Sea the spring bloom appears to have been more pronounced if not earlier in the period of high NAO indices since 1988. In a reverse sense, the higher colour levels

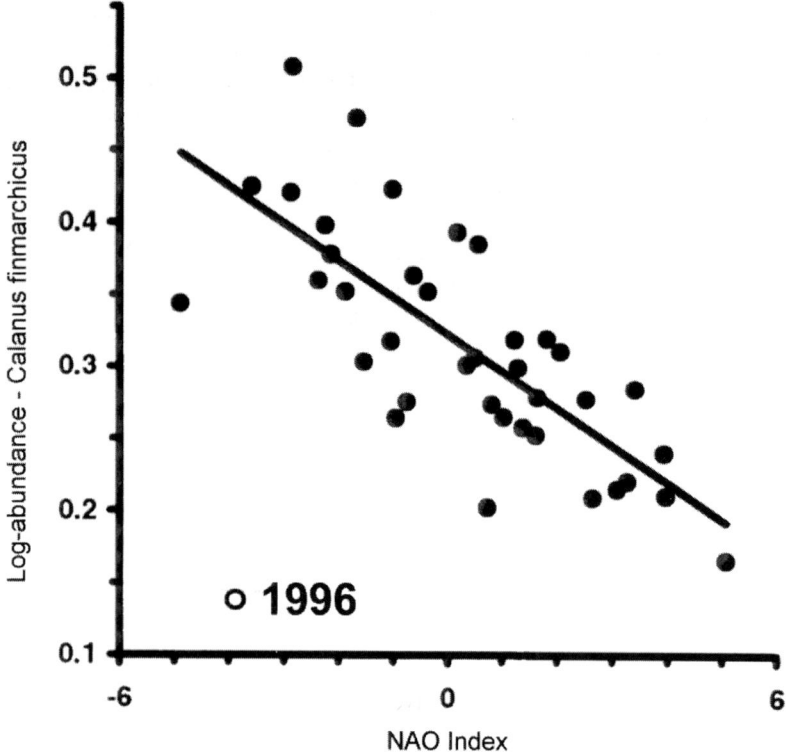

Fig. 12.6 Annual log-abundance of *Calanus finmarchicus* against the North Atlantic index. For the period 1958–95 the NAO explains 58% of the variance in the log-abundance of *C. finmarchicus*. The year 1996 clearly stands out from the regression line. From Planque & Reid 1998.

could be due to reduced grazing pressure from *C. finmarchicus*, but this seems unlikely on the basis of investigations to date.

Changes in the abundance of the population of *C. finmarchicus* located in the Gulf of Maine, western North Atlantic, show important similarities with those of the North Sea. While in the eastern Atlantic the NAO is associated with changes in wind and sea surface temperature (SST), it is associated only with wind in the Gulf of Maine (Enfield & Mestas-Nunez in press). Recent work by Marjorie Lambert (pers. comm.) shows that both NAO (related to wind) and winter SST signals have to be considered to explain the interannual variability in abundance of *C. finmarchicus* in the Gulf of Maine (Anon. 1998). Similarly to the opposition between *C. finmarchicus* and *C. helgolandicus* observed in the North Sea, there appear to be opposite trends in the abundance of *Centropages typicus* and *Calanus finmarchicus* in the Gulf of Maine.

The link between the NAO and *C. finmarchicus* in the eastern North Atlantic is so evident (Fig. 12.6) that it has been used in an attempt to predict the abundance of the copepod from the NAO index (Planque & Reid 1998). The predicted results compared well with observations for years 1993–95, but the relationship that held for more than 35 years broke down in 1996. This change coincided with a reversal in the NAO from one of its highest values to a low extreme, leading to intensive cooling of the North Sea in the early part of 1996 (Loewe 1996). There was however, no equivalent reversal in the fortunes of *C. finmarchicus* (Planque & Reid 1998) (Fig. 12.6). In 1997 the NAO index returned to a slightly positive value but the abundance in *C. finmarchicus* continued to decline, reaching the lowest level ever observed by the CPR survey (Reid *et al.* in press). Poor sampling in 1996 may have explained the continuing decline in abundance, but the results for 1997 suggest that the population has truly declined, possibly to such a low level that it will take several years to recover.

Plankton community

Colebrook, in a series of papers (1978, 1986, 1991), demonstrated for both the phyto- and zooplankton sampled by the CPR that the dominant and most abundant taxa showed similar patterns of change through time. Using Principal Component Analysis he showed that the main pattern of change for approximately 60% of the species/taxa was a declining trend. This trend was seen independently in approximately a dozen large statistical areas, for which averages were derived, covering the north-east Atlantic and north-west European shelf. By tradition, following early work by Colebrook, the first Principal Component in these analyses has been used as an index of the abundance of the plankton assemblage in the North Atlantic. The zooplankton data set started in 1948, but the time series for phytoplankton is shorter because a new method of counting was introduced in 1958. Reid & Hunt (1998) updated the PCA analyses to 1996, averaging data for the north-east Atlantic and North Sea, and showing that the decline in the first Principal Component was maintained as an almost continuous downward trend over approximately 50 years.

Recalculated Principal Component Analyses for both the north-east Atlantic and the North Sea up to 1996 are presented in Fig. 12.7. The first components for phyto- and zooplankton in both areas all show a downward trend, with the level for zooplankton in the Atlantic reaching the lowest on record in 1996. In contrast a marked increase is evident in the second phytoplankton and zooplankton components for the North Sea in the late 1980s, reflecting increases, particularly in species of copepods such as *Acartia* spp. and cladocerans such as *Podon* spp. (that have overwintering eggs), and, for the phytoplankton, dinoflagellates.

The changes reflected in the second components in the North Sea are part of the response to the North Sea regime shift reported by Reid *et al.* (in press), which includes an almost step-wise increase in Phytoplankton Colour after 1987 as described earlier. High catches of the horse mackerel (*Trachurus trachurus*) in the northern North Sea and changes in the distribution of other fish species such as the herring coincided with this event. The horse mackerel fishery took advantage of an exceptional northerly migration of this species along the shelf edge from the Bay of Biscay. With the presence of many other rare southerly plankton species in UK waters (Edwards et al. 1999), the horse mackerel changes provide some evidence to support an intensified eastern margin current. Reid et al. (in press) also described an increase in sea temperatures post 1987 and a major increase in the winter inflow of oceanic water into the North Sea after the same year (Fig. 12.8). All of these events are likely to be a reponse to measured changes in pressure distribution over the North Atlantic and followed a change to 9 years of high positive NAO indices.

Other copepods

The NAO is a Pan-Atlantic atmospheric oscillation but it has distinct regional hydrographic consequences. The effects on wind and heat and moisture transport vary regionally as well as the effects on oceanic circulation. It is most likely that the responses of zooplankton to the NAO may be region-specific. In the case of *C. finmarchicus*, the overwintering strategy of the species plays a crucial role in the relationship between the abundance of the population and the NAO. The responses of different taxa to the NAO will probably also depend upon their life cycle and their spatial distribution. Consequently, it is likely that the response of zooplankton to the NAO will be species dependent.

We have compared the regional changes in abundance of four copepod taxa during periods of low and high NAO. These taxa are found in European waters and are well sampled by the CPR. They are *Temora* spp., *Centropages typicus*, *Oithona* spp., and *Acartia* spp. The log-abundance of each taxa was calculated for years of high (1950, 57, 61, 67, 73–76, 81, 83–84, 89–95) and low NAO (1951, 55–56, 58, 60, 62–66, 68–71, 77, 79) in squares of 50 nautical miles around the British Isles. For each square the difference in abundance between years of high and low NAO was calculated and statistically tested using a Student *t*-test (Plate 12.4).

The responses of the selected taxa to the NAO are shown on Plate 12.4. All four taxa show significant positive and negative abundance anomalies during years of high NAO, depending on the region considered. The response of *Temora* spp. to the NAO is positive in the north-east North Sea and in Skagerrak waters, but negative in the eastern Celtic Sea. *Oithona* spp. shows a positive response to the NAO in most of the North Sea and to the south-east of Iceland, but the abundance of the taxa increases along the southern coast of Norway. The abundance of *Centropages typicus* increases in the North Sea but decreases in the Celtic Sea and Bay of Biscay. Finally, *Acartia* spp. displays an opposite response in the central north North Sea, where its abundance increases, and the southern North Sea and south-east of Iceland, where it decreases.

These contrasting results indicate that the life cycle of specific taxa as well as regional features may largely alter the potential effects of the NAO on copepod species abundance.

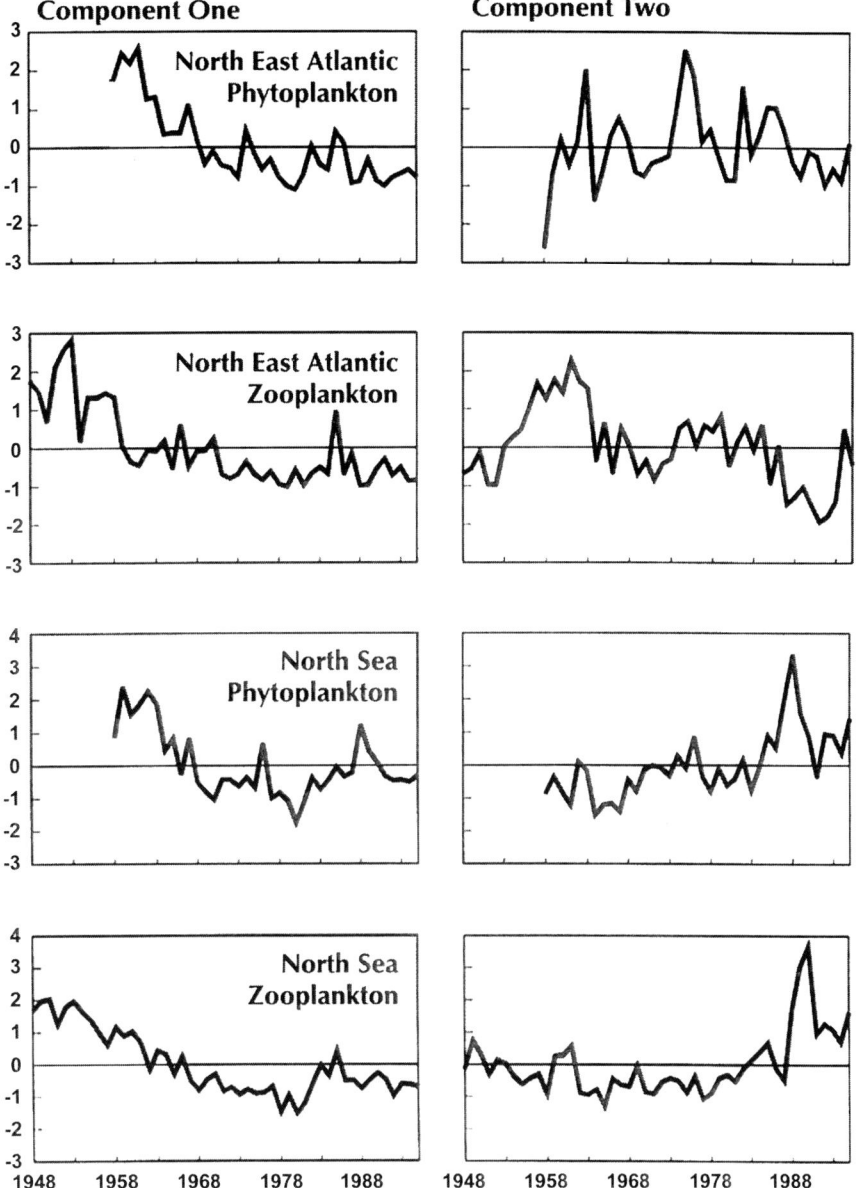

Fig. 12.7 Plots of the first and second components from a Principal Component Analysis of phytoplankton (1958–96) and zooplankton (1948–96) averaged for two region of the CPR Survey, the north-east Atlantic and the North Sea. The time series for phytoplankton is shorter because a new method of counting was introduced in 1958.

166   *The Ocean Life of Atlantic Salmon*

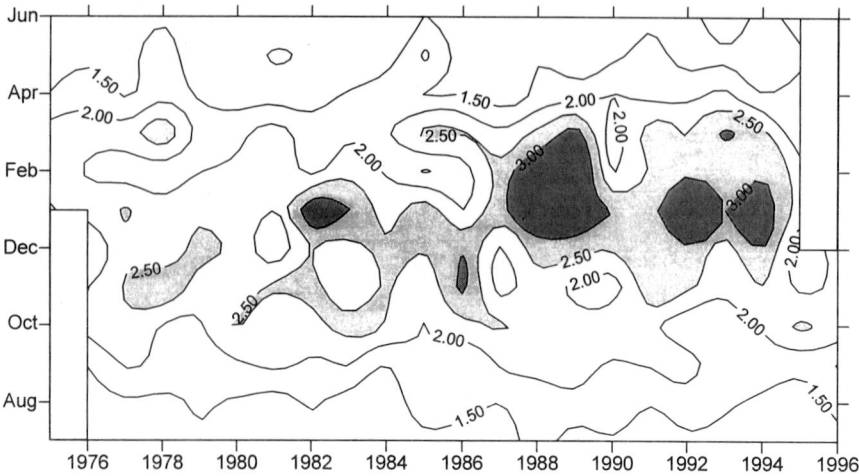

Fig. 12.8 A contour plot of monthly inflow into the North Sea, in sverdrups, across a section between Orkney, Shetland and Norway over the period January 1976 to December 1996. The plot has been centred on the winter months of the year. As in Reid *et al.* (in press) updated to 1996 with data kindly provided by Einar Svendsen, IMR, Bergen.

Conclusions

The main conclusions to arise from this study are the dominant role that the NAO has in oceanic forcing on North Atlantic and north-west European waters, and the very different responses that individual components of the ecosystem have to this forcing. For example, the copepods *C. finmarchicus, C. helgolandicus, Centropages typicus, Temora longicornis* and *Acartia* spp. all show clear and statistically significant spatial responses to different patterns of the NAO and each species appears to have a specific niche and complex response to environmental forcing mediated by the NAO. The NAO/GSI relationship suggests that there is a secondary or oscillatory and delayed response to the NAO that plays an important, if not yet fully understood, role of atmospheric control on spring hydrodynamic structure and subsequent plankton development. While primarily a winter phenomenon in controlling the strength of westerly winds, the effect of the NAO on sea surface temperature extends into the autumn. The response of the water column to the NAO also shows clear spatial patterns over extensive regions. The recently discovered link between ENSO and the GSI provides further evidence for teleconnections between the Pacific and Atlantic. Krovnin (1995) previously described teleconnection between sea surface temperature anomalies in these two oceans and some empirical evidence for climatic effects on some commercial fish stocks in the north-east Atlantic/North Sea. Narayanan *et al.* (1995) have demonstrated that Atlantic salmon are similarly influenced by environmental variability that may be climatically induced, but did not link their variability to ecosystem change.

A visual index of chlorophyll measured by the CPR provides evidence for a long-term progressive change in the phytoplankton of the north-east Atlantic and North Sea, that may be linked to major changes in the circulation of the North Atlantic, melting ice and permafrost in the Arctic, and global warming. Dickson & Turrell (1999) described evidence for a strengthened ocean circulation between the subtropical and subpolar gyres and concluded that this represents an increased flow in the North Atlantic Current in recent years with a linkage to the NAO. Reid *et al.* (1998a) have also recognised an increase in the circulation of the gyres, but have provided the opposite interpretation of the north-westerly directed flow of the North Atlantic Current.

The true response of the Gulf Stream/North Atlantic Current circulation and volume flow to the NAO is still far from clear. Understanding this relationship is likely to be of considerable importance to the stock management of salmon and other fish, and is also likely to affect the climate of north-west Europe. An augmentation in the south-north flow along the European shelf edge associated with the positive phase of the NAO has been more clearly recognised. It is likely that such increased flow acts adversely against salmon populations migrating across or against the current when returning from oceanic to UK waters. The regime shift in the North Sea described by Reid *et al.* (in press) is reflected in changes in oceanic inflow, sea surface temperature, an increase in abundance of mesoplanktonic crustaceans and dinoflagellates and a more intensive and longer phytoplankton growth period as reflected in Phytoplankton colour. These events coincide with a long period of high NAO indices after 1987 and a marked reduction in salmon stocks and catches (Figs 12.1, 12.2). What is not clear at present is whether the ecosystem has flipped into a new regime or the events are part of a more gradual change. The breakdown in the *C. finmarchicus*/ NAO relationships in 1996 and the continuing high levels of Colour in the North Sea, despite reduced inflow in that year measured from the Norwegian 3D model (E. Svendson pers. comm.) suggest that the system may have changed substantially. The possibility that the above change is linked to global warming may have substantial longer-term implications for the carrying capacity of the North Sea and salmon catches in north-west European rivers.

Acknowledgements

We thank the CPR analysis, logistic and administrative team for their contributions to this paper and the captains and crew of the ships that voluntarily tow CPRs. The CPR survey is operated by the Sir Alister Hardy Foundation for Ocean Science (SAHFOS), which gratefully acknowledges support from a consortium comprising the Intergovernmental Oceanographic Commission (IOC), the United Nations Industrial Development Organisation (UNIDO), the European Commission and agencies from Canada, Denmark, the Netherlands, Ireland, the United Kingdom and the USA.

References

Anon. (1998) Third ICES/GLOBEC Backward-Facing Workshop: Ocean climate of the NW Atlantic during the 1960s and 70s and consequences for gadoid populations. (Woods Hole, USA). International Council for the Exploration of the Sea, CM 1998/C:9.

Backhaus, J.O., Harms, I., Krause, M. & Heath, M.R. (1994) A hypothesis concerning the space-time succession of *Calanus finmarchicus* in the northern North Sea. *ICES Journal of Marine Science*, **51**, 169–80.

Beamish, R.J. & Bouillon, D.R. (1995) Marine fish production trends off the Pacific coast of Canada and the United States. In: *Climate Change and Northern Fish Populations* (ed. R.J. Beamish). *Canadian Special Publication of Fisheries and Aquatic Science*, **121**, 585–91.

Brodeur, R.D. & Ware, D.M. (1995) Interdecadal variability in distribution and catch rates of epipelagic nekton in the Northeast Pacific Ocean. In: *Climate Change and Northern Fish Populations* (ed. R.J. Beamish). *Canadian Special Publication of Fisheries and Aquatic Science*, **121**, 329–56.

Colebrook, J.M. (1978) Continuous plankton records: zooplankton and environment, north-east Atlantic and North Sea, 1948–1975. *Oceanologica Acta*, **16**, 9–23.

Colebrook, J.M. (1986) Environmental influences on long-term variability in marine plankton. *Hydrobiologia*, **142**, 309–25.

Colebrook, J.M. (1991) Continuous Plankton Records: from seasons to decades in the plankton of the north east. Atlantic. In: *Long Term Variability of Pelagic Fish Populations and their Environments* (eds T. Kawasaki, S. Tanaka, Y. Toba & A. Taniguchi), pp. 29–45. Pergamon Press, Oxford.

Dickson, R.R., Kelly, P.M., Colebrook, J.M., Wooster, W.S. & Cushing, D.H. (1988) North winds and production in the eastern North Atlantic. *Journal of Plankton Research*, **10**, 151–69.

Edwards, M., John, A.W.G., Hunt, H.G. & Lindley, J.A. (1999) Exceptional influx of oceanic species into the North Sea in late 1997. *Journal of the Marine Biological Association of the UK*, **79**, 737–39.

Ellett, D.J. (1993) The north-east Atlantic: A fan-assisted storage heater? *Weather*, **48**, 118–25.

Enfield, D.B. & Mestas-Nunez, A.M. (in press) Multiscale variabilities in global sea surface temperatures and their relationships with tropospheric climate patterns. *Journal of Climatology*.

Francis, R.C. & Hare, S.R. (1994) Decadal-scale regime shifts in the large marine ecosystems of the north-east Pacific: a case for historical science. *Fisheries Oceanography*, **3**, 279–91.

Frid, C.L.J. & Huliselan, N.V. (1996) Far-field control of long-term changes in Northumberland (NW North Sea) coastal zooplankton. *ICES Journal of Marine Science*, **53**, 972–7.

Fromentin, J.M. & Planque, B. (1996) *Calanus* and environment in the eastern North Atlantic. 2. Influence of the North Atlantic Oscillation on *C. finmarchicus* and *C. helgolandicus*. *Marine Ecology Progress Series*, **134**, 111–18.

George, D.G. & Taylor, A.H. (1995) UK lake plankton and the Gulf Stream. *Nature*, **378**, 139.

ICES (1995) Report of the working group on North Atlantic Salmon. International Council for the Exploration of the Sea CM 1995/Assess:14.

ICES (1997) Report of the ICES working group on North Atlantic Salmon. International Council for the Exploration of the Sea CM 1997/Assess:10.

Joint, I. & Pomroy, A. (1993) Phytoplankton biomass and production in the southern North Sea. *Marine Ecology Progress Series*, **99**, 169–82.

Loewe, P. (1996) Surface temperature of the North Sea in 1996. *Deutsche Hydrographische Zeitschrift*, **48**, 175–84.

Krovnin, A.S. (1995) A comparative study of climatic changes in the North Pacific and North Atlantic and their relation to the abundance of fish stocks. In: *Climate Change and Northern Fish Populations* (ed. R.J. Beamish). *Canadian Special Publication of Fisheries and Aquatic Science*, **121**, 181–98.

Myneni, R.B., Keeling, C.D., Tucker, C.J., Asrar, G. & Nemani, R.R. (1997) Increased plant growth in the northern high latitudes from 1981 to 1991. *Nature*, **386**, 698–702.

Narayanan, S., Carscadden, J., Dempson, J.B., O'Connell, M.F., Prinsenberg, S., Reddin, D.G. & Shackell, N. (1995) Marine climate off Newfoundland and its influence on salmon (*Salmo salar*) and capelin (*Mallotus villosus*). In: *Climate Change and Northern Fish Populations* (ed. R.J. Beamish). *Canadian Special Publication of Fisheries and Aquatic Science*, **121**, 461–74.

Planque, B. & Reid, P.C. (1998) Predicting *Calanus finmarchicus* abundance from climatic signal. *Journal of the Marine Biological Association of the UK*, **78**, 1015–18.

Planque, B. & Taylor, A.H. (1998) Long-term changes in zooplankton and the climate of the North Atlantic. *ICES Journal of Marine Science*, **55**, 644–54.

Reid, P.C. & Hunt, H.G. (1998) Are observed changes in the plankton of the North Atlantic and North Sea linked to climate change? In: *Pelagic Biogeography* (eds A.C. Pierrot-Bults & S. van der Spoel). *IcoPB II, Proceedings of the 2nd International Conference*. IOC/UNESCO, Paris, Workshop Report No. **142**, pp. 310–15.

Reid, P.C., Planque, B. & Edwards, M. (1998a) Is observed variability in the long-term results of the continuous plankton recorder survey a response to climate change? *Fisheries Oceanography*, **7**, 282–8.

Reid, P.C., Edwards, M.E., Hunt, H.G. & Warner, A.J. (1998b) Phytoplankton change in the North Atlantic. *Nature*, **391**, 546.

Reid, P.C., Borges, M. de F. & Svendsen, E. (in press) A regime shift in the North Sea *circa* 1988 linked to changes in the North Sea fishery. *Fisheries Research*.

Sathyendranath, S., Longhurst, A., Caverhill, C.M. & Platt, T. (1995) Regionally and seasonally differentiated primary production in the North Atlantic. *Deep Sea Research I. Oceanographic Research Papers*, **42**, 1773–802.

Stephens, J.A., Jordan, M.B., Taylor, A.H. & Proctor, R. (1998) The effects of fluctuations in North Sea flows on zooplankton abundance. *Journal of Plankton Research*, **20**, 943–56.

Taylor, A.H. (1995) North-south shifts of the Gulf Stream and their climatic connection with the abundance of zooplankton in the UK and its surrounding seas. *ICES Journal of Marine Science*, **52**, 711–21.

Taylor, A.H., Colebrook, J.M., Stephens, J.A. & Baker, N.G. (1992) Latitudinal displacements of the Gulf Stream and the abundance of plankton in the north-east Atlantic. *Journal of the Marine Biological Association of the UK*, **72**, 919–21.

Taylor, A.H. & Stephens, J.A. (1998) The North Atlantic Oscillation and the latitude of the Gulf Stream. *Tellus*, **50A**, 134–42.

Taylor, A.H., Jordan, M.B. & Stephens, J.A. (1998) Gulf Stream shifts following ENSO events. *Nature*, **393**, 638.

Warner, A.J. & Hays, G.C. (1994) Sampling by the Continuous Plankton Recorder Survey. *Progress in Oceanography*, **34**, 237–56.

Woodruff, S.D., Lubker, S.J., Wolter, K., Worley, S.J. & Elms, J.D. (1993) Comprehensive Ocean-Atmosphere Data Set (COADS) Release 1a: 1980–1992. *Earth System Monitor*, **4**(2), 4–10.

Chapter 13
Feeding Habits of Atlantic Salmon at Different Life Stages at Sea

JAN AARGE JACOBSEN and LARS PETTER HANSEN

Introduction

The abundance of Atlantic salmon has apparently decreased steadily in the last decade as inferred from the annual landings in the North Atlantic (ICES 1998a). The fishery, particularly the mixed-stock fishery, has for many years been thought to be the primary cause of the apparent decline. Recently, however, the focus has shifted from the high seas fishery to home water mixed-stock fisheries as the fishery in the ocean has virtually ceased. However, the continuing evidence of high mortality at sea, albeit with significantly reduced fishing efforts in recent years, has led researchers to look for biological and environmental factors likely to affect marine survival.

Probable causes to short- and long-term changes in abundance of salmon was dealt with in the International Council for the Exploration of the Sea (ICES 1996a, 1997), and included environmental factors, habitat limitation, diseases, pathogens and predation. Recently the effect of ocean environment on mortality in the early marine phase of salmon has received great attention (Friedland *et al.* 1993, 1998, in press, Reddin & Friedland 1993).

Although more information about Atlantic salmon in the marine phase has been generated in recent years (see Mills 1993), still little is known about the food and feeding habits of salmon in the ocean. Hislop & Shelton (1993) summarised the prey of Atlantic salmon in the sea, and Levings (1994) reviewed the food and feeding behaviour of post-smolts of Atlantic salmon in estuaries and fjords. Salmon probably spend most of their time pelagically in the ocean close to the sea surface, and prey on different pelagic animals such as fish, crustaceans and squid. Several authors have suggested that Atlantic salmon are opportunistic feeders (Hansen & Pethon 1985, Reddin 1988, Hislop & Shelton 1993). However, there is limited information available to compare the distribution of food organisms available to salmon with what they really eat.

Salmon post-smolts are defined as salmon the first few months after sea entry, pre-adults are salmon feeding in the open sea and offshore and adults are mature salmon on their homeward migration in coastal areas.

Our aim was to summarise what is known about the feeding habits of Atlantic salmon at different life stages in the North Atlantic (the Baltic area is not covered) including the most recent literature in the field. We also compare the available food organisms for salmon with the actual stomach contents observed in the area north of the Faroes (Jacobsen & Hansen 1996, 1999). Then we compare the food habits in the north-west and north-east Atlantic. Finally we discuss the possibility of food availability being a limiting factor for survival and abundance of salmon in the marine phase.

Food of post-smolts

Coastal areas

Morgan et al. (1986) reported on the stomach contents of 21 post-smolts of salmon caught shortly after emigration from the rivers into the Firth of Clyde, west Scotland. They found that the diet consisted principally of sandeels. In a larger study of post-smolts of Atlantic salmon in the northern Gulf of St Lawrence, Dutil & Coutu (1988) examined salmon stomachs from July till October. In July, the dominating food items were invertebrates (Insecta, mainly Chironomidae, and Crustacea, mainly Gammaridae). Later in the summer and autumn, the post-smolts consumed mainly small fish, *Ammodytes americanus* being most prevalent, which occurred in most of the stomachs. Capelin were also relatively abundant. Among the crustaceans, euphausiids were the most frequent.

Levings et al. (1994) and Hvidsten et al. (1995) caught post-smolts of Atlantic salmon by pair-trawling at different sites in Trondheimsfjord, Norway, in May and the beginning of June 1992–94. These fish may have been at sea for a few days to a few weeks. The fjord area was divided into zones, starting in the estuary, followed by the fjord area and ending in the outer coastal area. In the estuary, the stomach contents were dominated by prey taken in the river (Levings et al. 1994), whereas prey occurring in brackish and salt water were found in stomach samples collected farther out. The volume frequency of the main prey groups for the material collected in the fjord area just inside the open coastal zone showed great temporal variations, e.g. insects dominated the first 2 years and crustaceans the third year (Table 13.1). There were also spatial variations within a year, for example in 1994 euphausiids dominated in the fjord area, whereas farther out the importance of this group was somewhat reduced, as amphipods (Hyperiidae) and fish (herring larvae) became important as well (Table 13.2).

Table 13.1 Weight (volume) percentage of main prey groups in stomachs of Atlantic salmon post-smolts collected in the Trondheimsfjord area in May 1992, 1993 and 1994 (Levings et al. 1994, Hvidsten et al. 1995).

Prey group	1992 (N = 188)	1993 (N = 100)	1994 (N = 84)
Insects	72.0	45.0	3.4
Crustaceans	23.8	32.7	96.0
Fish	1.0	17.1	0.5
Other	—	5.1	—

Table 13.2 Food content of Atlantic salmon post-smolts collected in the Trondheimsfjord and on the coast in 1994. Percentage frequency of occurrence (%F) and weight (volume) percentages (%V) (Hvidsten et al. 1995).

Area:	Fjord area (N = 84)		Coastal area (N = 108)		Total	
Prey group	%F	%V	%F	%V	%F	%V
Insects						
Diptera larvae	3.7	3.44			1.4	0.01
Diptera adults	37.8	0.03	39.8	1.40	37.0	1.68
Coleoptera	2.4	2.40	0.9	0.02	1.4	0.02
Insecta remains	17.1	0.01	17.6	0.65	15.9	0.52
Crustaceans						
Calanus	6.1	0.03	14.8	0.34	10.1	0.17
Euphausiidae	64.6	92.34	38.0	39.78	49.5	63.05
Hyperiidae	3.7	0.11	63.0	8.74	35.1	4.20
Gammaridae			0.9	0.01	0.5	0.01
Decapoda (Zoea)	8.5	0.06	0.9	0.01	3.8	0.03
Corophium spp.	2.4		0.9	0.08	1.4	0.05
Isopoda	1.2	0.03			0.5	0.01
Crustacea remains	9.8	0.03	38.9	23.68	27.4	14.10
Fish						
Herring	+	0.92	7.4	15.77	4.3	7.93
Fish remains	3.7	+	24.1	9.42	15.9	6.40

Sturlaugsson (1994, 1995a) examined the food of ranched post-smolts of Atlantic salmon in coastal waters off west Iceland. He found that the most important prey of post-smolts were decapods (crab larvae), amphipods, copepods, dipterans (adult stage), euphausiids and fish larvae. The importance of some of the main prey types was area-restricted. For example, dipterans and epibenthic amphipods (Gammaridae) were important in near-shore areas, as opposed to the off-shore areas where pelagic amphipods (Hyperiidae) and fish larvae (*Ammodytes* spp.) became important prey.

Fjallstein (1987) studied runs of hatchery-reared salmon smolts in a Faroese fjord and found the smolts feeding on almost any object in the sea, including weed and inorganic material during the first 2 weeks, thereafter dipterans became the most important prey during the first month at sea.

The food of post-smolts seems to gradually shift from prey found in the freshwater system to marine prey such as crustaceans and fish larvae.

Oceanic areas

In recent years the feeding habits of post-smolts have been analysed in the area west of the British Isles and in the Norwegian Sea (Holst et al. 1993, 1996, Holm et al. 1996). The main fish species identified in the pelagic trawl catches beside salmon were herring, mackerel and lumpsucker. In total, 34 post-smolts were caught in the Norwegian Sea in 1991, 46 off the Hebrides in 1995, and 62 in the Norwegian Sea in

1995. The proportion of empty stomachs was relatively small (3–15%), which might indicate favourable feeding regimes in the sampled areas (Holst et al. 1996). In the Norwegian sea, the most common prey observed in post-smolts were hyperiid amphipods (*Themisto* spp.), euphausiids (Euphausiidae), O-group herring and redfish (*Sebastes* spp.) larvae, whereas off the Hebrides blue whiting (*Micromesistius poutassou*) larvae/0-group were the only food item eaten (Holst et al. 1996).

The total dominance of 0-group/larval blue whiting eaten by the post-smolts off the Hebrides in May–June 1995 coincided with the very strong 1995 year class of blue whiting, which was spawned earlier in the year, mainly on the Porcupine Bank area and along the slope west of Ireland and the Hebrides (ICES 1998b). The post-smolts seemed to take advantage of these abundant food resources (Holst et al. 1996). This could be considered as an example of opportunistic foraging of salmon post-smolts.

In some areas post-smolts probably compete with other fish species for food as they pass through various coastal and off-shore areas on their way to the high seas. One such example could be from the North Sea, where post-smolts from the west coast of Sweden, southern Norway, Denmark, south-eastern England and east Scotland are believed to travel through the northern North Sea in spring and summer before entry to the Norwegian Sea. The often abundant resource of Norway pout (*Trisopterus esmarkii*) in the North Sea, a fish of the same size as post-smolts (juvenile 8–13 cm and adults 14–21 cm), take food of the same size class as post-smolts, i.e. mainly pelagic zooplankton (Raitt & Adams 1965), including calanoid copepods, euphausiids, decapod larvae and fish larvae. Furthermore, both juvenile whiting and haddock are feeding in the same areas. Raitt & Adams (1965) considered only whiting as feeding on the same type of food as Norway pout. Probably the phase of the post-smolt stage in coastal areas and on the continental shelves could be considered food limited owing to competition from other pelagic predators and in some years by low marine productivity.

Food of pre-adults

There is some information about feeding of pre-adult salmon in the ocean, reviewed by Reddin (1988) and Hislop & Shelton (1993). In the north-west Atlantic area (Newfoundland, Labrador Sea, Davis Strait and West Greenland), salmon feed mainly on capelin, sandeel, euphasiids and amphipods (Templeman 1968, Lear 1980, Reddin 1985), whereas in the Irminger Sea the hyperiid amphipods *Themisto* spp., barracudinas (*Paralepis* sp.) and cephalopods (*Brachioteuthis riisei*) dominated (Jensen 1967, 1974). Jensen also noted the contrast in the food content between the Irminger Sea and West Greenland, where capelin and sandeel dominated in the food.

In the north-east Atlantic, salmon feed mainly on lantern fishes, sandeel, amphipods and euphausiids, and also some larger fish were consumed, mainly capelin and herring (Struthers 1970, 1971, Thurow 1973, Hislop & Youngson 1984, Hansen & Pethon 1985, Ólafsson 1987, Jacobsen & Hansen 1996).

A summary of the food of Atlantic salmon presented as the number of occurrences of each prey type for various geographic areas in the North Atlantic is presented in Table 13.3, which is a similar table to that of Hislop & Shelton (1993, their Table 5.1), but with percentage of occurrences instead of the number of occurrences. We also include the recent study by Jacobsen & Hansen (1996) for comparison, given as Faroese 1992–95 data (Table 13.3). The weight percentage data (Hislop & Shelton 1993) are also presented along with the new Faroese data (Table 13.4). The weight percentages from the Faroese area in 1983 (Hislop & Youngson 1984) have been recalculated in order to conform to the species list in the present data from the Faroes (Faroese data 1985 in Table 13.4).

It is interesting to note the difference in diet composition of salmon in oceanic areas in the north-west and north-east Atlantic. In the north-east Atlantic fish are found very frequently in the salmon stomachs, whereas crustaceans are contributing more in the north-east Atlantic, particularly in the oceanic areas (Tables 13.3, 13.4). This could be due to different abundances of various prey species in the two areas. This might affect the marine survival in the north-west and north-east Atlantic in different directions, apparently governed by a common environmental cause, owing to differential response in the abundance of the capelin and crustaceans in the north-west and north-east Atlantic from the anticipated common environmental cause. It might be speculated that the apparently less dependence on fish prey of salmon in the north-east Atlantic would act as a stabilising factor to the great fluctuations often observed in pelagic fish species, assuming less fluctuations in the zooplankton biomass. A prediction could, therefore, be that the salmon stocks in the north-east Atlantic would fluctuate less compared to north-west Atlantic stocks, mainly owing to the great fluctuations inherent in the short-lived capelin stock in that area (Frank *et al.* 1996), which are the main food for salmon off West Greenland and on the Canadian shelf area.

Faroese area

In October–November the first post-smolts/pre-adults appear in the catches in the Faroese long-line fishery north of the isles. The long-line is known to be size selective and the smallest pre-adult caught on the lines in November was 38 cm (J.A. Jacobsen, pers. comm.). The area north of the Faroes and in the southern Norwegian Sea seems to be a major feeding area for salmon in the north-east Atlantic during winter (Jákupsstovu 1988, Jacobsen & Hansen 1996).

Analysis of the stomach contents of 3848 salmon caught north of the Faroe Islands (Fig. 13.1) between November and March during the three fishing seasons 1992–93, 1993–94 and 1994–95 were analysed by Jacobsen & Hansen (1996). Crustaceans, including *Themisto* spp., euphausiids and pelagic shrimps, accounted for more than 80% of the food in number, but by weight more than 60% of the stomach content was fish, particularly lantern fishes, pearlsides, barracudinas, and silversides (mesopelagic fish). Some larger pelagic fish such as herring, blue whiting and mackerel were also present (0.2% by number and 13% by weight).

Table 13.3 Food of pre-adult and adult Atlantic salmon in the North Atlantic. Frequency of occurrence (%) of prey in each major geographic area. From Table 5.1 in Hislop & Shelton (1993) and from Jacobsen & Hansen (1996). A + means less than 1%.

Status of the salmon and period:	Adults	Pre-adults winter	Pre-adults	Pre-adults winter	Homing adults	Pre-adults winter	Pre-adults winter
Area:	Canadian shelf	Labrador Sea/ Davis Strait	West Greenland	Norwegian Sea	British Isles	Faroes 1983	Faroes 1992–95
No. examined	2267	231	1503	1293	3096	555	3848
No. empty or with only bait	1505	35	152	600	2719	258	1184
No. with food	762	196	1351	693	377	297	2664
Percentage with food	34	85	90	54	12	54	69
(1) Fish							
Clupeoids							
Clupea harengus	10	1	+	6	18	2	+
Clupea sprattus					15		
Alosa pseudoharengus	+						
Clupeoids					1		
Capelin							
Mallotus villosus	68	+	45	2		2	+
Fry (mostly *Mallotus villosus*)							4
Sandeel							
Ammodytids	21		48	2	34	3	+
Lantern fishes							
Benthosema glaciale		1		7			10
Notoscopelus kroeyeri	+	3		+			+
Lampanyctus macdonaldi		1		+			
Lampanyctus crocodilus							+
Myctophum punctatum							+
Protomyctophum arcticum		+		+			
Hierops arctica							
Myctophids			+	20		22	8

Contd.

Table 13.3 Contd.

Status of the salmon and period: Area:	Adults Canadian shelf	Pre-adults winter Labrador Sea/Davis Strait	Pre-adults West Greenland	Pre-adults winter Norwegian Sea	Homing adults British Isles	Pre-adults winter Faroes 1983	Pre-adults winter Faroes 1992–95
No. examined	2267	231	1503	1293	3096	555	3848
No. empty or with only bait	1505	35	152	600	2719	258	1184
No. with food	762	196	1351	693	377	297	2664
Percentage with food	34	85	90	54	12	54	69
Barracudinas							
Paralepis c. borealis							+
Notolepis rissoi kroyeri							+
Paralepids	+	30	2			1	3
Pearlside							
Maurolicus muelleri				+		3	18
Other fish							
Osmerus esperlanus					+		
Osmerus mordax	+						
Gasterosteus aculeatus	+						+
Gadus morhua	1		+				
Boreogadus saida	+		+				
Melanogrammus aeglefinus					1		
Merlangius merlangius					3		
Micromesistius poutassou				1		4	+
Pollachius pollachius					+		
Trisopterus esmarki							
Onogadus argentatus	+			2	+		+
Unidentified gadoid fry				+	+		
Belone belone				+			
Unidentified cottid fry	+		+	3	+		+
Anarhichas lupus							

Lumpenus maculatus						+	
Scomber scombrus	+						
Sebastes marinus			+				
Cottunculus microps			+				
Myxocephalus scorpioides			+				
Agonus decagonus			+				
Cyclopterus lumpus	+						
Reinharditus hippoglossoides	+						
Unidentified flatfish						+	
Unidentified fish	11	38	17	20	14	13	37
(2) Invertebrates							
Annelids	1		1		1		
Insects	+			+	1		
Crustaceans							
Copepods						+	+
Amphipods	3	37	24	12	6	61	66
Isopods				3			
Euphausiids	2	5	22	36	+	32	51
Other crustaceans	+	4		+	2	8	25
Unidentified crustaceans				2			19
Mollusca							
Spiratella helicina	1						
Squid							
Gonatus fabricii		33		7			1
Unidentified/other squid	+	+		5		+	
Unidentified invertebrates	+	+	+				
(3) Unidentified remains				13			5

Table 13.4 Food of pre-adult and adult Atlantic salmon in the North Atlantic. Percentage composition by weight in the diet in each major geographic area. From Table 5.2 in Hislop & Shelton (1993). Old Faroese data recalculated from Hislop & Youngson (1984) and new Faroese data from Jacobsen & Hansen (1996). A + means less than 1%.

Status of the salmon and period:	Adults	Pre-adults winter	Pre-adults	Homing adults	Pre-adults winter	Pre-adults winter
Area:	Canadian shelf	Labrador Sea/Davis Strait	West Greenland	British Isles	Faroes 1983	Faroes 1992–95
(1) Total stomach content						
Fish	98	83	82	99	66	66
Crustaceans	+	14	16	+	32	30
Molluscs	+	3				
Squid					2	2
Annelids	+	+	+	+		
Insects	+					
Unidentified	+	+	2			2
(2) Principal fish prey						
Clupeoids	13	+	+	57		18
Capelin	74	+	72		25	5
Sandeel	11	3	7	30		+
Lantern fishes	+	16			73	29
Barracudinas	+	59	2		2	22
Pearlside					1	16
Others	2	24	9	13		10
(3) Principal crustacean prey						
Amphipods	87	97	60	+	97	67
Euphausiids	+	+	40		1	17
Other	1	2		+	2	16

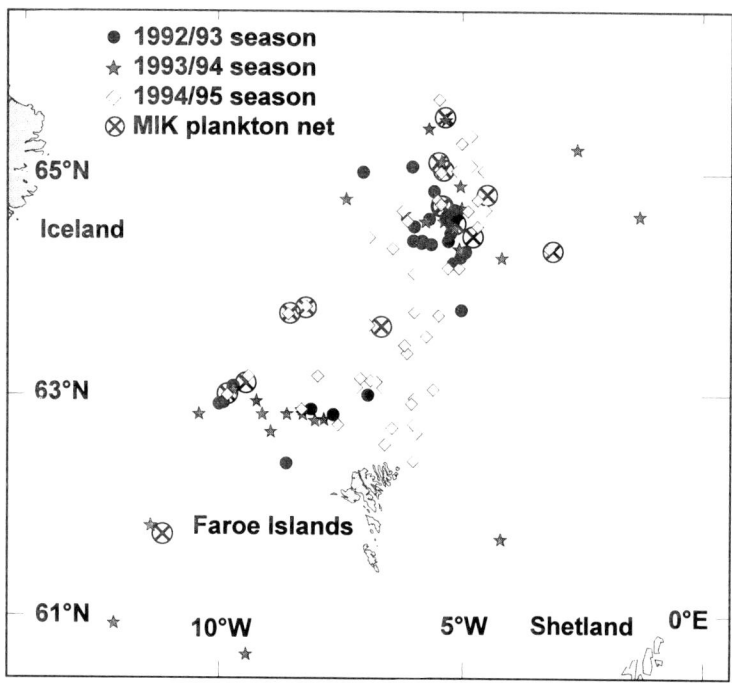

Fig. 13.1 Map showing long-line set locations where 3848 salmon stomachs were sampled during three consecutive fishing seasons 1992–1993 (circles), 1993–1994 (stars) and 1994–1995 (diamonds). The 13 plankton samples (MIK plankton net) taken in 1993–1994 and 1994–1995 are superimposed (crossed open circles), corresponding to 481 stomach samples at the same locations. From Jacobsen & Hansen (1996).

On the assumption that the proportion of empty stomachs is indicative of the intensity of feeding (Rae 1967), salmon feed more intensively in February–March (78% contained food) than in November–December (53% contained food) (Fig. 13.2a). Of the stomachs containing food, the average food content (g) per salmon also increased during the season (Fig. 13.2b), mainly owing to an increase in lantern fishes in the diet (Fig. 10.3), emphasising the lower feeding rate of salmon observed during autumn compared with winter. Salmon sampled in the Labrador Sea had less food in their stomachs in the autumn than in the spring (3.1 g and 5.7 g food per kg of salmon respectively) and were feeding less actively (28% and 8% empty stomachs respectively) (Lear 1980). Lower feeding rates during winter were also observed in the Baltic (Christensen 1961, Thurow 1966).

The low feeding rate observed in late autumn could be an indication of limited food availability in the sea, implying that salmon may have a hard time in the sea in this period. Mean sea surface temperature (SST $\pm$ s.d.) at the sampling locations was 7°C ($\pm$ 1.6°C) in November–December and 3°C ($\pm$ 1.3°C) in February–March. The growth rate in salmonids increases with temperature (Brett 1979) provided food (ration) is not limited. The higher SST during autumn combined with apparently less abundance of fish prey might result in sub-optimal growth rate in this period; it could

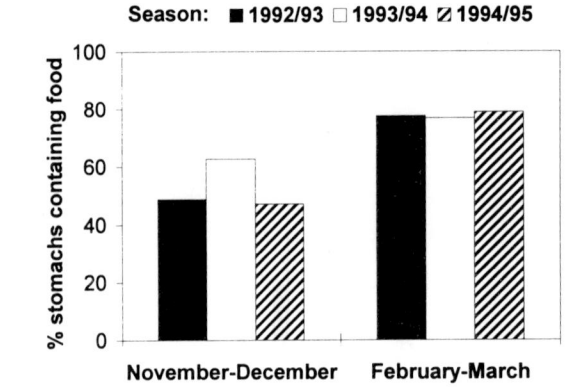

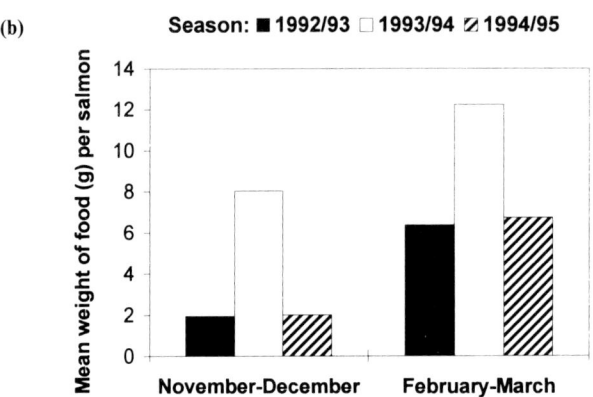

Fig. 13.2 (a) The proportion of salmon stomachs containing food caught north of the Faroes in the autumn (November–December) and winter (February–March) fishing seasons in 1992–93, 1993–94, and 1994–95: (b) Mean weight of prey (g) per salmon each period. From Jacobsen & Hansen (1996).

even be negative under conditions of starving. However, the lower temperature during winter would also slow down the turnover rate of food in the stomach owing to slower digestion (Jobling 1981), and the apparently higher feeding intensity suggested in the winter period might, therefore, not necessarily indicate higher feeding rate but slower digestion rate.

Owing to the increasing fish farming in the north-east Atlantic, fish farm escapees are mingled with the wild salmon on their feeding grounds (Hansen *et al.* 1993, 1998). The salmon caught were determined to be of wild or farmed origin by examining whether the fish showed external characters like fin erosion which is common on reared salmon (Lund *et al.* 1989). No significant difference in stomach content between wild and escaped farmed salmon in the high seas was observed, indicating that fish farm escapees fairly quickly adapt to the feeding habits in the wild environment (Jacobsen & Hansen 1996). Farmed salmon caught in Scottish coastal waters also fed on natural prey (Hislop & Webb 1992).

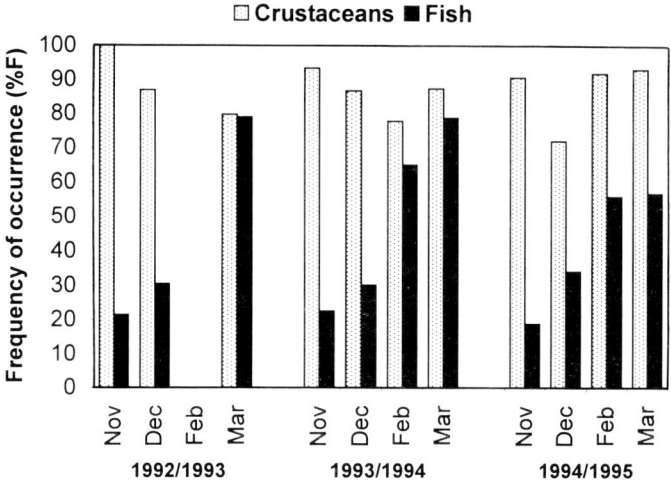

Fig. 13.3 Frequency of occurrence (%) of crustacean and fish prey per month during the three fishing seasons 1992–93, 1993–94, and 1994–95 north of the Faroes. From Jacobsen & Hansen (1996).

Hislop & Shelton (1993) suggested that the crustaceans were a much less important food than fish. This is generally so, but the extensive material studied by Jacobsen & Hansen (1996), including nearly 4000 stomachs sampled during three consecutive winter feeding seasons, places more emphasis on crustacean prey, particularly the hyperiid amphipods and to a lesser extent euphausiids as the typical prey in salmon beside mesopelagic fishes. Especially in the autumn salmon seemed to rely on amphipods as food (Fig. 13.3).

The diet of salmon compared with available prey

In the Atlantic there is limited information available that compares the diet of salmon with the potential food available. Although recent information from literature on plankton and micronekton distributions in the North Atlantic (Dalpadado *et al.* 1996, 1998, Johnson & Kitchell 1996) might give a clue to the potential prey of salmon, the non-overlapping temporal and spatial nature of the data prevents any conclusion on prey selection. Furthermore, the vertical distribution of salmon in the sea is poorly known and mainly inferred from catching (Templeman 1968, Reddin 1985, Reddin & Shearer 1987, Dutil & Coutu 1988). However, salmon are known to undertake deep dives, deeper than 150 m (Jákupsstovu 1988) for shorter or longer periods. The promising prospects from data storage tags (DST) might shed some light on the vertical distribution of salmon (Sturlaugsson 1995b, ICES 1998c). But so far the feeding behaviour and forage strategy of salmon in the ocean has only been inferred from indirect measures such as stomach analysis of salmon caught at the surface.

In the Faroese research programme corresponding plankton samples (0–50 m depth) were taken at 13 fishing locations in 1994–95 with an MIK plankton sampler (2 m diameter framed plankton net, 2.5 mm meshes) to compare the diet of salmon with available food in the surface layer (Jacobsen & Hansen 1996, 1999). The species found and their weight percentages in the salmon stomachs and in the corresponding plankton samples are listed in Table 13.5. The plankton tows generally included the same species as found in the stomachs, i.e. the hyperiid amphipods, euphausiids, the shrimp *Hymenodora glacialis*, lantern fishes and pearlsides (Jacobsen & Hansen 1996).

The apparent differences in the proportional weights of the species between the salmon stomachs and the MIK plankton samples (Table 13.5) have several probable causes, which could be categorised as sampling bias, selective feeding of salmon, and

Table 13.5 Species found and total wet weight percentage in the salmon stomachs and in the MIK plankton samples from 13 corresponding fishing locations (see Fig. 13.1). Table entries marked with a dash mean that prey species/group were not found in the sample. The entries marked *r* indicate that only remains of a specific group were found in the stomach content owing to advanced digestion, while in the MIK samples all material was fresh and identifiable to species or family. Total wet-weight of all samples is given at the bottom row in parenthesis. Data from Jacobsen & Hansen (in press).

Species/group		Stomachs[1] weight %	MIK samples[2] weight %
Scyphosoa	Jellyfish	—	2.17
Tomopteris sp.	Polychaeta	—	0.05
Bivalvia	Mussels	—	0.01
Sagitta sp.	Chaetognatha	—	0.98
Lepeophtheirus salmonis[3]	Crustaceans	—	0.01
Calanus finmarchicus	Crustaceans	—	0.10
Aristias tumidus	Crustaceans	0.003	—
Eusirus holmi	Crustaceans	0.25	—
Pareuchaeta norvegica	Crustaceans	0.002	1.25
Themisto spp.	Crustaceans	16.37	6.82
Ephausiidae	Crustaceans	8.72	87.54
Hymenodora glacialis	Crustaceans	5.15	0.05
Crustacea remains	Crustaceans	5.62	*r*
Gonatus fabricii	Squid	0.02	0.03
Maurolicus mülleri	Fish	4.74	0.54
Myctophidae	Fish	18.25	0.03
Mallotus villosus[4]	Fish	3.14	0.44
Paralepidae	Fish	9.45	—
Clupea harengus	Fish	15.90	—
Fish remains	Fish	12.38	*r*
Total wet-weight (g), all samples		(1719)	(837)

[1] Total weight % in 319 non-empty salmon stomachs (of 481 sampled) from 13 fishing locations.
[2] Total weight % in the MIK plankton samples taken on the 13 fishing locations as above.
[3] A few salmon lice have been found in salmon stomachs from other samples in the same area.
[4] Only capelin fry (< 50 mm) were caught in the MIK samples while both fry and adult specimens were found in the stomachs.

limited sampling range (depth). The most obvious bias in the material was considered to be avoidance of the plankton net by large animals. The large fish species such as herring, barracudinas and the larger capelin (> 50 mm) were not caught. Other researchers have reported avoidance of large zooplankton and small fish (> 4.5 cm) from the MIK plankton net sampler and similar sampling devices (Munk 1988, 1993, Dalpadado et al. 1996, 1998, ICES 1996b).

Since the abundance measure used was weight percentages, it is clear that if fish and larger crustaceans avoid the plankton net, the resulting percentage weight will be biased towards smaller individuals as compared with the true weight proportions of plankton and micronekton in the environment. Thus the few fish sampled in the plankton net represent only 1% in weight, whereas fish constituted about 63% in the salmon stomachs. Conversely the combined percentage weight of euphausiids and amphipods was 94% in the plankton samples and 37% in the stomachs.

Part of the observed difference could, however, also be due to selective feeding for fish, but the bias problem prevents any firm conclusions in that respect. Holst et al. (1996) noted that given that 0-group fish were available in the area (as observed from the trawl catches) and that amphipods also were available (from parallel plankton sampling), the post-smolts fed mainly on 0-group fish, indicating a selective feeding strategy. It appeared that 0-group fish were preferred to crustaceans as food. Lear (1972) found that salmon in the north-west Atlantic preferred herring to capelin, and that large salmon contained significantly more herring than small salmon. Pepin et al. (1987) stated that Atlantic mackerel positively selected fish larvae in preference to other food available (i.e. crustaceans).

An analysis of the crustaceans only, revealed opposite abundance indices between the two prey groups (amphipods and euphausiids) from the stomachs and the plankton samples, respectively. On average, the salmon had taken twice as many amphipods as euphausiids while the plankton sampler caught 13 times as many euphausiids as amphipods (Table 13.5). There were also great variations in the individual comparisons between the two prey groups and it seems as if salmon prefer amphipods to euphausiids (Fig. 13.4). In most cases it had taken more amphipods although there were apparently more euphausiids available in the upper 50 m (Fig. 13.4).

Holst et al. (1996) also found a tendency to size-selective feeding of salmon post-smolts on the hyperiid amphipods (*Themisto* spp.), as indicated by a positive relation between prey size in the stomachs and post-smolts size. Jacobsen & Hansen (1996) also found a significant but weak correlation between length of salmon and length of *Themisto libellula* prey.

Some species found in the plankton samples seemed to be missing from the stomachs, and the salmon appeared not to eat chaetognaths (*Sagitta* spp.) among the available prey observed in the plankton samples (Table 13.5). It might be speculated that *Sagitta* spp. either are too transparent or unpalatable, or are a low energy prey. However, chaetognaths are frequent food for Atlantic herring and mackerel in the north-east Atlantic (Fraser 1961, Gibson & Ezzi 1990, Melle et al. 1994, Dalpadado et

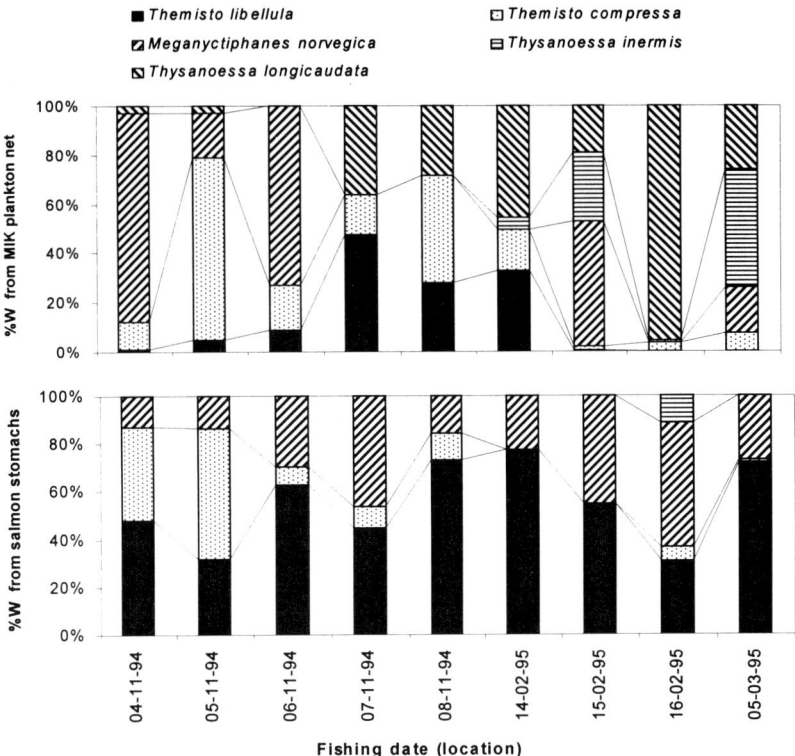

Fig. 13.4 Relative (%) weight distribution of two species of *Themisto* and three Euphausiid species compared from MIK plankton net samples, 0–50 m depth (upper panel) and from corresponding salmon stomachs (lower panel) at 9 fishing locations (with detailed species data) of the 13 fishing locations (dates). Date from Jacobsen & Hansen (in press).

al. 1996) and for salmon in the Pacific (Takeuchi 1972, Brodeur & Pearcy 1990, Tadokoro *et al.* 1996). Capelin north of Iceland have, however, been reported to prefer other crustacean prey to chaetognaths (Astthorsson & Gislason 1996).

The small copepod *Calanus finmarchicus* was captured by the plankton sampler, but was not found in the salmon stomachs (Table 13.5). Most of the copepods overwinter in deep water during autumn and winter, and hence the abundance of *C. finmarchicus* should be very small and only include sizes larger than 2 mm (stage V) during winter time (Hay *et al.* 1991, Hirche *et al.* 1994). Probably the fraction of *C. finmarchicus* of 2.5–3.5 mm available was too small to be eaten by salmon larger than 45 cm.

Even with paired data, it is still uncertain what proportion of the food was actually taken in the upper 50 m (maximum depth of plankton samples). However, the fact that most species found in the plankton samples, with the exception of larger fish and the largest zooplankton, probably lost owing to avoidance, were also found in the corresponding stomachs and *vice versa*, indicates that a significant spatial overlap in the data is likely. This implies an opportunistic foraging strategy of salmon, which is also reported elsewhere (Hansen & Pethon 1985, Reddin 1988, Hislop & Shelton 1993, Sturlaugsson 1994). Furthermore, the apparent species selection of amphipods to euphausiids and the selection of the larger *Meganyctiphanes norvegica* to the smaller *Thysanoessa* spp. (both euphausiids) indicates that the salmon is a selective feeder (Jacobsen & Hansen 1999).

Food of adults

During homeward migration the feeding activity of Atlantic salmon tends to cease. In samples collected from bag-net fisheries in Norwegian home waters only about 10% of the salmon stomachs on average contained food (L.P. Hansen, NINA, Norway, pers. comm.). Fish were by far the most abundant prey, with herring, sprat and sandeel as the dominant fish species. In a small number of adult salmon examined in a fjord in Northern Norway, Grønvik & Klemetsen (1987) found that herring were the most important prey. Neilson & Gillis (1979) found adult salmon feeding on the Labrador coast in August, and 86% contained food, mainly hyperiid amphipods (60% in weight) and *Ammodytes* sp. (25% in weight). Salmon caught in bag-nets on the Scottish west coast were shown to eat mainly sandeels (Fraser 1987). His results indicate that salmon were feeding up to a certain cut-off point in June or early July, after which all salmon he investigated contained no food. Lear (1972) also concluded that salmon cease to feed when they enter the estuary of their home river. Sturlaugsson *et al.* (1997) caught adult salmon returning to Icelandic rivers and found the majority of the stomachs empty (70%) and those with food almost invariably contained fish prey (sandeel and herring). Similarly Blair (1965) examined the stomachs of 919 grilse and older salmon in Bay of Islands (Gulf of St Lawrence) during summer and found 86% empty and the rest contained mainly capelin and sandeel, and Power (1969) examined stomachs of 300 adult salmon entering the rivers in Ungava Bay, none of which contained any food. Hislop & Webb (1992) examined stomachs of 54 salmon that had escaped from fish farms, and found that 35% contained food, predominantly juvenile whiting.

In summary, the food content of adult returning salmon consists entirely of fish if they take food at all. In this respect the feeding habits differ significantly from that of the pre-adult salmon in the open ocean feeding phase (Tables 13.3, 13.4), where up to one third of the stomach content weight is crustacean prey.

Is food a limiting factor for salmon?

Survival in the marine environment is a complex process, and Anderson (1988) summarised four major hypotheses that were used to explain marine mortality of fishes, namely starvation, predation, physical dispersal and disease. The effect of temperature and size on development, mortality and survival rates of the early pelagic life stages of marine fishes was reviewed by Pepin (1991).

Conditions that determine growth rate during the larval phase, such as food availability and temperature, are thought ultimately to determine survival (Ware 1975, Anderson 1988, Pepin 1991). It is likely that the growth-survival/mortality hypothesis put forward for other pelagic marine fishes (Anderson 1988) also apply to the early marine phase of Atlantic salmon. There is, however, limited knowledge of the factors influencing the growth and survival of post-smolts (Friedland & Reddin 1993, Friedland *et al.* 1998b), and it remains to be demonstrated that survival of salmon is a direct function of growth, mediated through size-dependent predation. Neither are the indirect effects of climate, such as food availability evident.

It has been shown that ocean climate conditions may affect marine survival of salmon (Friedland *et al.* 1993, 1998). There is at present no clear single factor that can be identified as causing mortality of salmon in the marine environment. Mortality is probably a combination of different factors as well as synergism between them. Traditionally it is assumed that the major mortality in salmon occurs at an early phase of the post-smolt phase.

Growth in salmonids is correlated with water temperature (Brett 1979), and Friedland *et al.* (in press) also observed that enhanced growth was associated with years having favourable temperature conditions, which in turn resulted in higher survival (return) rate to the rivers. The suggested relationship between growth of salmon and temperature does not help explain whether the reduced growth rates of Atlantic salmon in recent years were due to a reduced abundance of optimal food of high quality or a result of lower ocean temperatures. The slower growth rates may result in a higher predation pressure (Ware 1975, Folkvord & Hunter 1986, Pepin *et al.* 1987).

At which stages from smolt to returning adults are the various hypotheses in Anderson (1988) relevant to salmon? In marine fish species, the stages thought to be most vulnerable to mortality and where the year class strength most likely is determined, are the planktonic, egg and early larval stages. For salmon the anadromy gives rise to a special case of analogous critical life stages as in marine fishes. When salmon smolts enter the sea in spring and summer, they have to cope with the acclimatisation to salt water (Järvi 1989, Handeland *et al.* 1996), inducing stress and reduced tolerance of the fish to various external factors such as diseases and predation, a stage which could be analogous to the hatching process in marine fish. Furthermore, they have to switch from terrestrial to marine food, a stage which might be analogous to the first feeding in marine fish. Thus salmon smolts go through critical phases during early life in sea (post-smolt phase) analogous to those of marine fishes, with associated mortality at each stage.

If the abundance of prey is very low it is of course a limiting factor. However, the biomass of salmon is very low compared with other fish species such as herring, mackerel, blue whiting and capelin. Furthermore, studies suggest that marine mortality is density independent (Chadwick 1988, Chadwick & Claytor 1990, Crozier & Kennedy 1993, Jonsson *et al.* 1998). It is also suggested that salmon is an opportunistic feeder, and therefore is not dependent on one particular prey, and so food is expected to be a limiting factor only in extreme cases.

The optimal particle size, quality, quantity and diet composition is different at different life stages, and seems to be different in the north-east and north-west Atlantic. In the north-west Atlantic fish like capelin and sandeel are very important, while in the north-east Atlantic the crustaceans play a relative larger role as a food source. Are the large crustaceans as available in the north-west as in the north-east Atlantic? Salmon in the north-west Atlantic seem to suffer more in the ocean than do European salmon, as inferred from size distributions of North American and North European fish at West Greenland, and 2SW fish suffer more than 1SW fish (ICES 1998a). So what is the difference between the feeding grounds in the north-west and north-east Atlantic?

An external and man-made effect on food availability is the potential effect of the industrial fisheries (reduction fisheries) mainly operating in coastal areas and on continental shelves, which obviously involves the removal of potential salmon prey. A change in age and size structure of the prey populations indirectly affects the salmon by changing the predator-prey relationship. However, the likely effect might be less than expected owing to the presumed opportunistic forage behaviour of salmon. Thus, provided that the removal in absolute terms is less than the biomass needed as prey, the growth and survival should not be seriously impaired.

Limiting availability of suitable food, unfavourable temperatures and parasites are factors likely to influence growth of salmon and thus predator avoidance, swimming endurance, migration, and distribution in the sea. There is no simple relationship between the factors mentioned above and survival of salmon from smolt to returning adult. The effects of temperature on growth and on survival must be separated, in order to be able to access their relative contribution to mortality observed in salmon. Furthermore, we have to verify that salmon were taken by the presumed predators on salmon in the sea (Anthony 1996).

In the North Pacific several authors have indicated density-dependent survival due to food limitation for Pacific salmon (Peterman 1989, Bigler *et al.* 1996, Tadokoro *et al.* 1996, Welch *et al.* 1998). In the north-east Atlantic no such relationships have been found. There are differences between the Pacific and the Atlantic ecosystems, particularly concerning salmonids. In the Atlantic only one species inhabits the ocean, compared with several in the Pacific. Thus, the utilisation of potential prey must be different for the single species compared with more specialised niches reported in the Pacific. Furthermore, the abundance of salmonids in relation to other pelagic species is much lower in the North Atlantic than in the North Pacific.

Conclusions

Anadromous salmon move from fresh water to the ocean and gain weight. Although the biomass of salmon in the ocean is very small compared with the biomass of other pelagic marine species (e.g. herring, mackerel, blue whiting and capelin), Atlantic salmon are still part of the marine ecosystem and subjected to fluctuations and changes in that system.

The main feeding grounds for European salmon are in large areas in the north-east Atlantic and a large number of fish, in particular multi-sea-winter fish from south-west Europe move westwards in the Atlantic ocean, where they account for a significant component of the west Greenland fishery. North American salmon grow up in the north-west Atlantic, such as Labrador Sea and Greenland areas, although a few tagged Canadian salmon have been recovered in the east Atlantic and *vice versa*.

Comparison between plankton samples of available food organisms and corresponding stomach sampling of salmon suggests that salmon are opportunistic in the choice of food and that they also indicate that salmon selects prey in the upper end of the size spectrum available, and prefer fish to crustaceans if both are available. Salmon would furthermore select hyperiid amphipods to euphausiids as food in a mixture of both.

Salmon during its marine phase is thought to be an opportunistic pelagic and mid-water predator, supporting rapid growth rate by exploiting a wide range of invertebrates and fish prey. The wide variety of food in different areas and periods suggests that salmon abundance is unlikely to be very sensitive to annual changes in the availability of any particular prey.

The survival from smolt to adulthood is suggested to be limited mainly by density-independent factors during the first year after sea entry. These include fishing mortality, losses to predators and the availability of food. In recent years environmental factors such as sea surface temperatures have been shown to influence the marine survival of salmon. However, the underlying mechanisms linking ocean climate to survival are uncertain, but are thought to be governed by growth-mediated mortality, mainly through size-dependent predation. This, however, remains to be shown.

References

Anderson, J.T. (1988) A review of size dependent survival during pre-recruit stages of fishes in relation to recruitment. *Journal of Northwest Atlantic Fisheries Science*, **8**, 55–66.

Anthony, V.C. (1996) The predators of Atlantic salmon and their impact on salmon stocks. *NASCO Blue Book*, **59**, 3–58.

Astthorsson, O.S. & Gislason, A. (1996) Food of capelin in the subarctic waters north of Iceland. International Council for the Exploration of the Sea CM 1996/L: 32.

Bigler, B.S., Welch, D.W. & Helle, J.H. (1996) A review of size trends among North Pacific salmon (*Oncorhynchus* spp.). *Canadian Journal of Fisheries and Aquatic Sciences*, **53**, 455–65.

Blair, A.A. (1965) Bay of Islands and Humber River Atlantic salmon investigations. *Journal of the Fisheries Research Board of Canada*, **22**, 599–620.

Brett, J.R. (1979) Environmental factors and growth. *Fish Physiology*, **8**, 599–675.

Brandt, S.B. (1993) The effect of thermal fronts on fish growth: A bioenergetics evaluation of food and temperature. *Estuaries*, **16**(1), 142–59.

Brodeur, R.D. & Pearcy, W.G. (1990) Tophic relations of juvenile Pacific salmon off the Oregon and Washington coast. *Fishery Bulletin (US)*, **88**, 617–36.

Chadwick, E.M.P. (1988) Relationship between Atlantic salmon smolts and adults in Canadian rivers. In: *Atlantic Salmon: Planning for the Future* (eds D.H. Mills & D. Piggins), pp. 301–24. Croom Helm, London.

Chadwick, E.M.P. & Claytor, R.R. (1990) Predictability in a small commercial Atlantic salmon fishery in western Newfoundland. *Fisheries Research*, **10**, 15–27.
Christensen, O. (1961) Preliminary results of an investigation on the food of Baltic salmon. International Council for the Exploration of the Sea CM 1961/93 M:93.
Crozier, W.W. & Kennedy, G.J.A. (1993) Marine survival of wild and hatchery/reared Atlantic salmon (*Salmo salar* L.) from the River Bush, Northern Ireland. In: *Salmon in the Sea and New Enhancement Strategies* (ed. D.H. Mills), pp. 139–62. Fishing News Books, Oxford.
Dalpadado, P., Melle, W., Ellertsen, B. & Dommasnes, A. (1996) Food and feeding conditions of herring *Clupea harengus* in the Norwegian Sea. International Council for the Exploration of the Sea CM 1996/L: 20.
Dalpadado, P., Ellertsen, B., Melle, W. & Skjoldal, H.R. (1998) Summer distribution patterns and biomass estimates of macrozooplankton and micronekton in the Nordic Seas. *Sarsia*, **83**, 103–16.
Dutil, J.D. & Coutu, J.M. (1988) Early marine life of Atlantic salmon, *Salmo salar*, postmolts in the northern Gulf of St Lawrence. *Fishery Bulletin (US)*, **86**, 197–212.
Fjallstein, I. (1987) *Naturlig føde hos oppdrettet laksesmolt* (Salmo salar *L.*) *i indre kystområder på Færøyene*. Hovedfagsoppgave til cand. scient. graden i fiskeribiologi, Institutt for Fiskeribiologi, Universitetet i Bergen, Norge.
Folkvord, A. & Hunter, J.R. (1986) Size-specific vulnerability of northern anchovy *Engraulis mordax* larvae to predation by fishes. *Fishery Bulletin (US)*, **84**, 859–69.
Frank, K.T., Carscadden, J. & Simon, J.E. (1996) Recent excursions of capelin (*Mallotus villosus*) to the Scotian Shelf and Flemish Cap during anomalous hydrographic conditions. *Canadian Journal of Fisheries and Aquatic Sciences*, **53**, 1473–86.
Fraser, J.H. (1961) The oceanic and bathypelagic plankton of the North-East Atlantic. *Marine Research*, **4**, 1–48.
Fraser, P.J. (1987) Atlantic salmon, *Salmo salar* L., feed in Scottish coastal waters. *Aquaculture and Fisheries Management*, **18**, 243–7.
Friedland, K.D., Reddin, D.G. & Kocik, J.F. (1993) Marine survival of North American and European Atlantic salmon: effects of growth and environment. *ICES Journal of Marine Science*, **50**, 481–92.
Friedland, K.D., Hansen, L.P. & Dunkley, D.A. (1998) Marine temperatures experienced by postsmolts and the survival of Atlantic salmon, *Salmo salar* L., in the North Sea area. *Fisheries Oceanography*, **7**(1), 22–34.
Friedland, K.D., Hansen, L.P., Dunkley, D.A. & Maclean, J.C. (in press) Linkage between ocean climate, post-smolt growth, and survival of Atlantic salmon (*Salmo salar* L.) in the North Sea area. *ICES Journal of Marine Science*.
Gibson, R.N. & Ezzi, I.A. (1990) Relative importance of prey size and concentration in determining the feeding behaviour of the herring *Clupea harengus*. *Marine Biology*, **107**, 357–62.
Grønvik, S. & Klemetsen, A. (1987) Marine food and diet overlap of co-occurring Arctic charr *Salvelinus alpinus* (L.), brown trout *Salmo trutta* L. and Atlantic salmon *S. salar* L. off Senja, N. Norway. *Polar Biology*, **7**, 173–7.
Handeland, S.A., Järvi, T., Fernö, A. & Stefansson, S. (1996) Osmotic stress, antipredator behaviour, and mortality of Atlantic salmon (*Salmo salar*) smolts. *Canadian Journal of Fisheries and Aquatic Sciences*, **53**, 2673–80.
Hansen, L.P. & Pethon, P. (1985) The food of Atlantic salmon, *Salmo salar* L., caught by long-line in northern Norwegian waters. *Journal of Fish Biology*, **26**, 553–62.
Hansen, L.P., Jacobsen, J.A. & Lund, R.A. (1993) High numbers of farmed Atlantic salmon, *Salmo salar* L., observed in oceanic waters north of the Faroe Islands. *Aquaculture and Fisheries Management*, **24**, 777–81.
Hansen, L.P., Jacobsen, J.A. & Lund, R.A. (1998) The incidence of escaped farmed Atlantic salmon, *Salmo salar* L., in the Faroese fishery and estimates of catches of wild salmon. *ICES Journal of Marine Science*, **5**, 200–6.
Hay, S.J. Kiørboe, T. & Matthews, A. (1991) Zooplankton biomasses and production in the North Sea during the Autumn Circulation Experiment, October 1987–March 1988. *Continental Shelf Research*, **11**, 1453–76.
Hirche, H.-J., Hagen, W., Mumm, N. & Richter, C. (1994) Meso- and macrozooplankton distribution and production of dominant herbivorous copepods during spring. *Polar Biology*, **14**, 491–503.

Hislop, J.R.G. & Shelton, R.G.J. (1993) Marine predators and prey of Atlantic salmon (*Salmo salar* L.). In: *Salmon in the Sea and New Enhancement Strategies* (ed. D.H. Mills), pp. 104–18. Fishing News Books, Oxford.

Hislop, J.R.G. & Webb, J.H. (1992) Escaped farmed Atlantic salmon (*Salmo salar* L.) feeding in Scottish coastal waters. *Aquaculture and Fisheries Management*, **23**, 721–3.

Hislop, J.R.G. & Youngson, A.F. (1984) A note on the stomach contents of salmon caught by longline north of the Faroe Islands in March, 1983. International Council for the Exploration of the Sea CM 1984/M: 17.

Holm, M., Holst, J.C. & Hansen, L.P. (1996) Sampling Atlantic salmon in the NE Atlantic during summer: methods of capture and distribution of catches. International Council for the Exploration of the Sea. CM 1996/M: 12.

Holst, J.C., Nilsen, F., Hodneland, K. & Nylund, A. (1993) Observations of the biology and parasites of postsmolts Atlantic salmon, *Salmo salar*, from the Norwegian Sea. *Journal of Fish Biology*, **42**, 962–6.

Holst, J.C., Hansen, L.P. & Holm, M. (1996) Observations of abundance, stock composition, body size and food of postsmolts of Atlantic salmon in the NE Atlantic during summer. International Council for the Exploration of the Sea CM 1996/M: 4.

Hvidsten, N.A., Johnsen, B.O. & Levings, C.D. (1995) Vandring og ernæring hos laksesmolt i Trondheimsforden og på Frohavet. *NINA Oppdragsmelding*, **332**, 1–17.

ICES (1996a) Report of the Working Group on North Atlantic salmon. International Council for the Exploration of the Sea CM 1996/Assess: 11.

ICES (1996b) Report of the Study Group on Gulf III sampler efficiency calibrations. International Council for the Exploration of the Sea CM 1996/L: 8.

ICES (1997) Report of the Working Group on North Atlantic salmon. International Council for the Exploration of the Sea CM 1997/Assess: 10.

ICES (1998a) Report of the Working Group on North Atlantic salmon. International Council for the Exploration of the Sea CM 1998/ACFM: 15.

ICES (1998b) Report of the Northern pelagic and blue whiting fisheries Working Group. International Council for the Exploration of the Sea CM 1998/ACFM: 18.

ICES (1998c) Report of the Study Group on ocean salmon tagging experiments with data logging tags. International Council for the Exploration of the Sea CM 1998/M: 3.

Jacobsen, J.A. & Hansen, L.P. (1996) The food of Atlantic salmon, *Salmo salar* L., north of the Faroe Islands. International Council for the Exploration of the Sea CM 1996/M: 10.

Jacobsen, J.A. & Hansen, L.P. (in press) The food of wild and farmed Atlantic salmon, *Salmo salar* L., north of the Faroe Islands.

Jákupsstovu, S.H.í. (1988) Exploitation and migration of salmon in Faroese waters. In: *Atlantic Salmon: Planning for the Future* (eds D.H. Mills & D. Piggins), pp. 458–82. Croom Helm, London.

Jensen, J.M. (1967) Atlantic salmon caught in the Irminger Sea. *Journal of the Fisheries Research Board of Canada*, **24**, 2639–40.

Jensen, J.M. (1974) Salmon survey in the southern part of the Irminger Sea, 1974. International Council for the Exploration of the Sea CM 1974/M: 29.

Jobling, M. (1981) Mathematical models of gastric emptying and the estimation of daily rates of food consumption for fish. *Journal of Fish Biology*, **19**, 245–57.

Johnson, T.B. & Kitchell, J.F. (1996) Long-term changes in zooplanktivorous fish community composition: implications for food webs. *Canadian Journal of Fisheries and Aquatic Sciences*, **53**, 2792–803.

Jonsson, N., Jonsson, B. & Hansen, L.P. (1998) The relative role of density-dependent and density-independent survival in the life cycle of Atlantic salmon *Salmo salar*. *Journal of Animal Ecology*, **67**, 751–62.

Järvi, T. (1989) The effect of osmotic stress on the anti-predatory behaviour of Atlantic salmon smolts: a test of the 'Maladaptive Anti-predator Behaviour' hypothesis. *Nordic Journal of Freshwater Research*, **65**, 71–9.

Lear, W.H. (1972) Food and feeding of Atlantic salmon in coastal and over oceanic depths. *Research Bulletin of the International Commission for the Northwest Atlantic Fisheries*, **9**, 27–39.

Lear, W.H. (1980) Food of Atlantic salmon in the West Greenland-Labrador Sea area. *Rapports et Procès-verbaux des Réunions du Conseil International pour l'Exploration de la Mer*, **176**, 55–9.

Levings, C.D. (1994) Feeding behaviour of juvenile salmon and significance of habitat during estuary and early sea phase. *Nordic Journal of Freshwater Research*, **69**, 7–16.

Levings, C.D., Hvidsten, N.A. & Johnsen, B.O. (1994) Feeding of Atlantic salmon (*Salmo salar* L.) postsmolts in a fjord in Central Norway. *Canadian Journal of Zoology*, **72**, 834–39.

Lund, R.A., Hansen, L.P. & Järvi, T. (1989) Identification of reared and wild salmon by external morphology, size of fins and scale characteristics. *NINA Forskningsrapport*, **1**, 1–54. (In Norwegian with English abstract)

Melle, W., Røttingen, I. & Skjoldal, H.R. (1994) Feeding and migration of Norwegian spring spawning herring in the Norwegian Sea. International Council for the Exploration of the Sea CM 1994/R: 9.

Mills, D.H. (ed.) (1993) *Salmon in the Sea and New Enhancement Strategies*. Fishing News Books, Oxford.

Morgan, R.I.G., Greenstreet, S.P.R. & Thorpe, J.E. (1986) First observations on distribution, food and fish predators of post-smolt Atlantic salmon, *Salmo salar*, in the outer Firth of Clyde. International Council for the Exploration of the Sea CM 1986/M: 27.

Munk, P. (1988) Catching large herring larvae: gear applicability and larval distribution. *Journal du Conseil International pour l'Exploration de la Mer*, **45**, 97–104.

Munk, P. (1993) Describing the distribution and abundance of small 0-group cod using ring-net sampling and echo-integration. International Council for the Exploration of the Sea CM 1993/G: 40.

Neilson, J.D. & Gillis, D.J. (1979) A note on the stomach contents of adult Atlantic salmon (*Salmo salar*, Linnaeus) from Port Burwell, Northwest Territories. *Canadian Journal of Zoology*, **57**, 1502–3.

Pepin, P. (1991) The effect of temperature and size on development, mortality and survival rates of the pelagic early life stages of marine fishes. *Canadian Journal of Fisheries and Aquatic Sciences*, **48**, 503–18.

Pepin, P., Spearre, S., Jr. & Koslow, J.A. (1987) Predation on larval fish by Atlantic mackerel, *Scomber scombrus*, with comparison of predation by zooplankton. *Canadian Journal of Fisheries and Aquatic Sciences*, **44**, 2012–8.

Peterman, R.M. (1989) Application of statistical power analysis to the Oregon coho salmon (*Oncorhynchus kisutch*) problem. *Canadian Journal of Fisheries and Aquatic Sciences*, **46**, 1183–7.

Power, G. (1969) The salmon of Ungava Bay. Arctic Institute of North America, Montreal, Technical Paper, **22**, 1–72.

Rae, B.B. (1967) The food of cod on Faroese grounds. *Marine Research*, **6**, 1–23.

Raitt, D.F.S. & Adams, J.A. (1965) The food and feeding of *Trisopterus esmarkii* (Nilsson) in the northern North Sea. *Marine Research*, **3**, 1–28.

Reddin, D.G. (1985) Atlantic salmon (*Salmo salar*) on and east of the Grand Bank. *Journal of Northwest Atlantic Fisheries Science*, **6**, 157–64.

Reddin, D.G. (1988) Ocean life of Atlantic salmon (*Salmo salar* L.) in the Northwest Atlantic. In: *Atlantic salmon: Planning for the Future* (eds D.H. Mills & D. Piggins), pp. 483–511. Timber Press, Portland, Oregon.

Reddin, D.G. & Shearer, W.M. (1987) Sea-surface temperature and distribution of Atlantic salmon in the Northwest Atlantic ocean. *American Fisheries Society Symposium*, **1**, 262–75.

Struthers, G. (1970) A report on a salmon long lining cruise off the Faroes during April, 1970. *Freshwater Fisheries Laboratory, Pitlochry, Report, (54 FW 70)*, 1–8.

Struthers, G. (1971) A report on the 1971 salmon long lining cruise off the Faroes. *Freshwater Fisheries Laboratory, Pitlochry, Report, (33 FW 71)*, 1–6.

Sturlaugsson, J. (1994) Food of ranched Atlantic salmon (*Salmo salar* L.) postsmolts in coastal waters, W-Iceland. *Nordic Journal of Freshwater Research*, **69**, 43–57.

Sturlaugsson, J. (1995a) Postsmolts of ranched Atlantic salmon (*Salmo salar* L.) in Iceland: III. The first food of sea origin. International Council for the Exploration of the Sea CM 1995b/M: 16.

Sturlaugsson, J. (1995b) Migration study on homing of Atlantic salmon (*Salmo salar* L.) in coastal waters W-Iceland – depth movements and sea temperatures recorded at migration routes by data storage tags. International Council for the Exploration of the Sea CM 1995a/M: 17.

Sturlaugsson, J., þórisson, K. & Karlsson, H. (1997) Fæda laxa í hrygningargöngu um strandsævi. *Veidimálastofnun VMST-R*, **(97022)**, 1–13. (In Icelandic with English abstract)

Tadokoro, K., Ishida, Y., Davis, N.D., Ueyanagi, S. & Sugimoto, T. (1996) Change in chum salmon (*Oncorhynchus keta*) stomach contents associated with fluctuation of pink salmon (*O. gorbuscha*) abundance in the central Subarctic Pacific and Bering Sea. *Fisheries Oceanography*, **5**(2), 89–99.

Takeuchi, I. (1972) Food animals collected from the stomachs of three salmonid fishes (*Oncorhynchus*) and their distribution in the natural environment in the northern North Pacific. *Bulletin of the European*

Association of Fish Pathologists, **38**, 1–119. (In Japanese with English abstract, illustrations and tables)

Templeman, W. (1967) Atlantic salmon from the Labrador Sea and off West Greenland, taken during *A.T. Cameron* cruise, July–August 1965. *Research Bulletin of the International Commission for the Northwest Atlantic Fisheries*, **4**, 5–40.

Templeman, W. (1968) Distribution and characteristics of Atlantic salmon over oceanic depths and on the bank and shelf slope areas off Newfoundland, March–May, 1966. *Research Bulletin of the International Commission for the Northwest Atlantic Fisheries*, **5**, 62–85.

Thurow, F. (1966) Beträge zur Biologie und Bestandskunde des Atlantischen Lachses (*Salmo salar* L.) in der Ostsee. *Bericht Daten Wissenschaft Kommi ee Meeresforschungen*, **18**, 223–379.

Thurrow, F. (1973) Research vessel fishing on salmon off Norway. *Archives für Fischerei Wissenschaft*, **24**, 253–60.

Ware, D.M. (1975) Relation between egg size, growth and natural mortality of larval fish. *Journal of the Fisheries Research Board of Canada*, **32**, 2503–12.

Welch, D.W., Ishida, Y. & Nagasawa, K. (1998) Thermal limits and ocean migrations of sockeye salmon (*Oncorhynchus nerka*): long-term consequences of global warming. *Canadian Journal of Fisheries and Aquatic Sciences*, **55**, 937–48.

Chapter 14
The Food and Feeding of Atlantic Salmon (*Salmo salar* L.) During Feeding and Spawning Migrations in Icelandic Coastal Waters

JOHANNES STURLAUGSSON

Introduction

Research on food and feeding of salmon in the North Atlantic has mostly been directed towards the feeding in the oceanic area during post-smolt and the adult stages consisting of one-sea-winter (1SW), two- or multi-sea-winter fish (2SW; MSW). In addition, some work has been done in studying salmon foraging during their spawning migration along the coast.

Studies on the food and feeding of salmon post-smolts have been carried out in the North Atlantic and the Baltic Sea during the 1980s and especially during the 1990s. These research projects have been few and mainly covered the coastal areas, but recently considerable effort has been directed towards the oceanic area in the North Atlantic (Jacobsen & Hansen, Chapter 13). Sites of studies covering coastal areas range from the Gulf of St Lawrence (Dutil & Coutu 1988) in Canada north to Iceland (Sturlaugsson 1994a), and also include Norway (Hvidsten *et al.* 1993, Levings *et al.* 1994), Faroe Islands (Fjallstein 1987), west Scotland (Morgan *et al.* 1986) and the Baltic Sea (Lindroth 1961, Jutila & Toivonen 1985).

Studies on the food and/or feeding of 1SW and MSW have been carried out in the North Atlantic Ocean and the Baltic Sea since the 1960s. Studies on salmon food at their feeding grounds off West Greenland, Canada, Faroe Islands and Norway have shown that in general most of the captured salmon contained food. The food composition is very diverse but the main food of 1SW and MSW salmon are the following prey types: capelin (*Mallotus villosus* Müller), barracudinae species (Paralepidae), sandeel species (Ammodytidae), herring (*Clupea harengus* L.), lantern fish (Myctophidae), pearl side (*Maurolicus mülleri*), squid (Cephalopoda) and pelagic amphipods and euphausiids (Lear 1972, 1980, Hislop & Youngson 1984, Hansen & Pethon 1985, Reddin 1985, Jacobsen & Hansen 1996).

During the feeding migration the pelagic salmon are mostly eating pelagic prey, where they seem to select fish prey rather than crustaceans. However, the proportion of pelagic crustaceans in the food of salmon during the feeding migration can be considerable (Jacobsen & Hansen 1996). The proportion of these crustaceans is in general higher in oceanic areas such as off the Faroe Islands shelf than in coastal waters, as shown by the food composition of salmon within the coastal zone along Greenland and other areas (Lear 1972, Hislop & Webb 1992, Jacobsen & Hansen 1996).

Studies on salmon foraging during their spawning migration show that most of them are not eating when migrating through the coastal waters (Fraser 1987, Lear 1972). These results also indicated decreasing foraging towards the end of the spawning migration and showed that the salmon usually had ceased feeding when entering estuaries. Studies on salmon foraging during the end of their spawning migration in Scottish coastal waters showed that in June–July (1983–86), only 13% of the salmon contained food, which was in all instances sandeels (*Ammodytes marinus*) (Fraser 1987). In Canadian waters Lear (1972) studied the food of salmon in coastal waters in May–July (1969–71) and found a relatively high variation in prey types. His results also showed that the main prey species in these coastal areas was capelin, but herring and sandeels were also important. The feeding migration of Atlantic salmon (*Salmo salar*) from Iceland is widely spread over the North Atlantic, as shown by recaptures of tagged fish (Gudjonsson 1988).

In late 1980s and early 1990s the large scale sea ranching of Atlantic salmon in Iceland (Isaksson 1980,1994, Isaksson *et al.* 1997) facilitated research on the post-smolt salmon ecology in coastal waters and also made it much easier to study the returning adults during their spawning migration through coastal waters. Releases of up to several million smolts from a single release site gave the opportunity to capture post-smolts in large numbers with limited effort and with traditional fishing methods far from the release site.

Additionally, delayed release experiments made it possible to observe foraging under different environmental circumstances. The work referred to has already been partly described for post-smolts (Johannsson *et al.* 1991, Sturlaugsson 1994a,b, Sturlaugsson & Thorisson 1995a,b), and adults (Sturlaugsson 1994b, Sturlaugsson *et al.* 1997a,b). Further details on the study sites in Breidafjord, such as hydrography, plankton, competitors and predators are listed by Thorisson & Sturlaugsson 1995a,b).

The main aim of the research revealed here was to study the forage status of salmon during the beginning and end of their sea migration. The study sites used for sampling of post-smolts and adult salmon were distributed along the west coast (Fig. 14.1).

Catch

During the experimental fishing of post-smolts in 1987–94, approximately 1800 post-smolts were caught with fork lengths 10–36 cm and weights of 10–565 g. All these sample sets except from the near-shore fishing in 1994 are listed in Table 14.1 (a total of 1656 post-smolts). Stomach contents of 497 salmon were analysed during 1989–95 in Vogar, Hvalfjord, Hraunsfjord, Breidafjord and Adalvik and Jokulfjords (Tables 14.2, 14.3). The largest effort was in 1994–95, on which the forage is mainly based. The length distribution of adult salmon captured in Hvalfjord and Adalvik and Jokulfjords in 1994 and 1995 is given in Fig. 14.2. Table 14.4 provides information for microtagged salmon. Scale samples showed that these fish were of both wild and ranched origin, i.e. released as smolts in rivers or at salmon ranching stations (Table 14.4).

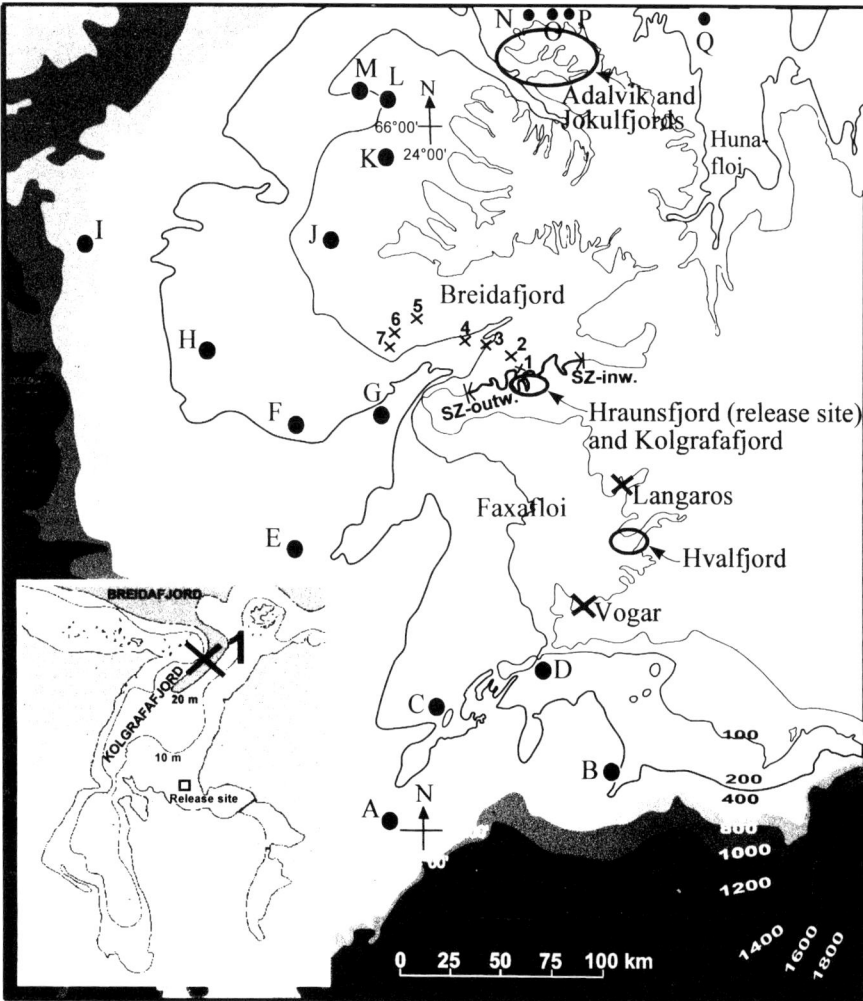

Fig. 14.1 Map of the study area. Sampling sites are symbolised (✘, ◯, |↔|). Post-smolt sampling was carried out in Hraunsfjord and Kolgrafafjord and on sites indicated with ✘ together with site name (Vogar & Langaros) or together with number (drift net stations). Additionally post-smolts were fished along shores zones in Breidafjord, both inshore (⟶|, SZ-inw.) and offshore (|⟵, SZ-outw.). Sites where adult salmon were sampled especially during spawning migration are indicated with open circles along with the name of the site. Capture sites of 1SW and 2SW salmon individuals derived from traditional fishing on sea fishes are listed as closed circles (●) along with letters (A–Q). The enlarged part shows Hraunsfjord and Kolgrafafjord and the release site of post-smolts is indicated as well as the site where homing adults are trapped in the head of the Hraunsfjord estuary. The depths are shown by depth isolines. Coastal waters (0–200 m bottom depth) have white background colour.

Table 14.1 Salmon post-smolt groups within fishing areas/sites and time, in relation to number captured. Forage status of the post-smolts is shown according to number of feeding fish (stomach content ≥ 0.01 g).

Fishing area (name and code)	Time of fishing (day/month year)	Number captured Total (no.)	Forage status of post-smolts Total with food (> 0.01 g) (no.)	(% of tot.)
Vogar	14–20/6 1989	271	1	0.4
Vogar	16–20/6 1990	110	1	0.9
Langaros (wild)	20/5–11/6 1987	15	14	93.3
Langaros (released)	20/5–11/6 1987	56	16	28.6
Hraunsfjord	3–4/7 1989	39	21	53.8
Kolgrafafjord	22–29/6 1990	65	46	70.8
Breidafjord[1] (SZ-outw.)	10/8 1993	61[1]	2	3.3
Breidafjord[1] (1)	10/8 1993	110[1]	3	2.7
Breidafjord (2)	10/8 1993	14	1	7.1
Breidafjord (3)	10/8 1993	20	14	70.0
Breidafjord (1)	22/6–28/8 1993	332	21	6.3
Breidafjord (SZ-outw.)	22/6–28/8 1993	200	23	11.5
Breidafjord (SZ-inw.)	22/6–28/8 1993	101	32	31.7
Breidafjord (4)	9/6 1994	430	120	28.0
Breidafjord[2] (5–7)	15/6; 8 & 18/7 1994	3	3	100.0

[1] This number of captured post-smolts serves as comparison with the capture on stations 2 & 3 the same day, but are also included in the total period given for this station in 1993.
[2] One of the post-smolts captured was of wild origin (captured 8 July at station number 5).

Post-smolts

Feeding migration and forage status

Salmon post-smolts of all sizes captured (10–34 cm) were feeding, although the proportion of post-smolts with empty stomachs was high (Table 14.1). Information on the digestion rate of salmonids shortly after release (Brodeur & Pearcy 1987, Johnsen & Ugeldal 1988), together with the short interval between visits to nets, indicate that the food amount observed is representative for the forage status of salmon post-smolts in the sampling area.

Simultaneous studies on wild smolts and released hatchery smolts in the productive Langaros estuary show well how effective the feeding of wild smolts can be in estuaries (Table 14.1) (Johannsson *et al.* 1991). On the other hand the forage status of released smolts was low, partly owing to the very short period from the time of release. Thus released smolts remaining in the area for some days did show low foraging compared with the wild smolts, which is likely to be due to lack of experience regarding feeding on living prey (Fjallstein 1987) and/or linked to their physiological status.

Table 14.2 Salmon (1SW & 2SW) groups within fishing areas and time, in relation to number captured. Forage status of the salmon is shown according to number of feeding fish (stomach content ≥ 0.1 g), shown as mean food weight per salmon and per 1 kg of salmon. The forage ratio (food weight/fish weight) is also shown.

Fishing area	Fishing time	Number of fish			Food weight (feeding fish)		Food weight/ Fish weight	
		Total (no.)	Total with food (no.)	% of total	Mean g/salmon	g/kg salmon	Mean (%)	Median (%)
Vogar	June 1989–90	14	2	14	20.3	9.9	1.04	
Hvalfjord	July 1993	9	5	56	9.8	9.8	0.98	1.13
Breidafjord	June–July 1993	26	7	27	41.6	1.4	1.11	0.12
Hraunsfjord	July 1993	49	5	10	Not measured			
Hvalfjord	June 1994	106	25	24	5.0	0.3	0.14	0.07
Hvalfjord	July 1994	118	40	34	7.8	0.8	0.21	0.07
Hraunsfjord	July 1994	100	16	16	1.8	0.1	0.07	0.05
Hraunsfjord	Sept. 1994	21	0	0				
Adalvik	June–July 1994	16	7	44	6.7	0.8	0.23	0.22
Jokulfjords	July 1995	38	30	79	6.9	1.9	0.22	0.14

Table 14.3 Salmon groups (i.e. total, 1SW, 2SW) within fishing areas and time, in relation to mean weight of salmon (S.D. in parentheses), total number of salmon and number of salmon containing food. The food weight of feeding salmon is the mean for each group (S.D. in parentheses).

Fishing area & time	Salmon groups			Food weight (g)
	Total & Sea age (1–2)	Mean weight kg (SD)	Number eating (no.) %	Mean (SD)
Hvalfjord June 1994	Total	3.7 (1.4)	25 24	5.0 (8.2)
	1SW	3.0 (0.6)	22 25	3.6 (5.0)
	2SW	6.0 (0.9)	3 13	15.6 (18.9)
Hvalfjord July 1994	Total	3.4 (1.1)	40 34	7.8 (14.0)
	1SW	3.0 (0.5)	35 29	4.2 (8.2)
	2SW	5.7 (1.2)	5 25	29.8 (25.5)
Hraunsfjord July 1994	Total	3.1 (1.1)	16 16	1.8 (1.5)
	1SW	2.9 (0.7)	16 16	1.8 (1.5)
	2SW	7.0 (1.3)		
Adalvik June–July 1994	Total	3.8 (1.8)	7 44	6.7 (2.9)
	1SW	3.1 (1.1)	6 50	7.2 (2.9)
	2SW	5.9 (2.0)	1 25	4.0
Jokulfjords July 1995	Total	2.9 (0.7)	30 79	6.9 (10.1)
	1SW	2.9 (0.7)	29 78	7.0 (10.2)
	2SW	3.7	1 100	2.6

The foraging of ranched post-smolts as observed through the proportion of feeding fish and their forage ratio (wet weight of food/wet weight of fish) reflected the strong urge of smolts to start seawards migration as soon as possible. This was shown repeatedly by immediate outward migration of the majority of smolts after their release. Forage status of post-smolts was in general low just hours after release owing to the swift outward migration of the post-smolts towards oceanic waters. This was most noticeable where the post-smolts were captured just after release, as was the case for most of the released smolts in Langaros and Vogar (Table 14.1). This was also observed among typical outwards migrating post-smolts sampled at site 1 where Kolgrafafjord opens into Breidafjord (approximately 4 km out from the release site (Fig. 14.1).

The post-smolts entered the mouth of Kolgrafafjord area as a large school within a few hours after release (down to roughly 2 hours) and had during that time already started to eat natural prey, although in small quantities (Table 14.1). This behaviour did not vary to any significant degree with changed circumstances such as different release dates, different prey density or changed light conditions. However, the forage status decreased through the summer, as expected, indicating a reduced availability of prey of suitable size later in the season but without altering the outward migration pattern. This might reflect that feeding is suppressed by the speedy outward migration.

The Food and Feeding in Icelandic Coastal Waters 199

Table 14.4 Number of captured Atlantic salmon in relation to fishing areas. Within fishing areas are given the proportion (%) of salmon of wild and ranched origin. For salmon of wild origin the smolt age intervals observed are shown. For salmon of ranched origin the sites where they were released as smolts are listed. For each release site is given the shortest sea route to capture site and the number of 1SW and 2SW salmon involved in the microtag data.

Fishing area & time	Total Number	Wild origin			Ranched origin				
		Number %	Smolt age Year	Number %	Release site	Information based on micro tags		Number	
						Distance (km)		1SW	2SW
Hvalfjord June 1994	106	18	3–5	82	Vogar	42		1	
					Kollafjord	19		1	
Hvalfjord July 1994	118	11	3–5	89	Vogar	42		8	4
					Kollafjord	19		8	1
					Laros	166		1	1
					Hraunsfjord	190		17	1
					River Laxá (NE Iceland)	582			1
Adalvik June–July 1994	16	31	3–4	69	River Gljúfurá (NW Iceland)	178			1
Jokulfjords July 1995	38	8	3–4	92	Hraunsfjord	222		1	

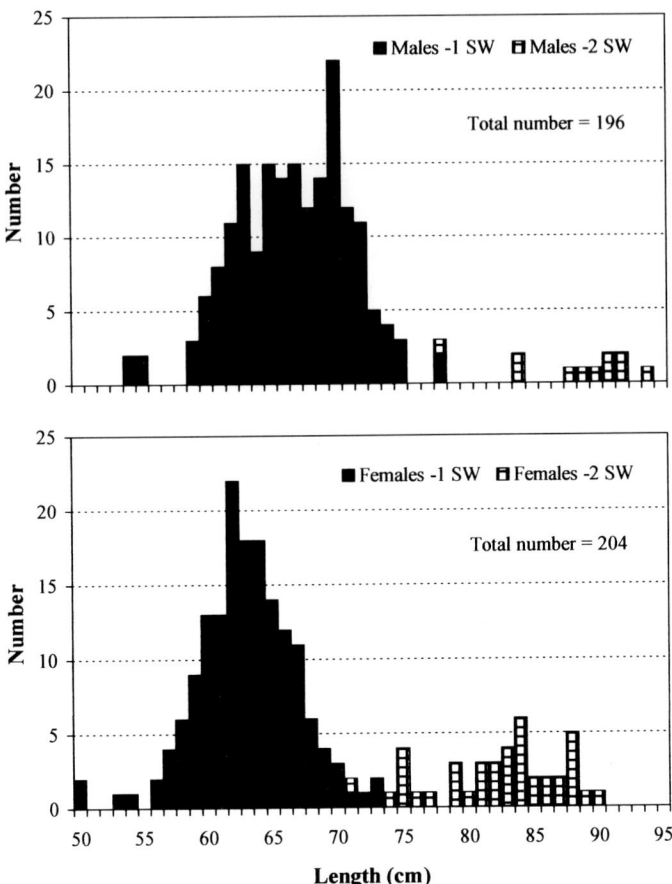

Fig. 14.2 Length distribution of 1SW and 2SW salmon sampled in 1994 and 1995. The number of salmon of each length is shown separately for males and females and for 1SW and 2SW salmon. The total number of sampled fish is shown for both sexes.

The feeding 'en route' that was observed during this fast outward migration would not be expected to be as effective as if the fish stopped and concentrated on feeding when experiencing favourable conditions. Despite this, the observed post-smolt feeding efficiency is enabling them to increase their foraging status considerably within the first days of their seawards migration (Table 14.1). The series of sampling from the shoal of outward migrating 1+ post-smolts during 10 August 1993 shown in Table 14.1 shows this well. Only 3% of the fish were feeding at station 1 and at the corresponding shore station (SZ-outward) but the proportion of feeding fish was found to have increased roughly twenty fold on the way to station 3 located 19 km (and 15 hours later) out in Breidafjord (Fig. 14.1, Table 14.1). The forage ratio expressed this well as food weight rose from being 0.09% of their body weight at the mouth of Kolgrafafjord to 0.30% at station 3. This increase of food content along the seawards migration axis was also observed a year later when a release of 2+ post-smolts was tracked with net fishing 40 km out from the release site (Table 14.1).

The most distant captures in the mouth of Breidafjord Bay (stations 5–7) included only three fish, one of them being wild (Table 14.1). This capture of wild post-smolts and released post-smolts after 70 km migration (shortest sea route) from Hraunsfjord, indicated that they were still being sampled on their migration route. These outermost captures were very few but the samples showed that the fish were feeding and had a higher forage ratio than in inner areas (mean = 0.87%, SD = 0.10). These findings suggest an increase in foraging of post-smolts as they move seawards. For comparison, information from feeding of salmon in West Greenland coastal waters has shown that during August–November (1968–70) 82–97% of the salmon were feeding and the mean food weight was in the range of 4.8–11.4 g/kg of salmon. Corresponding values for post-smolts in Icelandic coastal waters was 0.5–1.7 g/kg of salmon for near shore area (Langaros excluded) and up to 10.7 g/kg in offshore areas.

It should be mentioned that releases of smolts/post-smolts directly into estuaries or the sea always include some individuals that are not going to migrate oceanward that year. In some releases in the mouth of Hraunsfjord this fraction was dominated by maturing males that were going to spawn the following autumn. These fish, which in general were the largest ones, foraged along the shore. The majority of them migrated back into the shallow Hraunsfjord and were feeding there along with non-maturing post-smolts until starting their spawning migration. Additionally, part of this run of maturing males was going into Breidafjord and a further part out of the bay as shown by recaptures of microtagged individuals in rivers in Faxafloi the following autumn.

Prey types and their importance

In Langaros estuary simultaneous studies on post-smolts of wild and hatchery origin showed that their main food was amphipods and adult terrestrial dipterans that are drifting out through the estuary by wind and fresh water. In that area sand eels and polychaetes were also eaten. This diet is similar to that of anadromous arctic charr (*Salvelinus alpinus*) in the same area (Sturlaugsson *et al.* 1992, 1994).

In Vogar the post-smolts migrated quickly out through the narrow fishing zone, and only two fish were captured containing food, primarily amphipods. In Hraunsfjord the post-smolts were eating a considerable number of winged dipterans and gammarid amphipods, as in Langaros, but small copepods (cyclopoids) were also a dominant food item (Fig. 14.1). In addition to three-spined sticklebacks (*Gasterosteus aculeatus*), polychaetes (*Nereis* spp.) and crab larvae were of some importance volumetrically.

The pelagic megalopa larvae of *Hyas* spp. and *Pagurus bernhardus* were in later studies shown to be the main prey species of seawards migrating post-smolts both while migrating through Kolgrafafjord and in offshore areas (Fig. 14.3). In Kolgrafafjord the salmon were, in addition to crab larvae, mostly eating gammarids,

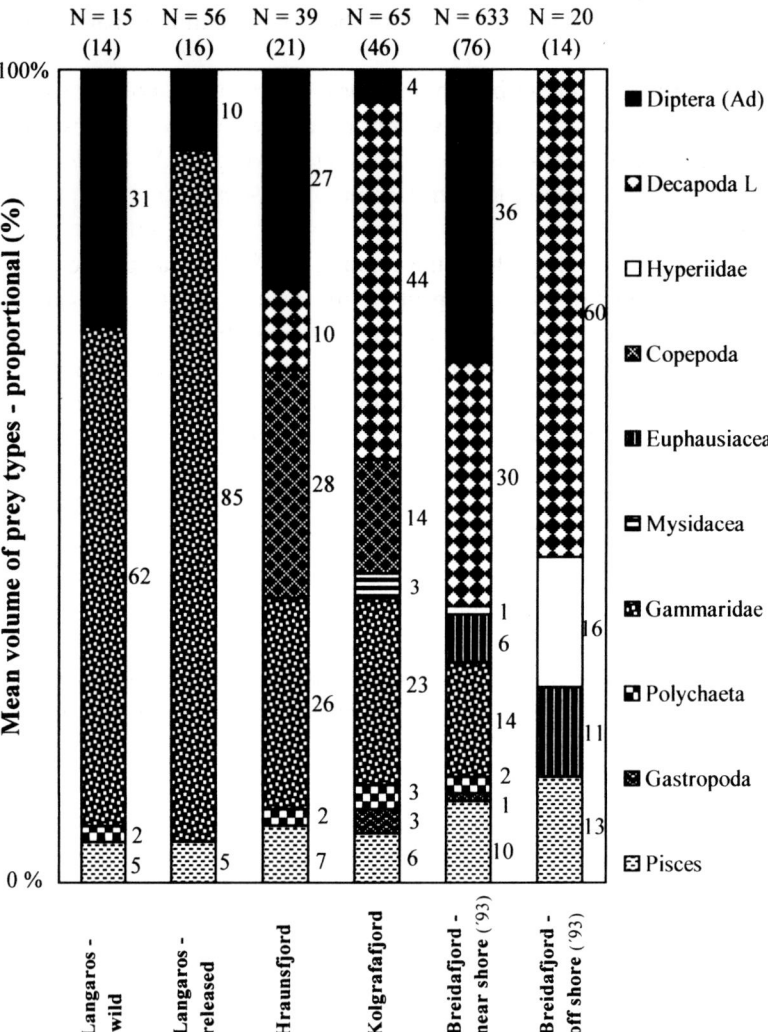

Fig. 14.3 The mean volume of prey types (%) in feeding salmon post-smolts. The mean volume composition of prey types is shown for each of the study sites involved. For Langaros the proportions of prey types eaten are shown separately for post-smolts of wild origin and of hatchery origin released. For Breidafjord the food composition is shown pooled for the shore zones and station 1 (nearshore) and separately for station 3 (offshore). The total number of anlaysed post-smolts is shown and in parentheses the number of fish that contained food. Values for the mean volume of each prey type are given.

copepods, sandeel larvae, winged flies, polychaetes, gastropods and mysids (Fig. 14.3). These observations also underlined the ability of post-smolts to eat epibenthic prey types in the beginning of their ocean-ward migration.

In the near shore area of Breidafjord (shore zones and station 1) the dipterans were still of considerable importance as food of the post-smolts along with crab larvae, amphipods and sandeels. (Figs 14.1, 14.3). While feeding in this area the pelagic euphausiids and hyperiids were added to the diet, an indication of coming changes in the diet.

In offshore areas, as shown by food composition of post-smolts at station 3, the importance of crab larvae still increased but now other planktonic crustaceans were eaten, namely hyperiid amphipods and euphausiids (Fig. 14.3). Sandeels were also one of the prey types taken in greater amounts. The same prey types were eaten in outer areas (sites 4–7).

Sampling post-smolts throughout the summer of 1993 revealed food composition in relation to the length of time spent in the nearshore area. In June decapod crustaceans dominated, but in July and August mainly flies.

Comparison of proportions of prey species in the environment with their proportions in the post-smolt stomachs using Ivlev's selectivity index indicated that postsmolts had active prey selection. It also showed post-smolts' predation on prey types not found in the zooplankton samples, which suggests that these results must be interpreted cautiously. The main reason, however, seemed to be related to the feeding of post-smolts near the surface and near-surface layers of the sea, where the zooplankton sampling was inadequate.

The diversity of prey types eaten was high in the nearshore areas (Sturlaugsson 1994b) and even the main prey types listed here were rather dispersed within the fjord and bay system. This prey diversity shows well the great flexibility of salmon as shown by their opportunistic feeding strategies. The strategy of a typical post-smolt migrant heading ocean-ward seems to be feeding 'en route' while travelling at an average speed of 1.6–1.7 km/hour through the coastal waters. At the beginning of the sea migration post-smolts were mainly eating benthic prey, as observed in Langaros estuary, and benthic species were also very important in the comparable Hraunsfjord area. Along the feeding migration axis the proportion of pelagic species increased ocean-ward and in the offshore area all prey types were pelagic and of marine origin. This observed proportion of benthic and pelagic prey, together with terrestrial flies, is similar to that observed among post-smolts in the Norwegian fjords and the Gulf of St Lawrence (Hvidsten *et al.* 1993, Dutil & Coutu 1988, Levings *et al.* 1994).

The typical forage behaviour observed in Icelandic waters among the salmon postsmolts migrating ocean-ward, indicates that the first feeding areas are of great importance on their migration route to the ocean.

Adult salmon

Spawning migration and forage status

When considering the spawning migration notice should be taken that salmon migrating from oceanic areas enter coastal waters over a large area. This is shown by individual captures of salmon that have been plotted in Fig. 14.1 (A–Q). These positions are from instances where salmon have been captured during traditional fishing as a by-catch and the fisherman has sent in information and/or samples. If sites H and I, which represent winter captures, are excluded all captures are within the summer months and therefore give an approximate view of the wide distribution of salmon, when homing towards the coast.

The majority of the 1SW and 2SW salmon had empty stomachs in all areas except in Hvalfjord in July 1993 and in the Jokulfjords (Table 4.3). The lowest forage status of salmon observed was in Hraunsfjord (no feeding) and the extreme was at the end of the salmon run in September when no feeding occurred despite very high densities of sticklebacks and considerable densities of one- and two-year-old pollack (*Pollachius virens*). Similarly, a small number were feeding at Vogar probably because of the phenomenon of feeding terminating when entering the home estuary, as was observed in Hraunsfjord and has been observed within the same migration phase in Scotland (Fraser 1987). Although the by-catch samples from Vogar and Breidafjord and Hvalfjord in 1993 were few in number they show well how effective these fish can be when they the enter areas of high abundance of food, such as sandeels and euphausiids, during their home migration.

Some feeding was also shown to be activated when salmon returning to Hraunsfjord encountered and preyed upon outward-migrating post-smolts from Hraunsfjord. This is probably the first documentation of cannibalism of salmon in the sea. In addition to a study in Scotland, the Icelandic results also have a resemblance to results from studies of Lear (1972) in Canadian waters based on sampling stations located in coastal waters. This is shown by the low proportion of feeding fish (73–98% empty) and low forage ratio with reference to low mean food weight (0.2–1.3 g/kg salmon), with exceptionally high forage status when capelin were abundant in the area.

Comparison of forage status between salmon in the southern Hvalfjord and northern Adalvik and Hraunsfjord in 1994 and 1995 showed that forage status was higher in the northern area (Table 14.2). This might be due to a number of reasons but one could speculate that the northern area is further away from the main salmon river systems and at the same time confronting the main current axis north-west of Iceland. Therefore it could be postulated that salmon migrating there are more likely to be at an earlier stage of the spawning migration than their 1SW and 2SW counterparts in Hvalfjord. Information was obtained on the origin of microtagged salmon and compared with known migration speed of homing salmon in Icelandic waters (Sturlaugsson 1995, Sturlaugsson & Thorisson 1995a, 1997). From that, one could deduce that most of the salmon captured in Hvalfjord and Adalvik and Jokulfjords had one or only a few days left of their spawning migration to their home estuary, with possible examples of continued migration in coastal waters for 3–4 weeks.

Comparison of foraging between 1SW and 2SW shows little difference in food consumption (Table 14.3), but the proportion of feeding 2SW fish is slightly less than 1SW but at the same time those large ones generally had a slightly higher forage ratio than 1SW counterparts (Sturlaugsson *et al.* 1997a,b).

As expected, the spawning migration of salmon in Icelandic coastal waters is relatively short compared with the total sea migration, ranging from a few days up to some weeks in duration. During these migrations in coastal waters off West Iceland the salmon spend little time feeding, resulting in low forage status. Thus no direct effect of feeding on survival is to be expected during this migration phase.

Prey and its importance

Based on the main sampling in 1994 and 1995, salmon in Jokulfjords had the most variable diet. In Hvalfjord the food was rather uniform, only sandeels and herring being eaten (Fig. 14.4). The main prey group of salmon was fish in all areas, but invertebrates were 49% of the food bulk in Jokulfjords according to mean food weight. The vast majority of the fish eaten were sandeels, but herring and fish of the family Lumpenidae were also eaten in considerable numbers (Fig. 14.4). The main invertebrates were decapods (megalopa larvae), euphausiids and amhipods. In addition to this the salmon were eating small amounts of polychaetes (*Nereis* spp.) and adult dipterans (Table 14.5).

Samples from 1989 to 1993 showed predation on sandeels in Vogar, but the lamellibranch *Mytulis edulis* was also found in small numbers. Detection of feeding of salmon in Hvalfjord in 1993 showed instances of highly effective predation on sandeels and euphausiids. Sampling in Hraunsfjord in 1993 showed that the main prey type was sandeels but one post-smolt (fork length 21.5 cm) was also found in one 2SW fish. In Breidafjord in 1993 the salmon captured were all of ranched origin and they were found to be considerably active in cannibalism. Out of 26 salmon captured at the inward shore zone (Fig. 14.1) four (15%) had eaten one or two post-smolts (fork length interval 12–21 cm). This was observed both among 1SW salmon (three) and 2SW salmon (one) 5–10 km out from the release site. The salmon 'cannibals' were the only fish that had a high forage ratio. Other main prey types eaten by salmon inshore in Breidafjord were sandeels, polychaetes, shrimps and gammarid amphipods. Megalopa larvae of crabs were also eaten occasionally.

Prey size and predation success

Laboratory studies have been carried out to establish information on how well prey of given type and stage suit feeding salmon in relation to their size (Wankowski & Thorpe 1979). The results showed that the maximum diameter of the prey is a good indicator of how successfully it can be eaten by fish of a given size, because that factor reflects both the fish's ability to swallow the prey, and to filter it through its gill rakers. This ratio of fish length to prey maximum diameter (Prey Forage Ratio – PFR) was calculated for prey types eaten by post-smolts in Hraunsfjord and Kolgrafafjord and confirmed the laboratory findings mentioned (Sturlaugsson 1994b). In addition it showed that post-smolts could also be very effective in foraging on prey that were as small as quarter of what the laboratory experiments calculated as maximum PFR (2–2.5%). The PFR observed among post-smolts in Hraunsfjord, Kolgrafafjord and Breidafjord bay ranged from 0.2–4.7%. The PFR for adult salmon was 0.2–5.7%, based on measurements of prey from captured adult salmon in 1994

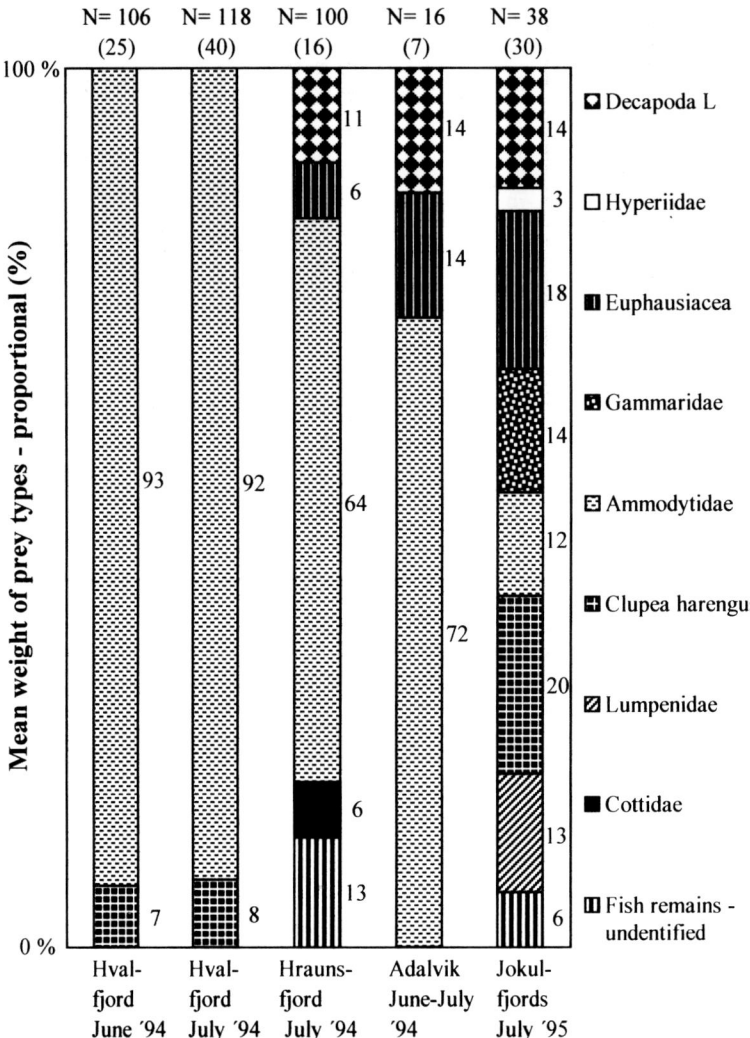

Fig. 14.4 The mean weight of prey types (%) in feeding adult salmon (1SW & 2SW) sampled in 1994 and 1995. The mean weight composition of prey types is shown for each of the study sites involved; sampling in June and July in Hvalfjord is shown separately. The total numbers of analysed post-smolts are shown and in parentheses the number of fish that contained food. Values for the mean weight of each prey type are given.

and 1995. When looking at the main prey species, their average value is close to the estimated optimum PFR. The decapod larvae eaten by post-smolts gave PFR in the interval 1.7–2.4% and the main prey type of adult salmon, sandeel and herring showed to have their PFR mean value fitted well to estimated optimum (Table 14.4).

In the context of the role of PFR then it must be noted that the prey density plays an important role because decreased density increases the influence of prey size on the foraging. When salmon at the post-smolt stage are observed containing hundreds of

Table 14.5 Size and number of prey types eaten by adult Atlantic salmon (1SW & 2SW) in relation to fish length based on 1SW and 2SW salmon sampled in 1994 and 1995. Sizes of prey types are given as total length and diameter (maximum width). Number of prey are given in relation to prey types, i.e. maximum (Max) and as mean (S.D. in parentheses). Number and length intervals (intv.) of feeding salmon are given for each prey type and for the salmon that were involved in the measurements on prey width (in parentheses). The proportion (%) between prey width and length of predating salmon are given, both as interval and as mean (S.D. in parentheses).

	Prey types				Fish that had eaten specified prey type		Prey width/Fish length (%)	
Type	Length intv. (cm)	Width intv. (cm)	Prey/stomach Max	Prey/stomach Mean (SD)	Number	Length intv. (cm)	Interval	Mean (SD)
Ammodytidae	5.2–18.6	0.6–1.9	10	1 (1)	81	54–84 (54–84)	0.9–3.0	1.8 (0.7)
Clupea harengus	6.1–23.5	0.6–3.7	13	2 (1)	13	63–84 (63–83)	0.8–4.4	2.4 (0.7)
Lumpenidae	4.9–5.4	0.3–0.4	52	18 (18)	6	59–75 (59–75)	0.4–0.6	0.5 (0.1)
Cottidae	13.0	3.5	1		1	61	5.7	
Diptera	0.8	0.2	1		1	63	0.3	
Decapoda larvae	0.4–0.6	0.2–0.3	782	97 (58)	13	59–74 (61–64)	0.3–0.5	0.3 (0)
Hyperiidae	0.7–3.4	0.2–0.7	22	8 (12)	3	62–67 (62–67)	0.5–0.8	0.6 (0.1)
Euphausiacea	2.3–6.0	0.3–0.6	104	21 (10)	13	59–69 (59–69)	0.5–0.9	0.6 (0.1)
Gammaridae	1.2–4.2	0.2–0.7	311	98 (135)	6	62–71 (63–71)	0.2–1.1	0.6 (0.3)
Polychaeta	11.2	0.7	1		2	63–69 (69)	1.0	

small cyclopoid copepods such as in Hraunsfjord, or when grilse are shown to have consumed hundreds of crab larvae, as in Adalvik (Table 14.5), then it is clear that these small prey species have been in swarming densities. Such stomach contents also indicate that filter feeding can partly be involved.

Although exceptional, one direct observation was made regarding grilse in the head of Hraunsfjord, which strongly indicated that the fish were filter feeding. On this occasion the fish larvae were very abundant within a thin vertical layer and the salmon seemed to be filtering them by swimming along the layer at a high speed with their jaws open.

During the stomach analysis in Vogar, Hvalfjord and Breidafjord the position of fish prey in stomachs was recorded. Such information was to find whether, in the majority of instances, the salmon ate their prey by swallowing them head first, seizing them from behind or crosswise. The results showed that the vast majority of sandeels and herring, as well as salmon post-smolts, were swallowed tail first.

Concluding remarks

The marine feeding migration of salmon is a survival combat, in which the feeding conditions experienced are known to play a significant role, and the opportunistic feeding of salmon makes the best out of these circumstances. During the feeding migration, especially in the beginning, the feeding success of salmon is directly linked to their body growth. Therefore the successful feeders grow more rapidly 'out of trouble' than their counterparts, because at a critical period their larger body size will enhance their chance to survive, e.g. to cope with predators, diseases, parasites and enable them to feed in colder water. The salmon that have for genetic or circumstantial reasons more successful feeding strategies and tactics have a head start in the race for survival. Here it must be remembered that post-smolts with diseases or parasitic infection above a certain level will, despite perfect feeding conditions, not survive.

The ecology of Atlantic salmon post-smolts is of special interest during the first summer owing to indications of high mortalities during early sea life, caused by localised predation on post-smolts (Larsson 1984, Hvidsten & Møkkelgjerd 1987, Hvidsten & Lund 1988, Montevecchi et al. 1988) and/or adverse sea conditions in the beginning of the feeding migration (Scarnecchia 1984).

It is concluded that the observed low forage efficiency among salmon post-smolts during their migration in Icelandic coastal waters is not likely to have any great effect on their survival owing to the short time spent in coastal waters. It is, however, believed that parasite infection during the coastal part of the feeding migration can have considerable effects on the survival rate of the post-smolts. This postulation is based on experiments on known lethal levels of sea lice infection (Grimnes & Jacobsen 1996) and on observed infected post-smolts in Breidafjord with examples of extreme infection of large post-smolts covered with 180 pre-adult stages of the copepod ectoparasite *Lepeoptheirus salmonis*.

The research on salmon during the feeding migration has focused not only on the quantity and quality of prey eaten by salmon and their corresponding fitness, but also on the relationship between the distribution of salmon in the sea and the environmental feeding conditions. With increasing knowledge of such a relationship there will be a correspondingly better understanding of the reasons behind the large fluctuations of stock size of Atlantic salmon.

References

Brodeur, R.D. & Pearcy, W.G. (1987) Diel feeding chronology, gastric evacuation and estimated daily ration of juvenile coho salmon, *Oncorhynchus kisutch* (Walbaum), in the coastal marine environment. *Journal of Fish Biology*, **31**, 465–77.

Dutil, J.D. & Coutu, J.M. (1988) Early marine life of Atlantic salmon, *Salmo salar*, Post-smolts in the Northern Gulf of St Lawrence. *US National Marine Service Fisheries Bulletin*, **86**(2), 197–212.

Fjallstein, I.S. (1987) *Naturlig føde hos oppdrettet laksesmolt* (Salmo salar L.) *i indre kystområder på Færøene*. Cand. Scient. thesis, University of Bergen.

Fraser, P.J. (1987) Atlantic salmon, *Salmo salar* L., feed in Scottish coastal waters. *Aquaculture and Fisheries Management*, **18**, 243–7.

Grimnes, A. & Jacobsen, P.J. (1995) Salmon lice (*Lepeoptheirus salmonis* Krøyer) infestation in Atlantic salmon (*Salmo salar* L.) postsmolt: physiological consequences and mortal impact. *Nordisk Jordbruksforskning*, **77**, 272.

Gudjonsson, T. (1988) Exploitation of salmon in Iceland. In: *Atlantic Salmon: Planning for the Future* (eds D. Mills & D. Piggins), pp. 162–78. Croom Helm, London.

Hansen, L.P. & Pethon, P. (1985) The food of the Atlantic salmon, *Salmo salar* L., caught by long line in northern Norwegian waters. *Journal of Fish Biology*, **26**, 553–62.

Hislop, J.R.G. & Youngson, A. (1984) A note on the stomach contents of salmon caught by long line north of the Faroe Islands in March, 1983. International Council for the Exploration of the Sea CM 1984/M:17.

Hislop, J.R.G. & Webb, J.H. (1992) Escaped farmed salmon, *Salmo salar* L. feeding in Scottish coastal waters. *Aquaculture and Fisheries Management*, **23**, 721–3.

Hvidsten, N.A. & Møkkelgjerd, P.I. (1987) Predation on salmon smolts, in the estuary of the river Surna, Norway. *Journal of Fish Biology*, **30**, 273–80.

Hvidsten, N.A. & Lund, R.A. (1988) Predation on hatchery-reared and wild smolts of Atlantic salmon, *Salmo salar* L., in the estuary of river Orkla, Norway. *Journal of Fish Biology*, **33**, 121–6.

Hvidsten, N.A., Sturlaugsson, J., Strand, R. & Johnsen, B.Ø. (1993) Næringsvalg hos fjordutsatt havbeitesmolt av laks pa Island og i Norge. *Norwegian Institute for Nature Research Report*, **187**, 1–16.

Isaksson, A. (1980) Salmon Ranching in Iceland. In: *Salmon Ranching* (ed. J. Thorpe), pp. 131–56. Academic Press, London.

Isaksson, A. (1994) Ocean ranching strategies, with a special focus on private ranching in Iceland. *Nordic Journal of Freshwater Research*, **69**, 17–31.

Isaksson, A, Oskarsson, S., Einarsson, S. & Jonasson, J. (1997) Atlantic salmon ranching: past problems and future management. *ICES Journal of Marine Science*, **54**, 1188–99.

Jacobsen, J.A. & Hansen, L.P. (1996) The food of Atlantic salmon, *Salmo salar* L., north of the Faroe Islands. International Council for the Exploration of the Sea CM 1996/M:10.

Johannsson, V., Sturlaugsson, J. & Einarsson, S.M. (1991) Faeda laxins i sjo (The food of the Atlantic salmon in the sea). *Veidimadurinn*, **136**, 100–106.

Johnsen, B.O. & Ugeldal, O. (1988) Naeringsopptak hos tosomrig settefisk utsatt i innsjø, vär, sommer og høst. *Direktoratet for Naturforvaltning, Rapport*, **3**.

Jutila, E. & Toivonen (1985) Food composition of salmon post-smolts (*Salmo salar*) in the northern part of the gulf of Bothnia. International Council for the Exploration of the Sea CM 1985/M21.

Larsson, P.O. (1984) Predation on migrating smolt as a regulating factor in Baltic salmon, *Salmo salar* L., populations. *Journal of Fish Biology*, **26**, 391–397.

Lear, W.H. (1972) Food and Feeding of Atlantic Salmon in Coastal Areas and over Oceanic Depths. *ICNAF Research Bulletin*, **9**, 27–39.

Lear, W.H. (1980) Food of Atlantic Salmon in the West Greenland-Labrador Sea Area. *Rapports et Procès Verbaux des Réunions, Conseil International pour l'Exploration de la Mer*, **176**, 55–9.

Levings, C.D., Hvidsten, N.A. & Johnsen, B.Ø. (1994) Feeding of Atlantic salmon (*Salmo salar* L.) in a fjord in central Norway. *Canadian Journal of Zoology*, **72**, 834–9.

Lindroth, A. (1961) Sea food of Baltic smolts. International Council for the Exploration of the Sea C.M. 196/No. 8.

Marcotte, B.M. & Browman, H.I. (1986) Foraging behaviour in fishes: perspectives on variance. In: *Contemporary Studies on Fish Feeding* (eds C.A. Simenstad & G.M. Calliet), pp. 25–34. Dr W. Junk Publishers, Dordrecht.

Montevecchi, W.A., Cairns, D.K. & Birt, V.L. (1988) Migration of postsmolt Atlantic salmon, *Salmo salar*, off northeastern Newfoundland, as inferred from tag recoveries in a seabird colony. *Canadian Journal of Fisheries and Aquatic Sciences*, **45**, 568–71.

Morgan, R.I.G., Greenstreet, S.P.R. & Thorpe, J.E. (1986) First observations on distribution, food and fish predators of post-smolt Atlantic salmon, *Salmo salar*, in the outer firth of Clyde. International Council for the Exploration of the Sea CM 1986:27.

Reddin, D.G. (1985) Atlantic salmon (*Salmo salar* L.) on and east of the Grand Bank. *Journal of Northwest Atlantic Fisheries Science*, **6**, 57–164.

Scarnechia, D.L. (1984) Climatic and oceanic variations affecting yields of Icelandic stocks of Atlantic salmon (*Salmo salar*). *Canadian Journal of Fisheries and Aquatic Science*, **41**, 917–35.

Sturlaugsson, J. (1994a) Vistfræða í Breiðafirði (Ecology of salmon post-smolts in Breidafjord Bay). *Ugginn*, **15**, 12–14.

Sturlaugsson, J. (1994b) Food of ranched Atlantic salmon (*Salmo salar* L.) post-smolts in coastal waters, West Iceland. *Nordic Journal of Freshwater Research*, **69**, 43–57.

Sturlaugsson, J. (1995) Migration study on homing of Atlantic salmon (*Salmo salar* L.) in coastal waters West Iceland: Depth movements and sea temperatures recorded at migration routes by data storage tags. International Council for the Exploration of the Sea CM 1995/M:17.

Sturlaugsson, J. & Thorisson, K. (1995a) Post-smolts of ranched Atlantic salmon (*Salmo salar* L.) in Iceland: II. The first days of the sea migration. International Council for the Exploration of the Sea CM 1995/M:15.

Sturlaugsson, J. & Thorisson, K. (1995b) Post-smolts of ranched Atlantic salmon (*Salmo salar* L.) in Iceland: III. The first food of sea origin. International Council for the Exploration of the Sea CM 1995/M:16.

Sturlaugsson, J. & Thorisson, K. (1997) Migratory pattern of homing Atlantic salmon (*Salmo salar* L.) in coastal waters West Iceland, recorded by data storage tags. International Council for the Exploration of the Sea CM 1997/CC:09.

Sturlaugsson, J., Johannsson, V. & Einarsson, S.M. (1992) Faeda sjóbleikju (*Salvelinus alpinus* L.) í Langárósi (Food of anadromous arctic char (*Salvelinus alpinus* L.) in Langaros estuary). *Institute of Freshwater Fisheries, Reykjavik, Research Report*, **VMST-R/92021**.

Sturlaugsson, J., Johannsson, V. & Einarsson, S.M. (1994) Forage of anadromous Arctic chars (*Salvelinus alpinus* L.) in estuary area, West Iceland. *Nordic Journal of Freshwater Research*, **69**, 96.

Sturlaugsson, J., Thorisson, K. & Karlsson, H. (1997a) Faeda laxa í hrygningargöngu um strandsaevi (Food of salmon during spawning migration in coastal waters. Fjölstofnarannsóknir (Multi-species research) 1992–1995. *Marine Research Institute, Reykjavik, Report*, **57**, 57–68.

Sturlaugsson, J., Thorisson, K. & Karlsson, H. (1997b) Fæða og far laxa á hrygningargöngu um strandsævi (Food and migration of salmon during spawning migration in coastal waters). *Institute of Freshwater Fisheries, Reykjavik, Research Report, VMST-R/97022*.

Thorisson, K. & Sturlaugsson, J. (1995a) Post-smolts of ranched Atlantic salmon (*Salmo salar* L.) in Iceland: I. Environmental condition. International Council for the Exploration of the Sea CM 1995/M:10.

Thorisson, K. & Sturlaugsson, J. (1995b) Post-smolts of ranched Atlantic salmon (*Salmo salar* L.) in Iceland: IV. Competitors and predators. International Council for the Exploration of the Sea CM 1995/M:12.

Gudjonsson, T. (1988) Laxagöngur í úthafinu (Salmon migration in the ocean). In: *Hafbeit (Salmon Ranching)* (ed. Valdimar Gunnarsson), pp. 44–61. Institute of Freshwater Fisheries Reykjavik.

Wankowski, J.W.J. & Thorpe, J.E. (1979) The role of food particle size in the growth of juvenile Atlantic salmon (*Salmo salar* L.) *Journal of Fish Biology*, **14**, 351–70.

Chapter 15
Problems Facing Salmon in the Sea – Summing Up

A.D. HAWKINS

The decline in salmon abundance

Throughout its geographical range the Atlantic salmon appears to be in a state of decline. There has been a steep fall in the recorded catches of Atlantic salmon in all countries around the North Atlantic, especially since 1990. The total nominal catch for the North Atlantic area in 1996 (3 128 t) was significantly lower than the 1995 catch (3 616 t), and indications are that the 1997 catch will be lower still.

A general 'health warning' must be attached to these catch trends. The catch data are mostly of poor quality. The dependence of catches on factors like weather and water conditions, the variable efficiency of catching methods, and changes in market conditions can all influence the effort put into fishing and affect catching efficiency. Changes in the behaviour of fish may result in variations in the numbers of fish entering the river at different times of the year. Some may enter outside the fishing season and may not be exposed to capture. The development of salmon ranching has augmented catches in some countries, while in others escapes of farmed fish have partially masked any decline in the abundance of wild fish. In some important fisheries, the catches are under reported, or simply not reported at all.

Trends in the catch records may also result from the closure of fisheries. Quotas for marine fisheries at Greenland and the Faroes have been reduced as a result of negotiations within the North Atlantic Salmon Conservation Organisation (NASCO), and in some years the limited quotas that have been set have been purchased by others and the fisheries closed. In many countries, including Scotland, Norway and Canada, there has been a significant decline in coastal netting, formerly the most important fisheries. Though these reductions may have been brought about in part as a response to the lower abundance of fish, a major contributory factor has also been a decline in the market price of fish, brought about by the great expansion in salmon farming.

Rod catches in fresh water have generally shown a less pronounced decline. Even here, however, the absence of any increase in rod catches following the demise of netting has been taken as evidence of a fall in salmon abundance.

Overall, despite our reservations about the quality and the interpretation of catch data, there has been a clear, widespread, downward trend in catches, which can only be interpreted as a fall in salmon abundance. The extent of the decline in the actual numbers of fish returning from the sea to spawn is difficult to measure, and may vary from country to country or from river to river. To take Scotland alone, the north-west coast has seen a major decline in the catches of both salmon and sea trout in the small rivers characteristic of this region. On the east coast, where the rivers are larger, there has been a strong decline in the catches and abundance of early running fish, but later running fish have been less badly affected, and may even be prospering.

One conclusion that we may draw from these different trends in different areas, or even in different parts of the same river, is that separate salmon populations behave in different ways and are influenced by different factors. It is now apparent from studies of the genetic make-up of salmon that they belong to more or less separate populations, with different inherited characteristics. Whatever the events that have led to the decline in salmon abundance, they have evidently affected different populations to a differing degree. Quite separate factors may be at work in different areas. Cross comparisons between rivers and between salmon populations may yield important information. The careful analysis of catch data and comparison with trends in other factors still has a place, and should be given a high priority by scientists. Catch data are often the only information we have on the state of salmon stocks.

The collective condition of salmon stocks in the North Atlantic is estimated every year by the International Council for Exploration of the Sea (ICES) through the application of models that estimate the pre-fishery abundance of North American and European salmon (see Chapter 3). A run-reconstruction approach has been adopted which back-calculates the stock size at points earlier in the life cycle from the numbers of fish returning to fresh water. These models are not ideal for the purpose. They are poorly documented, and they involve a number of questionable assumptions. For example, they assume that the mortality of salmon after the first year in the sea is low and stable, although this has not been clearly established. The data to support these models are also becoming increasingly fragmentary as more and more marine fisheries have closed. Nevertheless, although the interpretation of these models must be treated with care, they do point clearly to a significant decline in salmon abundance, and especially in the multi-sea-winter (MSW), early running components of the stocks.

Supporting information on trends in salmon populations comes from monitored rivers around the North Atlantic. Here, data on the return rates of fish are available from a combination of tag returns and fish counting methods. Recently, the great decline in netting effort has inevitably affected tag return rates and has made the interpretation of data from these monitored rivers more difficult. The return rates of one-sea-winter (1SW) fish have been very variable throughout the North Atlantic area. They showed a sudden decline at the end of the 1980s, and have remained low for some populations but recovered again for others. Return rates for two-sea-winter fish (2SW) have shown a more gradual decline but have been much lower in the 1990s than previously (see Chapters 3 and 8). Return rates for some individual rivers, the Western Arm Brook and the Conne River, both in Newfoundland, and the North Esk, in Scotland, have shown at least a two-fold decline in recent years compared with return rates earlier in the 1970s and 1980s (Chapter 8).

Changes have evidently taken place in the life history characteristics of fish in recent years as their numbers have declined. Reddin *et al.* (Chapter 8) have shown a progressive increase in the proportion of smolts returning as 1SW fish for the total North America population. Changes in the proportion of MSW fish and 1SW fish in the catches can be seen in the long-term catch records from many rivers. Detailed analyses of the catch statistics have led to speculation that climate change may alter the growth and development of fish, causing them to return at different sea ages (George 1982, Martin & Mitchell 1985). George, for example, refers to 'grilse cycles' and 'salmon cycles'. The current preponderance of late running, 1SW fish in the catches of many east coast Scottish rivers, and the dearth of early running, MSW fish, prompts the thought that fish that formerly remained at sea for two sea winters or longer may now be returning after one sea winter. There is an alternative view, however, that these large east coast rivers contain several different salmon populations, with different characteristics, which dominate at different times. Thus, under some conditions early running, MSW populations (which may also include a small proportion of early running, 1SW fish) may be especially strong and may dominate the catches. Under other conditions, later-running, mainly 1SW populations may dominate. Whether there is a developmental shift from one state to another, or whether different populations dominate at different times, the factors that lead to these changes may act upon the fish during their freshwater phase or during their marine phase.

Is the problem in the river or at sea?

There is quite strong evidence both from North American and European monitored rivers that freshwater production levels have generally been maintained in recent years. The main source of the decline in salmon abundance has been a reduction in marine survival. In the Western Arm Brook there has been a fall in marine survival from 14% in the 1970s and 1980s to 7% today. The North Esk has shown a similar fall. If this decline in marine survival is spread throughout the marine phase, then it will have a greater effect upon fish that remain longer at sea, resulting in a greater reduction in the abundance of MSW fish, as has been observed.

Potter & Crozier (Chapter 3) have made the point that we may, for some rivers, be placing too much emphasis on marine survival. In rivers like the Bush in Northern Ireland it appears to be variations in survival in the freshwater phase that are important. Of course, effects on the freshwater stages can subsequently affect the performance of salmon in the sea. Some environmental factors may act initially through the freshwater stage but later influence development in the sea. Thus, temperature changes, which affect growth, development rate and the time that fish leave the river, may subsequently affect growth and survival at sea.

Reddin *et al.* (Chapter 8) have reviewed the return rates of salmon in a number of monitored rivers and have concluded that they demonstrate an increase in mortality in the sea, though the causes are unknown. We are remarkably ignorant of the timing of this marine mortality. It is not clear whether it is highest soon after entry to the sea, whether it is spread over the whole marine phase, or whether there is a heavy loss rate when the adult fish return to coastal waters. MacLean *et al.* (Chapter 4) have drawn our attention to the summer scale checks which have recently affected fish in several rivers across Scotland. The presence of these checks indicates feeding or climatic problems in the middle of the salmon's stay in the sea.

It is remarkable how few of our rivers are being carefully monitored for variations in salmon and sea trout abundance and survival. Of the few that are being studied, some have significant inputs of hatchery fish, which may distort the picture for wild populations. There are especially few sites providing information on the return rates of MSW fish. Indeed, at a time when countries around the North Atlantic should be making a long-term commitment to further, new, monitored sites, there is a risk that because of their high operating costs the existing sites may no longer be maintained.

The life of salmon in the sea

Only recently have we begun to look at the life of salmon in the sea. Hansen & Jacobsen (Chapter 7) have summarised what is known of the distribution and migrations of the fish. Evidence suggests that the post-smolts do not linger near the coast, but move quickly into the open sea. Moore *et al.* (Chapter 5) describe the tracking of post-smolts in coastal waters and remark on their tendency to be carried by local currents. The swimming abilities of these tiny fish are of course very limited in relation to the speeds of water currents in the sea, and as they appear to stay close to the surface it is perhaps inevitable that they will be carried wherever those currents will take them.

In the open sea, the post-smolts again seem to be carried rapidly by the ocean currents, remaining close to the sea surface at least during the day (Holst *et al.*, Chapter 6). The speed of transport can be quite fast. The routes of the fish may not be well defined, as their movements may depend on variations in the ocean currents. Fish from particular rivers or regions may remain together in the initial stages of migration. Post-smolts captured in the Faroe–Shetland Channel and the western part of the Norwegian Sea originate mainly from southern European rivers (Spain, England, Wales and Ireland), carried there by the slope current running northwards along the continental shelf to the west of the British Isles. Post-smolts from northern European rivers (mainly Norway) are better represented in catches in the eastern part of the Norwegian Sea. Reddin & Short (1991) concluded that post-smolts in the Labrador Sea originated in rivers from Maine to Labrador.

Salmon from a large number of different rivers in North America and Europe may eventually come together in the same area of the ocean, and be harvested together in high seas fisheries. The salmon formerly caught in the waters off the western coast of Greenland originated in rivers from both North America and Europe. Those caught in the far north of the Norwegian Sea were mainly from Norway, but also included fish tagged in Great Britain, Sweden and Russia. Fish caught to the north of the Faroe Islands were from Scotland, Norway and Ireland and a few of the fish tagged and released in these waters were recaptured in Canada.

The distribution of the salmon of differing origin is still not well understood, but the limited information suggests that they are not evenly distributed. Salmon from North America seem to remain mainly in the western North Atlantic, although some may move into the eastern North Atlantic. It is also evident that a relatively large proportion of the older European salmon move into the western North Atlantic to feed. Salmon from particular rivers may differ in the length of time they spend at sea, and different age classes from the same populations may be present in different areas. As Hansen & Jacobsen (Chapter 7) have remarked, a better map of the distribution of the Atlantic salmon in space and time would greatly help us to understand the fluctuations in their survival and life history.

Though salmon from different countries of origin may gather together on the oceanic feeding grounds, the fish must subsequently make their separate ways homeward. It is now well established that a large majority of the fish returning to spawn in a particular river originated from that river. However, how they arrive in their home waters and whether their return migrations are precisely directed or result from trial and error searching is not yet clear. One model suggests that salmon migrate by means of random searching combined with a low degree of orientation to some outside stimulus (Saila & Shappy 1963). It may be unnecessary to invoke a very precise orientation of the fish, or the return of the fish through an area familiar to it, to explain the return migration. Adult fish reaching Scottish coastal waters have been found to be swimming on individual fixed compass headings, though their actual paths over the ground may be quite complex as a result of the deflecting action of surface currents (Hawkins *et al.* 1979, Smith *et al.* 1981).

There is the possibility that under some conditions a large proportion of the returning fish may be unable to find their home rivers, or may be unable to enter fresh water when they do find them. During this workshop Potter and others remarked on the apparent 'spontaneous generation' of fish in wet years, which may reflect greater ease of entry, or perhaps even better survival of the fish when entry conditions are easier. Low water conditions, low oxygen levels, high levels of pollution and perhaps high temperatures in estuaries may delay the entry of fish and may result in lower survival.

During the workshop we learned much about the further development of techniques which might enable us to look more closely at the behaviour and survival of salmon in the sea. The technology of tracking fish through the use of electronic tags has improved greatly. Tags are now smaller, and receivers more sophisticated. There are still severe limitations on the duration of active tracking sessions, however, and some of the earlier tracking studies by Holm *et al.* (1982) and others still set the standard. The long receiver lines and multiple receiver arrays described by Lacroix (see Moore *et al.* Chapter 5) have the potential to tell us more about marine mortality. The development of data storage tags also holds promise for the future. Indeed, the results reported by Reddin *et al.* (Chapter 8) on the temperatures experienced by tagged kelts, together with Sturlaugsson's results (Chapter 14) using depth recording tags, show what can now be achieved.

It has, in the past, been difficult to obtain good recapture rates from fish tagged in the high seas fisheries. The success encountered in recent trawling surveys specifically aimed at catching post-smolts suggests that this problem may be capable of being overcome. One of the great advances reported during the workshop by Holst (Chapter 6) was the development of a technique for catching salmon post-smolts alive and in good condition at sea. With this technique there is great potential for tagging fish and finding out much more about their behaviour and growth in the marine phase.

In particular, we need to find out more about the environmental preferences of salmon in the sea. Few attempts have been made to understand and model their behaviour. It should be possible to make quite simple assumptions about the way fish react to various stimuli and to predict their trajectories in the various current systems off the coast. It may prove possible to explain the patterns of distribution now being observed in the sea from trawling surveys, and to understand their relationship to temperature and currents, in terms of the fish obeying simple behavioural rules.

We clearly need to carry out more dedicated research vessel surveys, using the specialised trawls that have been developed to catch post-smolts in the sea. The major difficulty standing in the way of such surveys is the high cost of taking a large, well-equipped research ship into the inhospitable waters of the North Atlantic for periods of several weeks. However, if all the nations that are interested in the wellbeing of Atlantic salmon can join together in organising such surveys then it may be possible to make rapid progress.

The feeding of Atlantic salmon

The food of the salmon gradually changes as the fish move out from the river and into the sea (Chapters 13 and 14). In coastal waters their diet changes from insects to small and larval fish, including herring and sandeels and small marine crustaceans. In oceanic waters they feed on fish like capelin, sandeels, herring and lantern fishes, and on crustaceans like amphipods and euphausiids and on Arctic squid. They appear to be opportunistic predators, with a varied diet, preferring to select prey in the upper range of the size spectrum, and preferably preying upon fish rather than crustaceans. In productive marine waters, with such an opportunistic strategy, it seems unlikely that salmon will remain hungry for long. Indeed, the very rapid growth shown by salmon over their period in the sea suggests that death by starvation is unlikely. If food is limited, it may act by slowing growth rate, the slower growing fish perhaps being more vulnerable to predation or disease.

Ocean climate and the salmon

As we have seen, the survival of salmon in the sea can be measured by comparing the counts of outgoing smolts with the numbers of returning adults 1 or 2 years later. There is some coherence between the rates of return to the river for a group of North

American salmon populations from different rivers, and Reddin *et al.* (Chapter 8) have concluded that common events influence the production of North American salmon, suggesting that some widespread factor, like ocean climate, is a possible cause of the variation in survival.

Correlations between sea temperatures and salmon stocks have previously been reported by George (1982) and Martin & Mitchell (1985). Summers (1992, 1995) constructed a chronology of hydrographic and biological events in the North Atlantic over the last 200 years. He concluded that the high salmon and sea trout abundance in the 1960s more or less coincided with striking hydrographic changes across the whole North Atlantic, and was paralleled in the North Sea by an increased abundance of cod and haddock. Similar events had occurred almost 80 years earlier, in the 1880s, another period of increased salmon and sea trout abundance. Summers concluded that the return of salmon to Scottish rivers was influenced by natural marine factors, though the causative mechanisms could not readily be elucidated, and may have been different at different times in the past.

Reddin & Shearer (1987) suggested that the abundance of salmon west of Greenland could be directly related to the area of the north Atlantic enclosed by the 4°C and 10°C isotherms. These 'thermal habitat' ideas were developed further by Reddin & Friedland (1993) and Friedland *et al.* (1998), who have revealed correlations between salmon catches and salmon survival and an index of preferred spring habitat for post-smolts defined by sea surface temperature. At this workshop, Reddin *et al.* (Chapter 8) have suggested that the inshore sea surface temperatures around the coast of Newfoundland bear a relationship to survival at the post-smolt stage, and influence the numbers of adult salmon produced.

Bigg (Chapter 11) and Dickson & Turrell (Chapter 9) have drawn our attention to the large-scale changes taking place in the climate and ocean circulation of the North Atlantic. There are long-term trends in the North Atlantic Oscillation (NAO) index, with decadal wobbles superimposed on this. There may also be pronounced inter-annual changes in climate, which are difficult to account for, but which nevertheless may be very significant.

Drinkwater (Chapter 10) has shown how climate change can have major effects upon marine fish populations. It is well known that both the recruitment, and the movements and distribution of fish are affected by oceanic conditions.

Reid & Planque (Chapter 12) have shown that there have been major changes in the plankton colour index with time in the North Sea and other areas, indicating a progressive change in the plankton regime in these areas. Moreover, there is a strong correlation between the abundance of *Calanus finmarchicus*, a major prey item for many fish, and the NAO.

It is likely in the next few years that we will gain a better understanding of some of these correlations, for example between the NAO and the coupled oceanographic changes. We may also be able to relate the oceanographic changes to the biology of species like *Calanus* spp. With improved, more representative sampling methods for the pelagic prey organisms, and better coordinated surveys, we will be able to look at the effects of changes in ocean climate upon the food of salmon and the other important marine fishes. At the moment, however, we can only see a partial picture.

What we do not know is how climate change exerts its impact on the salmon stocks. Does it affect the availability of food, leading to death by starvation or reduced levels of feeding and slower growth? Does a reduction in growth and condition make fish more vulnerable to disease or to predation? Do the oceanographic changes affect the ability of fish to find their way home, or lead fish into parts of the ocean from which they cannot return?

Causes of mortality

Reid & Planque (Chapter 12) believe that the declining fortune of the salmon fisheries in the North Atlantic in the last decade is linked to changes in plankton productivity and in the food chain, brought about by variations in ocean climate. Long-term changes are undoubtedly taking place in the phytoplankton of the eastern North Atlantic and these may be linked to changes in the circulation of the North Atlantic. One fear we have to take away with us from this workshop is that climate change may now be accelerating. There may be the prospect of major changes taking place in the North Atlantic circulation in the next few years, including a diversion of the North Atlantic Drift towards the south. Such a possibility may make us pessimistic about the prospects for salmon. It is quite possible, however, that salmon may cope with such changes better than ourselves. The fate of salmon will not be the only thing on our minds if the North Atlantic becomes much colder.

What is not clear is how these ocean climate changes influence plankton production and then affect the salmon. There may have been a shrinking of the salmon's habitat and changes in water circulation patterns within the North Atlantic, and there may have been a reduction in the abundance of prey, but how has the salmon been affected?

Salmon migrate to the sea primarily to feed. Salmonids in fresh water have to cope with a very limited food supply and grow only slowly. If they mature in the river, they suffer the penalty of low female fecundity, for the eggs are large and yolky and cannot be produced in large numbers on a poor food supply. Salmonids that migrate to sea grow very rapidly, even by the standards of other marine fish. Food is more abundant, giving rise to rapid maturation and high fecundity.

The supply of food for salmon in the sea does vary, as we have seen, through natural variations in productivity. It can also be affected by marine fisheries. It has been argued that the so-called 'industrial' fisheries for forage species like sandeels and capelin may have brought about a reduction in the food available to salmon in the sea. With respect to sandeels, however, stocks of this species in the eastern North Atlantic have remained in a relatively healthy state, despite the large Danish and Norwegian fisheries (Hawkins et al. 1998). There has occasionally been poor recruitment of larval sandeels in some areas, like the waters around Shetland, but by and large the spawning stocks of sandeels have remained high. The heavy removal through human consumption fisheries of predatory species like cod, haddock, whiting, herring and mackerel, which consume sandeels, may have played a part in this.

The sandeels found alongside other prey species in the stomachs of post-smolts are mainly the larval stages, rather than adults. Larval sandeel production remains high across wide parts of the North Atlantic, and there is no evidence of any catastrophic depletion of the food supply of the emigrating post-smolts. The sandeel fishery is perhaps more likely to have an effect through depleting local grounds of adult sandeels, which may affect the ability of returning adult salmon and sea trout to find food.

Fisheries for capelin, a small arctic fish species, take place around Iceland, Greenland, and the northern part of the Norwegian Sea, where feeding salmon are to be found. The Barents Sea stock of capelin is currently in a state of severe decline, and the fishery has been closed. The capelin stock in the waters between Iceland, East Greenland and Jan Mayen is currently in a healthy state, though it is subject to high variability. We do not know what effect the fishery for capelin is having upon salmon in the sea.

If the food supply for salmon in the sea has been drastically reduced in the last decade we would surely have noticed the effects upon the salmon themselves. Perhaps we have not been observant enough. In the future we should be looking more closely at the growth made by salmon in the sea. As MacLean and his colleagues have shown (Chapter 4), we can investigate the relationship between growth, feeding and ocean climate by examining the scales of returning fish.

Little was said at the workshop about two other major potential causes of death in the sea; disease and predation. It is known that diseases and parasites affect marine fish but we are remarkably ignorant of their effects. Fish which have died of disease are simply lost and not recovered. One particular infectious agent of fish which has been especially well studied is *Ichthyophonus*, a protistan parasite that infects a wide range of species, including salmonids (McVicar 1998). Epidemics of *Ichthyophonus* have occurred in herring (*Clupea harengus*) in both the eastern and western North Atlantic in the last decade. Indeed, massive mortalities of herring associated with the presence of this parasite have been recorded in the Gulf of Maine, the Gulf of St Lawrence, and the Kattegat. The same parasite has also been linked with high mortalities of plaice (*Pleuronectes platessa*) in waters to the north of Scotland, and it appears to be a particular problem where fish are gathered together in great numbers. However, *Ichthyophonus* is just one example of the many infectious agents that can afflict fish. There is a wide range of other parasites and bacterial and viral diseases which may cause significant mortalities. On the west coast of Scotland it is considered by many that diseases and parasites from fish farms have played a significant role in the decline of wild salmon and sea trout. The problem for wild fish is that unlike outbreaks at fish farms epidemics of disease at sea are difficult to observe and may be missed altogether. We simply do not know what part disease plays in the survival of salmon in the wild, and whether the incidence of disease may be affected by climatic conditions.

A wide range of species prey upon the salmon. On initial entry to the sea, birds may be especially important, together with fish like the cod (*Gadus morhua*) and the pollack (*Pollachius pollachius*). There has been a massive increase in the numbers of a particular marine predator, the grey seal (*Halichoerus grypus*) in North Atlantic waters in recent years. Indeed, grey seal numbers at Scottish breeding sites have doubled since the 1970s. Common seals (*Phoca vitulina*) have not shown the same expansion in numbers but are still important predators of salmon. Both species are common in coastal waters, and are especially common in the estuaries of east coast Scottish rivers in the months September–February when predation may have a strong effect upon the numbers of spring fish (Pierce *et al.* 1997). However, seals are not confined to estuaries or the coast, and they may prey upon salmon throughout their life in the sea. The level of predation may be affected by changes in climate, especially if salmon are weakened by poor food supplies, or constrained in their movements.

Another major predator is man. Holst *et al.* (Chapter 6) have remarked that salmon post-smolts may be especially vulnerable to purse seines and to pelagic trawls, rigged to fish close to the sea surface. No large scale purse-seine fisheries appear to be taking place in areas where post-smolts are believed to be concentrated. There is, however, a trawl fishery for mackerel in the international zone of the Norwegian Sea where post-smolts are found, pursued mainly by Russian vessels. There is a need to carry out systematic sampling of this and other fisheries for evidence of any salmon by-catch.

Conclusions

This workshop has shown that rapid progress is now being made in understanding the problems facing salmon in the sea. Our knowledge of the physical factors affecting salmon is growing all the time, and our knowledge of the biology of the fish is gradually improving. Collectively, quite a lot of effort is now going into the study of salmon in the sea, despite the expense and the difficulty of doing such work. Nevertheless, we are still largely ignorant of many of the influences acting upon salmon once they leave their rivers as smolts. We must put more effort into studying the marine life of salmon.

The workshop has been especially valuable in bringing together scientists from many different disciplines to discuss what is known of salmon in the sea. It has been a very worthwhile initiative by the Atlantic Salmon Trust, and we owe them our thanks for their foresight. We must congratulate Derek Mills, Jeremy Read and Dick Shelton for all the effort they have put into the organisation of the workshop.

References

Friedland, K.D., Hausen, L.P. & Dunkley, D.A. (1998) Marine temperatures experienced by postsmolts and the survival of Atlantic salmon in the North Sea area. *Fisheries Oceanography*, **7**(1), 22–34.

George, A.F. (1982) *Cyclical variations in the return migration of Scottish salmon by sea-age c. 1790 to 1976.* M.Phil. thesis, Open University, Milton Keynes.

Hawkins, A.D., Christie, J. & Coule, K. (1998) *The Industrial Fishery for Sandeels.* Atlantic Salmon Trust, Pitlochry.

Hawkins, A.D., Urquhart, G.G. & Shearer, W.M. (1979) The coastal movements of returning Atlantic salmon. *Scottish Fisheries Research Report*, **15**, 1–14.

Holm, M., Huse, I., Waatevik, E., Døving, K.B. & Aure, J. (1982) Behaviour of Atlantic salmon smolts during seaward migration. International Council for the Exploration of the Sea CM 1995/M:7.

McVicar, A.H. (1998) *Ichthyophonus* and related organisms. In: *Fish Diseases and Disorders*, Vol. 3. *Viral Bacterial and Fungal Infections* (eds P.T.K. Woo & D.W. Bruno), pp. 661–87. CAB International, Wallingford, Oxon.

Martin, J.H.A. & Mitchell, K.A. (1985) Influence of sea temperature upon the numbers of grilse and multi-sea-winter Atlantic salmon caught in the vicinity of the River Dee (Aberdeenshire). *Canadian Journal of Fisheries and Aquatic Science*, **42**, 1513–21.

Pierce, G.J., Hislop, J.R.G. & Carter, T.J. (1997) Interactions between seals and salmon in northeast Scotland. *The Salmon Net*, **28**, 49–55.

Reddin, D.G. & Friedland, K.D. (1993) Marine environmental factors influencing the movement and survival of Atlantic salmon. In: *Salmon in the Sea and new Enhancement Strategies* (ed. D.H. Mills), pp. 79–103. Fishing News Books, Oxford.

Reddin, D.G. & Shearer, W.M. (1987) Sea-surface temperature and distribution of Atlantic salmon in the Northwest Atlantic Ocean. *American Fisheries Society Symposium*, **1**, 262–75.

Reddin, D.G. & Short, P.B. (1991) Postsmolt Atlantic salmon (*Salmo salar*) in the Labrador Sea. *Canadian Journal of Fisheries and Aquatic Sciences*, **48**, 2–6.

Saila, S.B. & Shappy, R.A. (1963) Random movement and orientation in salmon migration. *Journal du Conseil International pour l'Exploration de la Mer*, **28**, 153–66.

Smith, G.W., Hawkins, A.D., Urquhart, G.G. & Shearer, W.M. (1981) Orientation and energetic efficiency in the offshore movements of returning Atlantic salmon. *Scottish Fisheries Research Report*, **21**, 1–22.

Summers, D.W. (1992) *Studies of the Atlantic Salmon in Scotland.* PhD. thesis, University of Aberdeen.

Summers, D.W. (1995) Long-term changes in the sea-age at maturity and seasonal time of return of salmon, *Salmo salar* L., to Scottish rivers. *Fisheries Management and Ecology*, **2**, 147–56.

Index

Acartia spp., 163–4, 166
acidification, 24,
acoustic tags (pingers), 10, 50, 54, 58–9, 62–3.
Adalvik, 194–9, 204, 206, 208
Advanced Very High Resolution Radiometer (AVHRR), 88
American lobster, *Homarus americanus*, 122–3, 126
American plaice, *Hippoglossoides platessoides*, 118, 122, 130
Amlach, 99–100, 102
Ammodytes, see sandeels
amphipods, 4, 15, 171, 173–4, 177, 182, 184, 188, 192, 201–203, 205–207, 216
annelids, 177–8
Arctic charr, *Salvelinus alpinus*, 201
Arctic cod, *Boreogadus saida*, 121, 176
Atlantic Centre for Remote Sensing, 2
Atlantic salmon, *Salmo salar*, (pre- and maturing adults)
  cannibalism, 205
  countries of origin, 80
  diet (pre-adults), 173–81
  diet (adults), 185, 193–4
  distribution, 75–84
  exploitation rate, 80
  feeding behaviour, 208
  feeding habits, 170–88
  food preference, 183, 188
  foraging strategy, 181
  homing migration, 82–4, 202, 204
  river entry, 127
  tag recoveries, 77–81
  vertical distribution, 181
Atlantic salmon catches, 7, 26, 37, 109, 153, 155, 167, 211
Atlantic Salmon Federation, 1, 2, 49, 91
Atlantic Salmon Trust, 1, 2, 49, 91, 220

archival tags (*see* data storage tags)
aromatic hydrocarbons (PAHs), 4
Atlantic menhaden, *Brevoortia tyrannus*, 122
automatic listening stations, 55–6, 59–61
automated receivers, 51–2
availability, 126

Baltic salmon, 37
Baltic Sea, 16, 83, 179, 193
Barents Sea, 72, 139, 219
Barents Sea–Fram Strait, 101
barracudinas, *Paralepis* spp., 4, 173–4, 176, 178, 182–3, 193
Bay of Biscay, 164
Bay of Fundy, 16, 52, 56, 118
Bay of Islands, 185
Bear Island, 72
benzofurans, 4
blue whiting, *Micromesistius poutassou*, 4, 173–4, 176, 186–7
Brachioteuthis riisei (cephalopod), 173
Breidafjord, 194–7, 200–202, 204–205, 208
British Isles, 175–6, 178

Cabot Strait, 122
Canadian Shelf, 174–5, 178
catchability, 3, 126
Calanus spp. 157
Calanus finmarchicus, 15, 161–3, 166–7, 182, 184, 217
Calanus glacialis, 161
Calanus helgolandicus, 161–2, 166
Calanus hyperboreus, 161
capelin, *Mallotus villosus*, 2, 4, 16, 118–21, 130, 171, 173–5, 178, 182–4, 186–7, 192, 204, 216, 218–19
Carlin tags, 25

cataracts, 16
catchability, 3, 126
catch rates, 90–91
CEFAS Laboratory, Lowestoft, 57, 60
Centropages typicus, 162, 164, 166
cephalopods, *see* squid
chaetognaths (*Sagitta* spp.), 182–4
chironomids, 171–2
climatic change, 2, 14, 112, 137–50, 220
closed circuit television (CCTV), 73
cod, *Gadus morhua*, 2, 16, 116–26, 130–31, 176, 217, 219–20
coho salmon, *Oncorhynchus kisutch*, 16, 20
Communicating History Acoustic Transponder (CHAT) tags, 55, 61
common seal, *Phoca vitulina*, 220
Comprehensive Ocean Data Set (COADS), 156
Continuous Plankton Recorder (CPR), 15, 153–6, 159, 163, 165, 167
counters, 34
crab larvae, 172, 201–202, 205–208
Cromarty Firth, 66
crustaceans, 15, 17, 172, 174, 178, 181–2

data storage tags (DSTs), 12, 17, 49, 56–8, 60–61, 181, 215
Davis Strait, 4, 173, 175–6, 178
decapods, *see* crab larvae
Denmark Strait, 139
depth recording tags, 215
dinoflagellates, 163, 167
dioxins, 4
dipterans, 201–202, 207
drift-nets, 3, 10, 53

El Niño, 13, 153
El Niño Southern Oscillation (ENSO), 13, 159, 166
ERS-2 earth observation satellite, 2
euphausids, 4, 15, 171–2, 174, 177–8, 181–3, 185, 188, 193, 202–203, 206–207, 216

Fair Isle, 99–100, 102
Fair Isle Channel, 71
farmed salmon, 12, 77–8, 81, 180, 185, 211, 219
Faroe Islands, 1, 4, 5, 8, 11, 15, 49, 77–9, 84, 174–5, 180, 193, 214
Faroe–Shetland Channel, 56, 67, 71–2, 148, 161, 214
Faroese long-line fishery, 174
Firth of Clyde, 171
fishing moratoria, 128, 211
Fram Strait, 106, 158
Freshwater Fisheries laboratory, Pitlochry, 38, 65
freshwater survival, 31–2
Fyllas Bank, 130

gannets, 16
geo-locating system (GLS), 56–7, 60
Georges Bank, 116–18, 124–5
giant scallop, *Placopecten magellanicus*, 119
glacial termination, 143
Global Positioning System (GPS), 50
global warming, 14, 144–7, 161, 167
Globigerinoides ruber, 141
Grand Banks, 89–91, 117–19, 121, 126, 130
Great Salinity Anomaly, 106
greater black-backed gull, 16
green crab, *Carcinus maenas*, 122
greenhouse warming, *see* global warming
Greenland, 1–5, 8, 11, 24, 49, 75–7, 83, 106, 125, 174, 187–8, 214
grey seal, *Halichoerus grypus*, 220
Gulf of Maine, 122–3, 125, 162
Gulf of St. Lawrence, 11, 75–6, 119, 121–3, 171, 193, 203
Gulf Stream, 89, 124, 139, 167

Gulf Stream Index (GSI), 153, 156–9, 166
Gulf Stream rings, 124
gyres, 2, 4, 107–109, 138–9, 141, 144, 148

habitat degradation, 9, 31
halibut, 120
Hadley Centre Climate model, 146
haddock, *Melanogrammus aeglefinus*, 118, 124–5, 176, 217, 219
herring, *Clupea harengus*, 4, 16, 65, 118, 120, 122, 171–5, 182–3, 185–7, 193, 206–207, 216, 219
herring gull, 16
horse mackerel, *Trachurus trachurus*, 163–4
Hraunsfjord, 194–9, 201, 205–206, 208
Hvalfjord, 198–9, 204–206, 208
hydroacoustic gear, 1

icebergs, 2
Iceland, 193–209, 219
Ichthyophonus, 219
index rivers, 33, 35, 109, 113
industrial fisheries, 218–19
Institute of Marine Research, Bergen, 65
International Council for the Exploration of the Sea (ICES), 2, 7–8, 33, 111, 212
inverse weight hypothesis, 9, 20–21, 34
Iridium satellite system, 61
Irminger Current, 120, 139
Irminger Sea, 173

Jan Mayen, 219
Jokulfjords, 194–9, 204–206

Kattegat, 219
kelts, 17, 57, 84, 215
Kolgrafjord, 195–6, 198, 201, 205

Labrador Current, 91

Labrador Sea, 4, 13, 16, 76, 89–90, 93, 96–107, 110, 116, 125–30, 138, 144, 149, 173, 175, 178–9, 188, 214
Labrador Sea Water, 107
Langaros estuary, 186, 198, 201–203
lantern fishes, 4, 173–5, 178–9, 182, 215
Last Glacial Maximum, 142
Laurentian Channel, 122
Lerwick, 97–8, 101
Little Ice Age, 143
long-lines, 56, 77
lumpsucker, *Cyclopterus lumpus*, 172, 177, 205–207

mackerel, *Scomber scombrus*, 4, 16, 120, 122, 172, 174, 183, 186–7, 219
mackerel fishery, 11, 70, 220
Magdalen Shallows, 122
magnetic field, 82–3
marine climate, 88–91
marine habitat index, 33
Marine Laboratory, Aberdeen, 65
marine survival, 24, 29–30, 33, 37–8, 213
marine survival indices, 32, 47
Meganyctiphanes norvegica, 185
micronekton, 181
micro tags, 10, 22, 25, 199
MIK plankton net, 179, 182–4
Millport, 99–100, 102
Ministry of Agriculture, Fisheries and Food, 2, 91
molluscs, 177–8, 202, 205
Molson Family Foundation, 2, 91
monitored rivers, 24, 33–4, 212, 213
Moray Firth, 66
Multi-Channel Sea Surface Temperature (MCSST) data, 88–90
mutagens, 4
myctophids, *see* lantern fishes

nonylphenol (4-nonylphenol), 9
North Atlantic Current, 107–8
North Atlantic Deep Water, 137, 142

North Atlantic Drift, 137–9, 143–4, 146, 148, 218
North Atlantic Ocean, 14, 33–4, 37, 49, 65, 70, 73, 78, 82, 89, 101, 140, 149, 150, 161–2, 170, 174, 183, 187, 193, 217
North Atlantic Oscillation, 13, 92–113, 127–8, 139, 140, 144, 149, 153, 156, 158, 160, 161–2, 166–7
North Atlantic Oscillation Index, 13, 15, 104, 105, 110, 112, 127, 129, 156–7, 164, 217
North Atlantic Salmon Conservation Organisation (NASCO), 211
North Atlantic Salmon Working Group, 80, 111, 156
North Sea, 66, 70, 93, 95, 101, 109, 111, 161, 164, 173, 217
Northern Minch, 71
Norwegian Atlantic Current, 103, 104, 105
Norwegian Coastal Current, 72, 94, 138–9
Norway pout, *Trisopterus esmarkii*, 173, 176
Norwegian Sea, 1, 4–5, 11, 49, 65, 67, 70, 71–2, 76–7, 80, 94–5, 104, 138, 143, 157, 172–3, 175, 214, 220

ocean climate, 127–8
ocean habitat index, 24
Oithona spp., 164
orientation, 82
organochlorines, 4
overfishing, 123
oxygen isotope, 140, 141

Pacific salmon, 13, 16, 18, 21, 37, 45, 75, 82–3, 153, 187
Passamaquoddy Bay, 52
pearlsides, *Maurolicus muelleri*, 174, 182, 193
pelagic trawls, 11, 66, 70

Peterhead, 99, 100, 102
pheromones, 83
phosphates, 4
phytoplankton, 2–4, 123, 147, 159–61, 163
phytoplankton colour, 159, 160, 161, 163, 167, 217
plaice, *Pleuronectes platessa*, 219
Podon spp., 163
pollack, *Pollachius pollachius*, 16, 124–5, 204, 220
polychaetes, 201, 202, 205, 207
polychlorobiphenyls (PCBs), 4
'pop-up' tags, 12, 61–2
Porcupine Bank, 67, 173
post-smolt
  cataracts, 16
  distribution, 65–73
  feeding behaviour, 170, 171–3, 183, 193–4, 196, 198, 200, 205, 206, 216
  forage status, 196, 198, 200, 201
  migration, 11, 12, 51, 65–73, 90, 196, 198, 200
  mortality, 73, 208
  recaptures, 67–70, 84, 194
  surveys, 4, 65–73
  swimming ability, 214
  tracking, 4, 10, 49–63
pre-fishery abundance (PFA), 22–3, 28–9, 35, 154, 212
Prey Forage Ratio (PFR), 197, 204, 205
Principal Component Analysis, 163, 165
purse seine, 70

radio telemetry, 61
recapture rates, 29, 39, 78–80
redfish, *Sebastes marinus*, 120, 124–5, 130, 173, 177
research vessel surveys, 215
return rates, 20, 26–8, 212
rivers
  Canada
    Conne, 9, 88, 212

Fraser, 82
Kouchibouguac, 80
Magaguadavic, 52
Middle, 76
Miramichi, 78, 80
Sandhill, 20
St. Croix, 52
Western Arm Brook, 8–9, 23, 76, 88, 212–13
England
　Avon, 51
　Test, 51
　Wye, 9
France
　Bresle, 25, 27
　Nivelle, 25, 27
　Oir, 25, 27
Iceland
　Ellidaar, 25–7
　Gljúfúra, 199
　Laxá, 199
　Vesturdalsa, 25–9
Ireland
　Ballynahinch, 67
　Bundorragha, 67–8
　Burrishoole, 30, 32–3, 67–9
　Corrib, 25–9
　Delphi, 69
　Lee, 69
　Shannon, 67–9
Northern Ireland
　Bush, 8–9, 19, 20, 26–7, 29, 30–32, 213
　Erne, 67, 69
Norway
　Drammen, 8
　Figgio, 10, 29, 37, 49, 94
　Imsa, 25–9, 30, 84
Scotland
　Dee, 23
　Don, 42
　Girnock, 14
　North Esk, 9, 10, 15, 21, 24–5, 27–9, 37, 39, 40, 41–3, 49, 88, 94, 212–13
Spain
　Asón, 67, 69
USA
　Penobscot, 8, 75
Wales
　Conwy, 51
　Ogmore, 57–8
　Tawe, 51
run-reconstruction model, 7
Russia, 77, 79, 90

salmon ranching, 194, 198, 205, 211
sand eel, *Ammodytes americanus*, 171, 178
sandeels, 4, 171, 178, 185, 187, 193–4, 202, 203, 206, 207, 216, 218
Sargasso Sea, 124
satellites, 2, 12, 88–9, 93, 159, 161
scales, 10, 16, 17, 37–48
Scarborough, 99, 100, 102
scorpion fish, *Myxocephalus scorpioides*, 177
Scotian Shelf, 116–18, 120, 121, 125–6
sea lice, *Lepeophtheirus salmonis*, 16, 73, 182, 208
sea surface temperature, 2–4, 12, 88–93, 95, 101, 108, 109, 110, 139, 148, 156, 162, 179, 217
sea trout, *Salmo trutta*, 219
seals, 16, 220
sequential imprinting, 83
sharks, 5, 61
silicates, 4
silver hake, *Merluccius bilinearis*, 122
Skagerrak, 164
skate, 130
Slope Current, 70, 71–2, 94, 138, 148
Slope Water, 91
smelt, *Osmerus eperlanus*, 176
snow crab, *Chionoecetes opilio*, 121
sockeye salmon, *Oncorhynchus nerka*, 21, 82

sonar buoys, 51, 55
spawning gravels, 31
Spitsbergen, 72
sprat, *Clupea sprattus*, 175
squid, 5, 173, 177–8, 182, 193, 216
St. Johns, Newfoundland, 128–9
stock-recruitment curve, 31
storm index, 105

tags, 76–9
telemetry acoustic transmitters, 51
telemetry tags, 54, 59
Temora longicornis, 166
Temora spp., 164
Themisto spp., *see* amphipods
thermal habitat, 13, 93–4, 109–11, 217
thermohaline circulation, 148–9
three-spined stickleback, *Gasterosteus aculeatus*, 201
Thysanoessa spp., 185
Trondheimsfjord, 171–2

Ungava Bay, 75, 185

Viking Bank, 66
volcanic eruptions, 150
Vogar, 194–9, 204, 205, 208

West Greenland Current, 91
Weymouth, 99–100, 102
white butterfish, *Peprilus triancanthus*, 122
white whales, 120
whiting, *Merlangius merlangius*, 176, 185, 219
winter flounder, *Pseudopleuronectes americanus*, 118
wolf fish, *Anarhichas lupus*, 130, 176
Wyville–Thomson Ridge, 71

Yellowtail flounder, *Limanda ferruginea*, 122, 124–5, 130
Younger Dryas event, 143

zooplankton, 2, 15, 44, 97, 123, 159, 161–6, 173–4, 182